The Bacterial Cell Wall

Springer

Berlin
Heidelberg
New York
Barcelona
Hong Kong
London
Milan
Paris
Tokyo

Guntram Seltmann · Otto Holst

The Bacterial Cell Wall

With 114 Figures and 23 Tables

 Springer

Dr. rer. nat. habil. Guntram Seltmann
Karl-Marx-Str. 9
38855 Wernigerode
Germany

Prof. Dr. rer. nat. Otto Holst
Forschungszentrum Borstel
Zentrum für Medizin und Biowissenschaften
Parkallee 1-40
23845 Borstel
Germany

ISBN 3-540-42608-6   Springer-Verlag Berlin Heidelberg New York

Library of Congress Cataloging-in-Publication Data
Seltmann, Guntram. The bacterial cell wall / Guntram Seltmann, Otto Holst.
p. cm. Includes bibliographical references and index.
ISBN 3540426086
1. Bacterial cell walls. I. Holst, Otto. II. Title.
QR77.3 .S45 2001   571.6'8293--dc21

Springer-Verlag Berlin Heidelberg New York
a member of BertelsmannSpringer Science+Business Media GmbH
http://www.springer.de
© Springer-Verlag Berlin Heidelberg 2002
Printed in Germany

The use of general descriptive names, registered names, trademarks, etc. in this publication does not imply, even in the absence of a specific statement, that such names are exempt from the relevant protective laws and regulations and therefore free general use.

Cover Design: *design & production*, Heidelberg
Camera-ready by the authors

SPIN: 10785602   31/3130 – 5 4 3 2 1 0 – Printed on acid free paper

Contents

Preface

The bacterial cell wall represents a very complex structure disconnecting the interior of single-cell organisms from the environment, thus protecting, but also enabling, them to interact with the surrounding milieu and to exchange both substances and information. Knowledge of the biochemistry of the cell wall (components) and the genetic background helps to understand their significance with regard to microbiology and immunology of bacteria.

This book represents the second edition of a publication which was presented nearly 20 years ago in the German language (*Die bakterielle Zellwand*). Since that time our knowledge in this field has been significantly enlarged. Therefore, the manuscript had to be completely revised and updated. To maintain both the size and the introductory character of the book at least to a great extent, the authors had to restrict the presented material to that which appears basic and most important. This requirement must inevitably bring about many subjective factors.

As pointed out in the first edition, the term cell wall was not taken too strictly. Since the constituents located outside the cytoplasmic membrane are frequently difficult to divide in structure, localisation, and/or function into true cell wall components and supplementary substances, they are all at least briefly mentioned. For this reason capsules, S-layers and appendices like flagellae, fimbriae and pili are discussed. In addition, sometimes deeper regions of the bacteria like cytoplasmic membrane or cytoplasm are crucial for the course of processes necessary for the cell wall (components), e.g. for their biosynthesis or transport. In such cases, these also are discussed.

The progress of natural sciences depends mainly on the development and introduction of modern methods. Therefore in the text important methods are mentioned and briefly discussed. In some cases it appeared useful, by historical and didactic reasons, to discuss older approved methods.

Science does not only live in the present. Current developments are mostly based on results obtained in the past. Thus, it was both tempting and attractive to sketch or outline historical developments in particular cases.

In most cases the binary taxonomic designation of the species is used unabbreviated. In the case of well-known bacteria (e.g. *Escherichia coli*) the first term is abbreviated. In agreement with the commonly used practice for Salmonella serovars of subspecies I (*Salmonella enterica* subsp. *enterica*), the old names were maintained, however, not italicised, and the second term capitalised (e.g. Salmonella Typhi).

Some decades ago (1977) archaebacteria (archaea) were recognised as a third form of life besides (eu-)bacteria and eukaryotes. They are prokaryotes whose cell walls in many cases resemble those of the "classical" bacteria, but which may show very interesting modifications in structure-function relationship. Therefore they are discussed in an individual chapter.

Our writing was supported by several people, whose advice and encouragement were very helpful. We are indebted to Prof. Dr. E. Th. Rietschel and Prof. Dr. H. Tschäpe for their very generous help, to Prof. Dr. K. Jann and Dr. Barbara Jann for their very valuable and extensive advice for a revised version of Chapter 6, to Prof. Dr. mult. W. Köhler, Prof. Dr. U. Seydel, Prof. Dr. M. Wagner, Prof. Dr. H. Labischinski, Prof. Dr. U. B. Sleytr, Prof. Dr. T. J. Beveridge, Prof. Dr. J. Wecke and Dr. H. Engelhardt for providing us with literature and/or photos. Mrs. Dorothea Eitze prepared the majority of the figures. We also express our gratitude to the Spektrum Akademischer Verlag, Heidelberg, as legal successor of the Gustav-Fischer Verlag for the re-assignment of the copyright and to Mrs. Christiane Glier of the Springer-Verlag for her complaisance and help during the preparation of this Volume.

Wernigerode/Borstel   Guntram Seltmann
August 2001   Otto Holst

1 Introduction

The cell wall represents the outermost boundary of bacterial cells. Because of its exposed location it possesses manifold functions, some of which are apparently contradictory such as:

- the separatory functions: the cell wall separates the interior of the bacteria from the environment, while protecting the cell from harmful influences exerted by the surrounding milieu;

- the connecting functions: the cell wall enables transport of substances and information from outside to inside and vice versa, and enables the bacteria to contact (communicate with) the environment.

Furthermore, the cell wall provides the bacteria with sufficient solidity (e.g. shape stability, insensitivity to osmotic shock) and enables metabolism, growth and multiplication of the cells. However, it must be elastic enough to withstand considerable expansion.

The structures of bacterial cell walls are adapted to these functions, but solving the problem is absolutely different in individual bacteria. Thus, the cell walls appear to be rather complex and may vary greatly in details. However, on comparing the cell walls of different bacteria, one also can find structural principles common to many or most of them.

More than 100 years ago (1884), Gram succeeded in differentiating into two large groups all bacteria known at that time, using a simple staining method that was later named after him. After heat fixation, the bacteria are stained successively by solutions of carbolgentiana violet and of iodine (Lugol solution). Then ethanol is added until no further dye is eluted from the layer. Subsequently, the bacteria are restained with a diluted ethanolic fuchsine solution. Bacteria appearing darkblue to violet under the microscope are designated Gram-positive, red ones Gram-negative. The difference in stainability between typical representatives of both groups depends on the structure of their cell walls. Both types of cell walls contain the so-called peptidoglycan (murein, see below) as rigidity-causing component; however, that of the Gram-positive bacteria is much thicker than that of the Gram-negative. In addition, both groups contain different accessory cell wall components. Gram-positive cell walls can differ much more from each other in composition and structure than the Gram-negative ones, which follow a more general structural format.

Later, two additional groups of microorganisms were described: mycoplasma, containing no cell wall and no peptidoglycan; and archaea, possessing a differently composed rigid layer instead of peptidoglycan in or on the cell wall, e.g. a pseudomurein or an S-layer, respectively.

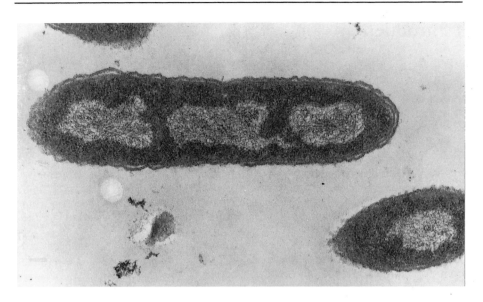

Fig. 1.1. Electron micrograph of an ultrathin section of a Gram-negative bacterium (*Pseudomonas aeruginosa*, F. Mach)

Not in all cases does the Gram stain yield clear and unambiguous results. Several bacteria, e.g. *Clostridium tetani*, may react in a Gram-positive or Gram-negative way depending on cultivation conditions; these are called Gram-variable bacteria. In addition, untypical Gram-negative and Gram-positive bacteria have been described: the Gram-negative representative of cyanobacteria containing a thick peptidoglycan and the Gram-positive mycolata, the cell wall of which resemble that of Gram-negative bacteria. Likewise, in the case of archaea the results are equivocal and therefore without taxonomic importance. However, because of its simplicity and high reliability, the Gram stain is still commonly used.

In general, microorganisms of the four groups mentioned above differ fundamentally from each other in many details of cell wall composition and structure. In Fig. 1.1 the electron micrograph of an ultrathin section from a Gram-negative bacterium is shown. Such pictures may look different depending on the cultivation conditions and the fixation procedure. However, close to the cell surface three layers are clearly visible. The innermost layer represents the inner or cytoplasmic membrane. Although most authors do not regard this membrane as a cell wall component, it represents the structural unit in which many components of the cell wall are biosynthesised. To point out this relationship the term cell envelope is defined, which comprises both the cell wall and the cytoplasmic membrane.

Further outwards is situated the periplasmic space, in which a rigid layer, the peptidoglycan, is located. This layer is present in all vegetative Gram-positive and

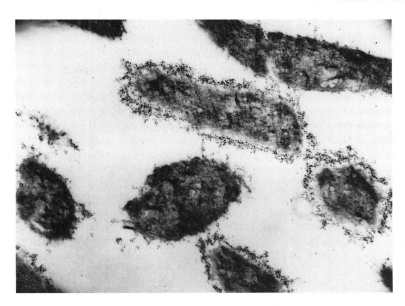

Fig. 1.2. Electron micrograph of an ultrathin section of Gram-negative bacteria after reaction with ferritin-labelled antibodies directed against surface O-antigens (*Shigella sonnei*, B. and M. Wagner)

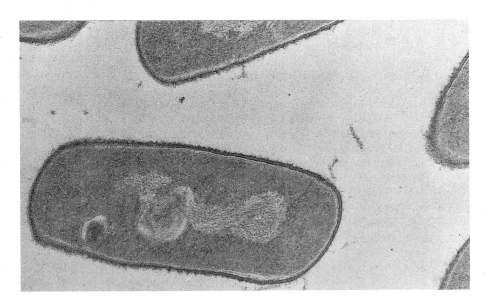

Fig. 1.3. Electron micrograph of an ultrathin section of a Gram-positive bacterium (*Bacillus subtilis*, F. Mach)

Gram-negative cells, although not in L-forms (penicillin-induced stable forms that have completely lost their cell wall) and, as mentioned above, in mycoplasma, in which it is totally missimg, and in archaea, in which it is replaced by other rigid structures. The peptidoglycan brings about the stability of the bacteria, and is therefore common to practically all bacterial species. However, in many Gram-negative bacteria it is not visible in electron micrographs. This is explained by the recently developed hypothesis that the peptidoglycan forms part of a periplasmic gel. The structural principle of the rigid layer is identical among Gram-positive and Gram-negative bacteria, though this layer is much thicker in the case of the former. In addition, the periplasmic space contains in solution oligosaccharides and proteins of different functions.

The outermost of the three layers is represented by the outer membrane. This structural unit is unique to the Gram-negative bacteria. Though in electron micrographs the outer membrane shows the typical trilaminar structure of common biological membranes, it is differently composed and much more stable than the latter. As the outermost layer it represents an effective penetration barrier.

In Fig. 1.1 no additional material is detectable outside the outer membrane. However, using special techniques, i.e. utilisation of ferritin-labelled specific antibodies, in many Enterobacteriaceae one can show fibrous material being firmly connected with the outer membrane and protruding into the surrounding medium (Fig. 1.2). From results of chemical analyses it can be concluded that in most cases this material consists of the polysaccharide chains of the lipopolysaccharides (LPS, see Sect. 2.2), a substance specific to the Gram-negative cell wall.

The cell wall of Gram-positive bacteria (Fig. 1.3) is distinctly different. It is much thicker than that of the Gram-negative bacteria and appears in many cases relatively structureless. However, it also contains a thick electron-dense rigid layer (20-80 nm compared with 10-15 nm in Gram-negative bacteria). In some Gram-positive bacteria, e.g. *Bacillus polymyxa*, distinct cell wall structures have been found electronmicroscopically. Likewise, using special techniques one can show that, for instance, staphylococcal cell walls also reveal a highly sophisticated architecture. Although an outer membrane is not detectable, the existence of a region comparable with the periplasmic space of the Gram-negative bacteria is postulated by some authors (Dijkstra and Keck 1996). Because of the missing outer membrane the Gram-positive cell wall is much more penetrable than the Gram-negative.

The Gram-positive cell walls, like the Gram-negative ones contain substantial amounts of polysaccharides, mainly the (lipo-)teichoic acids (see Sect. 4.1.1). Without previous specific treatment of the cells they cannot be detected in electron micrographs; however, they become visible after reacting with ferritin-labelled specific antisera.

Mycolata are classified as Gram-positive bacteria, although their staining is rather Gram-variable. Their cell wall structure is highly complex and unique (see Sect. 4.3) and, thus, differs from that of other Gram-positive species. A peptidoglycan layer is present, on top of which a thick layer composed of various lipoglycans, glyco(peptido)lipids, and proteins is present. Evidence for the presence of a

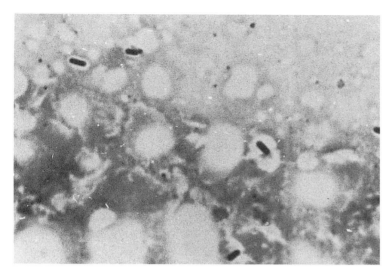

Fig. 1.4. Microscopical picture of a capsule-carrying bacterium (*Clostridium perfringens*, E. Dinger)

capsule as outer surface layer that is composed of proteins and polysaccharides has been obtained in recent years.

Finally, using specific techniques, one can frequently demonstrate additional layers above the cell wall, of which the longest known are the so-called capsules. An appropriate procedure for their demonstration is to treat bacteria spread out on a slide with Indian ink and then stain with fuchsin. Under the microscope the colourless capsules are readily identified between the reddish bacteria and the black background (Fig. 1.4).

Further examples are slime layers (slime walls), S-layers and sheaths. They are all not actual components of the cell walls, but situated at the cell surface, thus providing the bacteria with significant advantages in the struggle with the environment (see Chap. 6).

Finally, filamentous proteinaceous appendages of different length and diameter (flagellae, fimbriae, pili) must be mentioned. They are anchored with one end in the cell wall, while the other end extends into the surrounding medium.

The scientific investigation of the cell walls and the interpretation of the results are impaired by the fact that most investigations are carried out on laboratory cultures of bacteria. It is well known that bacteria grown in vitro are composed differently to those grown in natural habitats and are, therefore, not identical in behaviour and reactions. In addition, even under in vitro conditions, composition and structure of the bacteria are strongly dependent on the medium, the growth temperature, and the age of the culture. Thus, for instance, carbon-starved *E. coli* cells are more resistant to heat shock, osmotic challenge or oxidative stress than exponential-phase cells. On the other hand, under in vivo conditions the bacteria need not be stable in composition and structure. For instance, the different growth situations during an infection cycle require fast adaptations to new situations by the expressions of different factors.

Bibliography

Beveridge TJ, Graham LL (1991) Surface layers of bacteria. Microbiol Rev 55:684-705

Dijkstra AJ, Keck W (1996) Peptidoglycan as a barrier to transenvelope transport. J Bacteriol 178:5555-5562

Dmitriev BA, Ehlers S, Rietschel ET (1999) Layered murein revisited: a fundamentally new concept of bacterial cell wall structure, biogenesis and function. Med Microbiol Immunol 187:173-181

Gupta RS (1998) Protein phylogenies and signature sequences: a reappraisal of evolutionary relationship among archaebacteria, eubacteria, and eukaryotes. Microbiol Mol Biol Rev 62:1435-1491

Noll KM (1992) Archaebacteria (archae). In: Lederberg J (ed) Encyclopedia of microbiology, vol 1. Academic Press, San Diego, pp 149-160

Siegele DA, Kolter RR (1992) Life after log. J Bacteriol 174:345-348

2 The Outer Membrane of the Gram-Negative Bacteria and their Components

2.1 Lipids

2.1.1 Classification, Significance

Unlike proteins and polysaccharides, the lipids are not characterised by similarities in structure but by a commonly high hydrophobicity: they are therefore more or less insoluble in water, but exhibit good solubilities in hydrophobic organic solvents such as benzene, chloroform or ether. Thus, significant structural differences exist between the individual lipid groups.

Most lipids of bacterial origin represent derivatives of long-chain fatty acids, they are mainly located in the outer cell layers. Two groups of lipids are predominant in bacterial cell walls: *i.* neutral lipids, mostly esters of fatty acid with polyols, mainly glycerol, and *i.i.* complex lipids, containing a hydrophilic moiety (head group) besides the hydrophobic fatty acids and thus representing amphiphilic compounds. The latter, mainly the phospho- and glycolipids and lipoglycans (including lipopolysaccharides), are the most important for bacterial cell walls. Their amphiphilicity represents a crucial prerequisite for their ability to form membrane or membrane-like structures.

Lipids in bacteria perform the following functions:
- they represent important structural elements (penetration barriers),
- they are protective components,
- they form biologically active materials, which are essential in biosynthetic and transport processes such as, for instance, substrate activation, enzyme activation, organisation of multienzyme complexes or carrier function,
- they serve, though to a lesser extent, as energy sources or function as reservoirs.

The lipid composition of a given bacterium is not stable under all circumstances, but varies in part very drastically depending on cultivation conditions such as temperature, composition of the medium or age of the culture.

The presence of lipids proves to be essential for the bacteria. Lipid-rich groups of bacteria in general are more resistant to environmental influences such as action of detergents, antibiotics and others. Addition of sublethal doses of antibiotics often results in an increase in lipid production and in resistance caused thereby. Very often not only the amount of lipids but also their composition varies in a manner to modify the cell wall permeability.

In part, bacterial lipids differ drastically from lipids of plants and animals. Their unusual variety is astonishing; in a given bacterial species up to 20 different fatty acids and even more different kinds of lipids can be detected (see below). They all are imperative for bacteria in their struggle for survival. The variety of fatty acids and lipids implies a variety of biosynthetic pathways, among them reactions that are unknown in higher organisms.

2.1.2 Isolation and Analysis

Aspects of isolation and structural investigations of lipids are described in detail in biochemistry textbooks. Therefore only some information about the analysis of phospho- and glycolipids is given here.

Lipids can be extracted from the bacterial cell wall by (mixtures of) organic solvents, if necessary after mechanical disintegration. Because of the differences in solubility of the individual lipids in a given solvent, it is possible to achieve a preseparation by the appropriate choice of the extraction conditions. The extracts obtained can then be further separated by chromatography on hydrophobic sorbents.

To determine their qualitative and quantitative composition, the lipids are hydrolysed and the mixtures obtained are analysed using chromatographic and photometric methods. For the separation of less volatile components adsorption-chomatographic procedures are preferred, more volatile ones are in most cases separated by gas-liquid chromatography, if necessary after increasing volatility by derivatisation. Also, high-performance liquid chromatography (HPLC) on reversed-phases represents an important tool. Separated lipids can be identified by (coupled) mass spectrometry and NMR (nuclear magnetic resonance) spectroscopy (see Sect. 2.2.3.2). Using these methods, it is possible to determine the chain length of fatty acids or the position of double bounds. Information on the structure can also be obtained by defined chemical or enzymatic hydrolyses.

Bacteria are known to contain significant amounts of phospholipids. For their separation thin layer chromatographic systems (mainly high performance thin layer chromatography = HPTLC), in part two-dimensionally, have been developed.

The analysis of the oligosaccharide moiety of glycolipids and lipoglycans is described in detail for lipopolysaccharides (Sect. 2.2.3.2).

2.1.3 Chemical Structure

2.1.3.1 Fatty Acids

Fatty acids play a fundamental role in the bacterial cell wall. Since they are bactericidal in higher concentrations, only minute amounts of free acids are present in bacteria. However, as hydrophobic components they are bound in larger quantities in phospho- and glycolipids as well as in glycerides and waxes.

All fatty acids consist of a hydrocarbon chain and a terminal carboxylic group. The hydrophilic carboxylic group represents the far more reactive part of the molecule, through which the binding of fatty acids in lipids is mediated.

The hydrocarbon chain represents the hydrophobic moiety of the fatty acid molecule which in bacteria comprises mainly unbranched hydrocarbon chains, but branched chains were also found. The chains occur in most cases in the saturated or mono-unsaturated (both *cis* and *trans* linkage), while polyunsaturated fatty acids are usually less frequent in bacterial lipids. In addition, substitutes such as cyclopropane rings or hydroxyl groups have been identified.

The majority of chain lengths varies between 12 and 28 C-atoms, sporadically shorter or (even drastically) longer chains have been found. These disparities are not a whim of nature, but are necessary for the regulation of the lipid properties which is connected with bacterial regulation in general and contributes to the molecular geometry of the fatty acids.

Figure 2.1 shows molecular models of fatty acids with a saturated, a *trans*- and a *cis*-unsaturated, and a branched hydrocarbon chain. The models are idealised and exist, strictly speaking, exclusively at absolute zero. At temperatures above this, deviations from the ideal structure arise which increase with rising temperatures due to thermal movements. Because of their almost linear zig-zag structure, the saturated and the *trans*-unsaturated fatty acids tend to form arrangements in which their are packed parallel to each other. They thus resemble the structure of matches in a matchbox, the carboxylic groups forming the heads. On the other hand, hydrocarbon chains containing *cis*-double bonds, branches or cyclopropane rings appear much more bulky, disturb the parallel packing and result in a less dense structure. The effectiveness of these groups depends on their location in the chain: the nearer to the middle, the stronger the effect. Several such groups in one chain result in summation of the effects. These properties are very important for the formation of biological membranes. At ambient temperature, only about three quarters of the C-C-linkages are in the *trans*-conformation. The resulting deviations from the stretched form of the hydrocarbon chains, however, either are not too serious or affect larger areas as a whole. In total, the parallelity of the chains remains maintained to a high degree. The results of X-ray diffraction investigations performed on crystalline saturated fatty acids are in agreement with this assumption.

The extent of disturbances in the molecular arrangements of fatty acids manifests itself in their melting point. To illustrate these connections, the melting points of individual C_{18}-fatty acids are presented in Table 2.1.

Table 2.2 presents some examples of fatty acids which were found in bacteria. In general, even-numbered saturated and unsaturated fatty acids predominate in Gram-negative bacteria as well as uneven-numbered ones containing a cyclopropane ring. Gram-positive bacteria, on the other hand, mainly contain uneven-numbered branched fatty acids. The unsaturated fatty acids predominantly contain *cis*-double bounds.

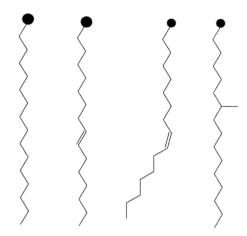

Fig. 2.1. Molecular model of fatty acids containing each a saturated, a *trans*- and a *cis*-unsaturated, and a methyl-branched chain (from left to right)

Table 2.1. Melting points of some C_{18} fatty acids

n-Octadecanoic acid (stearic acid)	+7
16-Methyl heptadecanoic acid	+6
cis-Δ^2-Octadecenoic acid	+5
trans-Δ^9-Octadecenoic acid (elaidic acid)	+4
10-Methyl heptadecanoic acid	+3
cis-Δ^9-Octadecenoic acid (oleic acid)	+1
cis, *cis*-$\Delta^{9,12}$-Octadecadienoic acid (α-linoleic acid)	-5
cis, *cis*, *cis*-$\Delta^{9,12,15}$-Octadecatrienoic acid (α-linolenic acid)	-1.

2.1.3.2 Waxes and Glycerides

In both classes of substances the fatty acids are esterified with an alcohol, in waxes mostly with a monovalent one and in glycerides with the trivalent glycerol. Depending on the number of fatty acids bound in the molecule, mono-, di- and triacylglycerols can be distinguished. The triacylglycerols are also called (neutral) fats.

In the natural glycerides, all three hydroxyl groups of the glycerol are esterified either with the same fatty acids (simple glycerides) or with different ones (mixed glycerides). In this way, their molecular geometry, and therefore their melting points, are varied almost continuously.

Table 2.2. Selection of fatty acids found in bacteria

Trivial name	C-Atoms	Structure
Lauric acid	12	$CH_3-(CH_2)_{10}-COOH$
Myristic acid	14	$CH_3-(CH_2)_{12}-COOH$
Palmitic acid	16	$CH_3-(CH_2)_{14}-COOH$
Stearic acid	18	$CH_3-(CH_2)_{16}-COOH$
Arachidic acid	20	$CH_3-(CH_2)_{18}-COOH$
Behenic acid	22	$CH_3-(CH_2)_{20}-COOH$
Lignoceric acid	24	$CH_3-(CH_2)_{22}-COOH$
Montanic acid	28	$CH_3-(CH_2)_{24}-COOH$
Sarcinic acid	14	$CH_3-CH-(CH_2)_{10}-COOH$ $\mid$ CH_3
	15	$CH_3-CH-(CH_2)_{11}-COOH$ $\mid$ CH_3
	15	$CH_3-CH_2-CH-(CH_2)_{10}-COOH$ $\mid$ CH_3
	16	$CH_3-CH-(CH_2)_{12}-COOH$ $\mid$ CH_3
	17	$CH_3-CH-(CH_2)_{13}-COOH$ $\mid$ CH_3
Tuberculostearic acid	19	$CH_3-(CH_2)_7-CH-(CH_2)_8-COOH$ $\mid$ CH_3
Lauroleic acid	12	$CH_3-CH_2-CH=CH-(CH_2)_7-COOH$
Myristoleic acid	14	$CH_3-(CH_2)_3-CH=CH-(CH_2)_7-COOH$
Palmitoleic acid	16	$CH_3-(CH_2)_5-CH=CH-(CH_2)_7-COOH$
Palmitvaccenic acid	16	$CH_3-(CH_2)_3-CH=CH-(CH_2)_9-COOH$
Oleic acid	18	$CH_3-(CH_2)_7-CH=CH-(CH_2)_7-COOH$
cis-Vaccenic acid	18	$CH_3-(CH_2)_5-CH=CH-(CH_2)_9-COOH$
	24	$CH_3-(CH_2)_{13}-CH=CH-(CH_2)_7-COOH$
C_{27}-Phthionic acid	27	$CH_3-(CH_2)_{17}-CH-CH_2-CH-CH=C-COOH$ $\quad\quad \mid \quad\quad \mid \quad\quad \mid$ $\quad\quad CH_3 \quad CH_3 \quad CH_3$
Corynomycolenic acid	32	$CH_3-(CH_2)_5-CH=CH-(CH_2)_7-CH-CH-COOH$ $\quad\quad\quad\quad\quad\quad\quad\quad \mid \quad \mid$ $\quad\quad\quad\quad\quad\quad\quad OH\ CH_2-(CH_2)_{12}-CH_3$

Table 2.2, continued.

	17	$CH_3-(CH_2)_5-CH-CH-(CH_2)_7-COOH$ \ / CH_2
Lactobacillic acid	19	$CH_3-(CH_2)_5-CH-CH-(CH_2)_9-COOH$ \ / CH_2
	19	$CH_3-(CH_2)_5-CH-CH-(CH_2)_8-CH-COOH$ \ / | CH_2 OH
Mevalonic acid	6	CH_3 | $CH_2-CH_2-C-CH_2-COOH$ | | OH OH
β-Hydroxymyristic acid	14	$CH_3-(CH_2)_{10}-CH-CH_2-COOH$ | OH

The importance of waxes and glycerides for the bacterial cell wall is far less than that of the phospho- and glycolipids. As neutral components, however, they can be deposited in the lipid double layer of the outer membrane or on the cell wall. This is often paralleled by a significant increase in the bacteria´s resistance. Likewise, the poly-β-hydroxybutyric acid is named wax, and in a range of bacteria (among others *Pseudomonas*) serves as a storage lipid.

2.1.3.3 Phospholipids and Glycolipids

Although these two classes of lipids possess different chemical compositions, they are both characterised by their amphiphily. Their lipophilic regions are formed by long-chain fatty acid residues being ester-linked to glycerol or amide-linked to the long-chain amino alcohol sphingosine. The sphingolipids formed in the latter case play a minor role in bacteria.

Phosphoric acid (phospholipids) or carbohydrates (glycolipids) can be found as the hydrophilic component. Because of the pronounced amphiphily of these lipid classes they are of great importance for bacterial cell walls. The antigen carrier lipid (ACL) mentioned in Sections 2.2.5.3 and 3.2.2.2, which is very important for the biosynthesis of LPS and the peptidoglycan, also belongs to this group. Roughly, it represents the phosphate of a C_{55}-polyisoprenoid alcohol (undecaprenol).

2.1.3.3.1 Glycerolphospholipids. These lipids are of crucial importance for the cell wall of Gram-negative bacteria in particular, as already indicated by the fact that about one quarter of the outer membrane consists of glycerol-phospholipids. Chemically, they derive from glycerol, in which one of the two primary hydroxyl groups is esterified with phosphoric acid. To overcome the ambiguity caused by the asymmetrical C-atom of the substituted glycerol (the substance can be named either D-glycerol-1-phosphate or L-glycerol-3-phosphate), the stereospecific numbering (= sn) signifies that the secondary hydroxyl group is located on the left side of the projection formula. The compound is thus sn-glycerol-3-phosphate and has the L-configuration.

Both the other hydroxyl groups also carry long-chain fatty acids in ester linkage. In the case of bacterial lipids the C-1 atom generally carries a saturated (in *E. coli* mainly palmitic acid) and the C-2 atom a monounsaturated, branched-chain or cyclopropane fatty acid. This is in accordance with the rule that the fatty acids with the highest melting point possess the highest affinity for position C-1. The resulting compound is named phosphatidic acid (structure see Fig. 2.3) which represents a key intermediate in glycerolphospholipid biosynthesis. The phosphoric acid in the phosphatidic acid usually is further substituted by other ester-linked components. Figure 2.2 lists the most important ones that occur in bacteria.

The name of the resulting compound is composed of phosphatidyl and the name of the ester-linked alcohol, e.g. phosphatidyl ethanolamine (formerly called kephalin) or phosphatidyl choline (formerly lecithin). Depending on the nature of the group bound to the phosphatidic acid, anionic and zwitterionic glycerolphospholipids are distinguished. The anionic glycerolphospholipids contain neutral groups (glycerol, inositol), the zwitterionic ones basic groups (ethanolamine and its derivatives). To the former belongs also cardiolipin (Fig. 2.3) in which two

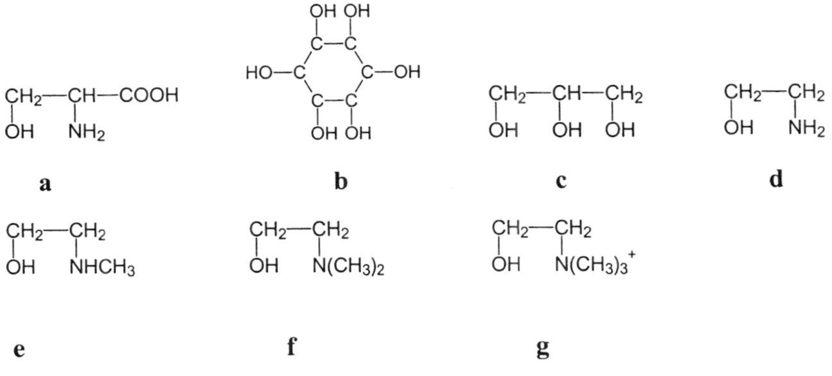

Fig. 2.2. Examples of substituents of phosphatidic acid in bacteria. *a*, serine; *b*, inositol; *c*, glycerol; *d*, ethanolamine; *e*, monomethylethanolamine; *f*, diemethylethanolamine; *g*, choline. The last three are less frequent

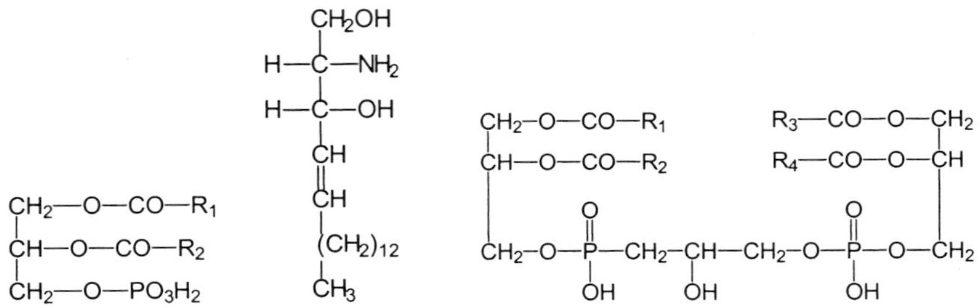

$$
\begin{array}{ccc}
& \mathrm{CH_2OH} & \\
& | & \\
& \mathrm{H-C-NH_2} & \\
& | & \\
& \mathrm{H-C-OH} & \\
& | & \\
\mathrm{CH_2-O-CO-R_1} & \mathrm{CH} & \\
| & || & \\
\mathrm{CH-O-CO-R_2} & \mathrm{CH} & \\
| & | & \\
\mathrm{CH_2-O-PO_3H_2} & \mathrm{(CH_2)_{12}} & \\
& | & \\
& \mathrm{CH_3} &
\end{array}
$$

$$
\begin{array}{cc}
\mathrm{CH_2-O-CO-R_1} & \mathrm{R_3-CO-O-CH_2} \\
\mathrm{CH-O-CO-R_2} & \mathrm{R_4-CO-O-CH} \\
\underset{\mathrm{OH}}{\mathrm{CH_2-O-\overset{O}{\overset{||}{P}}-CH_2-\underset{OH}{CH}-CH_2-O-}} & \underset{\mathrm{OH}}{\overset{O}{\overset{||}{P}}-O-CH_2}
\end{array}
$$

Fig. 2.3. Structure of phosphatidic acid, sphingosin and cardiolipin (from left to right). R_1, R_2, R_3, R_4 Long-chain fatty acids

molecules phosphatidic acid are bound to each other via a glycerol molecule. This substance plays a role, for example, in the biosynthesis of the membrane-derived oligosaccharides of the periplasmatic space (Sect. 3.1) as well as being diglyceride donor for the Braun´s lipoprotein (Sect. 2.3.2.4.1). The cardiolipin content of bacterial membranes increases in stress situations.

The Gram-negative cell wall contains both anionic and zwitterionic glycerolphospholipids. Under all conditions examined, the ratio of both groups is highly constant, whereas variation occurs within the groups. For example, the cardiolipin content in the cell wall of *E. coli* can be reduced to 1/15 and be replaced by phosphoglycerol without leading to phenotypic modifications.

Instead of a fatty acid ester, glycerolphospholipids may contain a long aliphatic chain bound as enolether, i. e. via a vinyl group. These lipids are called plasmalogens. In contrast to ester linkages, such linkages benefit from stability to (enzymatic) hydrolysis.

As integral components of the outer membrane, phospholipids are essential for the bacterial cell wall. They crucially codetermine its structure and flexibility and are essential in transport processes as well as for the regulation of enzymes. Their presence is necessary for the resistance and adaptation of bacteria to the particular environment. Variation in the composition of head group and fatty acid residues results in the formation of hundreds of different lipid species adapted to their specific functions. About 25% of the enterobacterial cell wall consists of phospholipids, of which phosphatidyl ethanolamine makes up about 75%, phosphatidyl glycerol about 20% and cardiolipin 1-5% (Raetz 1986).

2.1.3.3.2 Glycolipids. Glycolipids are glycosyl derivatives of lipids such as acylglycerols, ceramides and prenols. Thus, they represent substances in which a mono- or an oligosaccharide is substituted either directly by long-chain fatty acids or long-chain fatty alcohols or by glycerol that, in turn, carries fatty acids. In the last case, glycolipids are called glyceroglycolipids. Bacterial glycolipids can be sub-divided into two groups: the glycosyldiacylglycerols (glycosyldiglyceride) and the acylated sugar derivatives.

Glycosyldiacylglycerols consist of a glycerol moiety which carries two ester-linked long-chain fatty acids and glycosidically substitutes a sugar. Thus, structurally they resemble phosphatidic acid derivatives. Mannose, glucose, galactose and rhamnose, but also glucuronic acid and galactose sulfate, have been found as sugar components, present as monosaccharides or as di- to tetrasaccharides comprising one or two different sugars. The linkage of the carbohydrate moiety is always located at the sn-3-hydroxyl group of the glycerol. As in the case of phospholipids, the fatty acids show an asymmetrical positional distribution but, in contrast to them, the chain length of the fatty acids is more important than chain branches or the number of double bonds. Glycosyldiacylglycerols have been found in a wide variety of bacteria; they in some cases seem to be of a certain taxonomic importance. They are obviously essential for the supply of the bacteria with carbohydrates.

In the case of acylated sugar derivatives, the acylation of one, several or all hydroxyl groups of a saccharide with either long-chain or also short-chain (e.g. acetic acid) acyl residues may occur. Acylated sugar derivatives are widespread in bacteria. They may be intermediate stages during biosynthesis of phosphoglycolipids.

2.1.3.3.3 Sphingolipids.

Sphingolipids derive from sphingosine, an unsaturated amino alcohol with a long aliphatic side chain (Fig. 2.3), or from dihydrosphingosine in which the double bond of sphingosine is hydrated. Long-chain fatty acids are linked to the amino group.

The primary alcohol group of sphingosin is either esterified with a phosphoric acid molecule carrying a second component, as in the case of phosphoglycerides, or it is glycosidically linked to a mono- or an oligosaccharide. Compounds of the first kind are called phosphosphingolipids, those of the second glycosphingolipids, and comprise cerebrosides and gangliosides. Up to now, both phosphosphingolipids and glycosphingolipids have only rarely been detected in bacteria. Only the ceramides are rather significant. Here, the sphingosine is esterified with long-chain fatty acids (mostly C_{16}, but also C_{18}, C_{22} or C_{24}) and other polar head groups are not present.

2.1.3.3.4 Hopanoids.

Hopanoids (Fig. 2.4) are pentacyclic triterpenoids and represent a new class of lipids detected rather recently. They are sterol analogues with regard to both structure and function. Both molecule classes possess a rigid and hydrophobic polycyclic skeleton and a polar headgroup. Their molecular similarities allow both the insertion into (phospholipid) bilayers and van der Waals interactions with acyl side chains. Numerous structural variations of hopanoids have been detected in bacteria which occur mainly at the C-35 functional group and comprise, e.g. its substitution by amino acids or sugars and their derivatives, or its methylation.

2.1.4 Spatial Structure and Physical Properties

Amphiphilic lipids consist of a hydrophilic (polar) head group and a hydrophobic tail. This arrangement represents a prerequisite for the membrane-forming properties especially of phospho and glycolipids. When dispersed in aqueous systems, they assemble in such a manner that the hydrophobic regions of the molecules are hidden and the hydrophilic ends are turned towards the water. Such an organization is lower in energy than any other possible one and therefore stable. However, several spatial arrangements are possible. They depend on the ratio of the areas occupied by the hydrophobic acyl chains and by the polar head groups. In the most simple arrangements, namely, the vesicles (Fig. 2.5) the polar head points outwards, whereas the apolar tails are directed inwards and thus mutually create a hydrophobic milieu. Such an arrangement, for geometrical reasons, needs bulky head groups.

At phase interfaces, the molecules may be orientated in such a manner that the hydrophilic heads dip into the aqueous phase while the hydrophobic tails protrude into the air. The molecules thus create a monomolecular film (Fig. 2.5).

The most preferred arrangement, however, contains molecules in the form of a double layer by positioning their hydrophobic tails against each other. Spherically shaped, they are named liposomes, whereas in large two-dimensional arrangements (Fig. 2.5) they represent the simplest model of biological membranes. In the presence of lipids with relatively small head groups in the latter structure non-bilayer tubular regions termed inverted hexagonal (H_{II}) phases occur, which may be essential for the biological membrane functions.

Basically, two different arrangements of lipids may occur in biological membranes, i.e. a solid, often hexagonally packed gel state and the fluid-crystalline (liquid) state. In the crystalline state, two kinds of packing are possible. In the first and tighter one strong specific chain-chain interactions exist. In the second, the chains are more loosely packed and the specific chain-chain interactions are partially lost due to partial rotation of the chain or of its -CH_2- groups.

In cells, lipid aggregates are not typically crystalline, but rather liquid crystal-like. This state combines the properties of both liquids and crystals. Because it combines order and fluidity, it is particularly suited to cellular functions, especially to the formation of membranes. Such liquid crystals are classified depending on their degree of long-range order. When having three-dimensional (3D) long-range order they are packed in a cubic lattice (see Fig. 2.21) and could be considered to be fully crystalline. However, their chain packing is liquid. Two-dimensional (2D) liquid crystals are ordered in rectangular or hexagonal lattices. The rectangularlattice is also called the ribbon-like structure (in Fig. 2.21 lamellar), the hexagonal states are composed of long rods or cylinders packed in a hexagonal array. Finally, in the 1D state the acyl chains may be either quasicrystalline (gel state) or liquid.

The liquid states of lipids show the general properties of liquids, e.g. fluidity and rapid molecular motions.

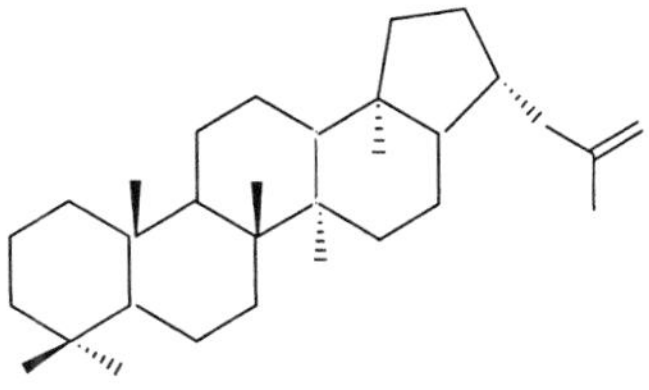

Hopene

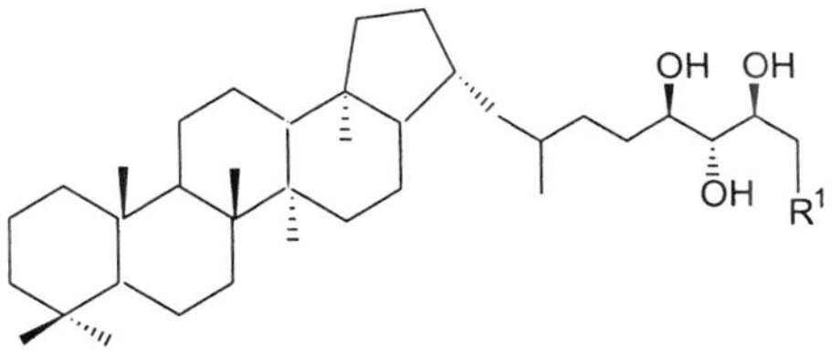

R^1 = OH, Bacteriohopanetetrol; R^1 = NH$_2$, Aminohopanetriol

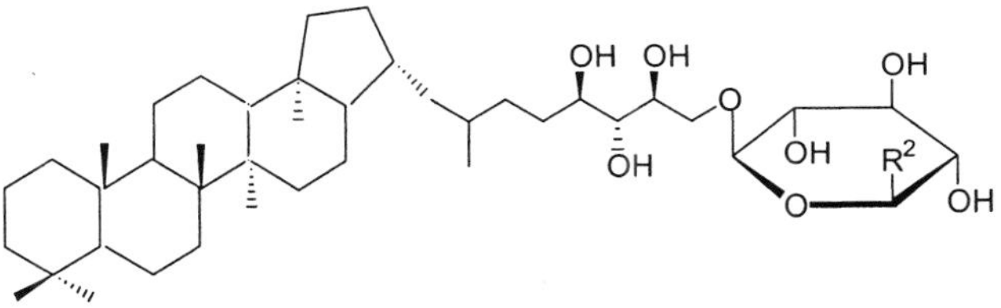

Hopane glycolipids

Fig. 2.4. Examples of non-elongated and elongated hopanoids and of hopane glycolipids. R^2 is either –CH$_2$OH (in *Rhodospirillum acidophila*) or –CH$_2$NH$_2$ (in the cyanobacterium *Synechocystis*) or –COOH (in *Rhodospirillum rubrum*)

A phase transition occurs between both states at a definite temperature T_t. Depending on the structure of the double layer, the phase transition can be clear-cut or comprise a larger temperature range, Δt. In the latter case, the midpoint of the transition is defined as T_t. Both T_t and Δt depend on the fatty acid composition of the membrane-forming lipids since

1. T_t is higher if the membrane-forming lipids exclusively contain saturated fatty acids.
2. T_t Rises with increasing chain-length of the lipid fatty acids.

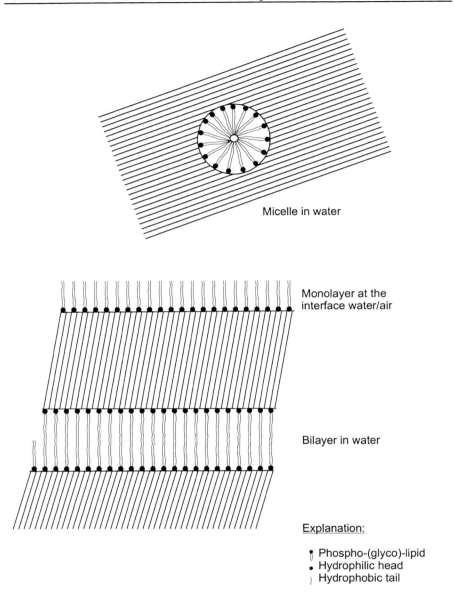

Micelle in water

Monolayer at the
interface water/air

Bilayer in water

Explanation:

Phospho-(glyco)-lipid
Hydrophilic head
Hydrophobic tail

Fig. 2.5. Schematic representation of a micelle, a mono- and a bilayer, each formed by amphiphilic molecules

3. *cis*-Unsaturated fatty acids reduce T_t, the effect increases with the number of double bonds.
4. *trans*-Monoenoic fatty acids reduce T_t far less than *cis*-monoenoic fatty acids.
5. Branched chain and cyclopropane fatty acids reduce T_t.

6. If the lipids contain only one single type of fatty acids, Δt is very small, increasing the heterogeneity of chain length and structure enlarges Δt.

All this is connected with the molecular geometry of the membrane lipids. According to the molecular models shown in Fig. 2.1, one can easily imagine that two phospholipids containing only saturated fatty acids aggregate more close and in more orderly fashion than phospholipids, which contain branched or unsaturated fatty acids. For this reason, the energy necessary for the disturbance of the ordered solid state is higher in the case of phospholipids of the first type, i.e. the transition from the solid to the liquid state occurs at higher temperatures.

Of course, this statement does not mean that the saturated fatty acids in the liquid interior of the membrane contain only stretched hydrocarbon chains. On the contrary, during the transition from the solid to the liquid state, distinct deviations from the linear form occurs (transition from stretched to crumbled geometry). However, only such deviations appear in which the chains in principle remain in parallel, termed an ordered double layer with unordered chains. At physiological temperatures (about 30 °C), the chains rotate mainly around their longitudinal axes, with a frequency of about 10^{10} Hz. Besides this, the lipids diffuse within the membrane plain whereas diffusion vertical to this plain (flip-flop) only rarely occurs, at least in the case of amphiphilic lipids.

The main regulating pathway for membrane liquidity adjusted by the melting point of the incorporated fatty acids consists of the incorporation of unsaturated fatty acids. Up to now, it remains unclear why bacteria also incorporate branched-chain and cyclopropane fatty acids, although the regulation of the melting point could be sufficiently maintained via the main pathway. It is supposed that phospholipids containing branched-chain and cyclopropane fatty acids are more stable against degradation and that their incorporation represents a protection or stabilisation of the membranes. This is supported by the fact that the ratio of cyclopropane to unsaturated fatty acids increases drastically during the transition of a bacterial culture to the stationary growth phase.

If a membrane consists of a mixture of phospholipids, if their homogenisation is perfect in all regions, and if the interior of the membrane is in the solid state, the cis-unsaturated fatty acids are the first escape from the ordered state at a slow rise in temperature. In comparison, the saturated fatty acids are still in the ordered state. Macroscopically, the membrane becomes softer without yet turning into the fluid state. As temperatures continue to rise, the trans-unsaturated and finally the saturated fatty acids leave the ordered state, i.e. the fluidity increases more and more until at least the whole membrane interior is in the liquid state. Therefore the starting and the end point of the phase transition as well as its range (Δt) are determined by the fatty acid composition of the lipids.

T_t also depends on the nature of the polar groups: it is about 20 °C higher in the case of dipalmitoyl phosphatidylethanolamine compared to the homologous choline lipid; these relations, however, are less distinct than in the case of the influence of fatty acids.

If there are proteins with larger hydrophobic surface regions (see Sect. 2.3.1) in the double layer, strong lipid-protein interactions take place. The fatty acids adjacent to the proteins adapt to the shape of the latter (Fig. 2.6), resulting in a higher degree of disorder in such regions.

2.1.5 Biosynthesis of the Lipids

2.1.5.1 Biosynthesis of Fatty Acids

At least in *E. coli* the biosynthesis of the fatty acids takes place by means of a type II fatty acid synthase system. Each enzyme necessary for the individual synthesis stages is a separate protein which can be easily separated from the other enzymes and then purified. All intermediate stages are bound to an acyl-carrier protein (ACP) which is necessary to pass on the growing fatty acid chain from one enzyme to the next. ACP (molecular mass 8.86 kDa) is a protein containing one single thiol group; it is found in relatively large quantities (about 0.25% of the total soluble protein content).

The biosynthesis starts with the reaction of acetyl-CoA and malonyl-CoA (synthesised from acetyl-CoA and carboxy-biotin, the active form of CO_2) with ACP, which results in binding both to the thiol group of ACP as the prosthetic group.

After that, according to

$$CH_3\text{-}CO\text{-}S\text{-}ACP + HOOC\text{-}CH_2\text{-}CO\text{-}S\text{-}ACP \rightarrow CH_3\text{-}CO\text{-}CH_2\text{-}CO\text{-}S\text{-}ACP +$$
$$HS\text{-}ACP + CO_2$$

acetoacetyl-S-ACP is formed.

The ketonic CO-group of acetoacetyl-S-ACP is transformed into $CH_3\text{-}CH_2\text{-}CH_2\text{-}CO\text{-}S\text{-}ACP$ (butyryl-S-ACP) by enzymatic reduction with NADPH, enzymatic elimination of water to remove oxygen and repeated enzymatic reduction. Like acetyl-S-ACP, butyryl-S-ACP serves as an acceptor for the renewed condensation with malonyl-CoA, the further reaction cycle is repeated, now forming caproyl-S-ACP. Except for the irreversible condensation reaction, all stages of the cycle are reversible.

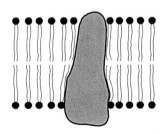

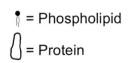

= Phospholipid

= Protein

Fig. 2.6. Spatial adaptation of adjacent fatty acid chains to the shape of proteins embedded in a membrane

The metabolic pathway of higher organisms for unsaturated fatty acids, i.e. the enzymatic dehydration of saturated CoA-fatty acids in the presence of an H-acceptor, is only widespread in obligate aerobic bacteria. In the majority of bacteria mainly an elongation of already existing unsaturated fatty acids takes place. Two important key substances are the *cis*-Δ3- and *trans*-Δ2-decenoic acids, the latter of which is formed from 3-hydroxydecanoic acid by dehydration to the *trans*-Δ2 compound and its isomerization. The chains of both compounds are elongated by the introduction of C_2-building blocks. In bacteria, however, also monoenoic fatty acids with shorter chains (Δ2- and Δ3-octenoic acids) have been found. Since 3-hydroxydecanoic acid represents the starting molecule, the mono-unsaturated fatty acids of bacteria have their double bond mainly between C-7 and C-8 taken from the chain´s CH_3-terminus (*cis*-vaccenate series), whereas in those found in higher organisms it is present between C-9 and C-10.

Unsaturated fatty acids represent the starting substances for cycloprane fatty acids. To these, a C_1 fragment is added, which mostly originates from the CH_3 group of an S-adenylated methionine. To make this reaction possible, the fatty acids in question must already be bound in a phospholipid.

Branched-chain fatty acids are also often obtained by chain elongation. The starting substance is e.g. isovaleric acid derived from leucine. For the biosynthesis of tuberculostearic acid in the case of the Gram-positive mycobacteria (Table 2.2), however, the reaction starts from stearic acid, which is dehydrogenated to oleic acid and subsequently transferred by means of the CH_3-group of methionine into the desired compound. Hydroxy fatty acids are taken directly from fatty acid biosynthesis (D-stereoisomers) or from fatty acid degradation (L-stereoisomers).

The regulation of the chain length depends on the substrate specificity of the condensing enzymes, on the availability of malonyl-CoA as well as the use of the formed acyl-ACP by (glycerolphosphate-)acyltransferase, and on the feedback inhibition by long chain acyl-CoA and -ACP.

2.1.5.2 Biosynthesis of Waxes and Glycerides

The biosynthesis of waxes is carried out by a reaction of a fatty alcohol, formed by reduction of a fatty acid, with an activated fatty acid.

The formation of triacyl glycerides mostly takes place via phosphatidic acid (see below). It is dephosphorylated by means of a phosphatase and subsequently transformed to the triester by a molecule of CoA-activated fatty acid. This pathway is not very common in bacteria.

2.1.5.3 Biosynthesis of Phospholipids and Glycolipids

2.1.5.3.1 Biosynthesis of Glycerolphospholipids. The starting-material for this biosynthesis is L-glycerol-3-phosphate. This compound represents an important key substance in the general metabolism. It is formed either from dihydroxyacetone phosphate, an important product of the carbohydrate metabolism, by reduction with NADPH or to a lesser extent by phosphorylation of glycerol with ATP. For the formation of phosphatidic acid, substitution takes place of both free OH groups of glycerol by fatty acids (saturated ones to C-1, unsaturated ones to C-2) which have been activated by binding to CoA or to ACP. The order of attachment of both fatty acids demonstrated in Fig. 2.7 can also be reversed.

Both reactions are catalysed by a transferase system specific to the individual fatty acid. It is known, however, that the ratio of saturated to unsaturated fatty acids in the cell wall depends on the cultivation temperature of the bacteria. The contradiction is only apparent. Investigations demonstrated that both transferases are specific only in the case of a sufficient quantity of the respective fatty acid being available, otherwise they become unspecific. This fact is of great importance for the maintenance of the membrane fluidity at different temperatures.

The formed phosphatidic acid is activated with CTP under formation of CDP-diacylglycerol which then can further react to the individual glycerolphospholipids according to the scheme shown in Fig. 2.8. The molar ratio between the different glycerolphospholipids is regulated by feedback inhibition of the individual biosynthetic enzymes.

2.1.5.3.2 Biosynthesis of Glycolipids. The starting substance for the biosynthesis of glycosyldiacylglycerol is 1,2-diacylglycerol which is formed by enzymatic dephosphorylation of phosphatidic acid. The incorporation of the sugars occurs gradually starting with their dinucleotide phosphates. Figure 2.9 shows the biosynthesis of 1-[α-D-Manp-(1→3)-α-D-Manp-(1→)]-2,3-diacylglycerol.

The biosynthesis of acylated sugar derivatives is exemplified for a rhamnolipid of *Pseudomonas aeruginosa*. Two molecules of D-3-hydroxy-decanoyl-CoA react under formation of D-3-hydroxy-decanoyl-D-3-hydroxydecanoyl-CoA. Its free hydroxyl group is then substituted with L-rhamnose or α-L-Rhap-(1→2)-α-L-Rhap (as TDP compound each).

2.1.5.3.3 Biosynthesis of Sphingolipids. As mentioned above, only the ceramides play a role in bacteria. The first stage of their biosynthesis consists of the reaction of palmitoyl-CoA and serine under formation of 3-keto-sphinganine (Fig. 2.10). This compound is dehydrated to sphinganine and transformed to ceramide by one molecule of fatty acid-CoA. The incorporation of polar head groups occurs as

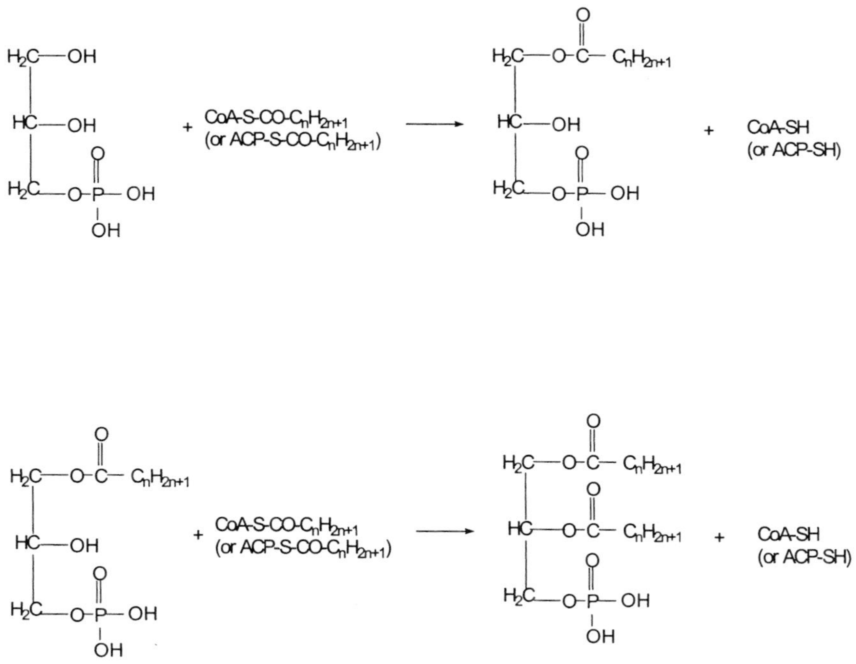

H$_2$C—OH
HC—OH   + CoA-S-CO-C$_n$H$_{2n+1}$
H$_2$C—O-P—OH   (or ACP-S-CO-C$_n$H$_{2n+1}$)
        OH

$\longrightarrow$

H$_2$C—O-C—C$_n$H$_{2n+1}$
           O
HC—OH   + CoA-SH
H$_2$C—O-P—OH   (or ACP-SH)
        OH

H$_2$C—O-C—C$_n$H$_{2n+1}$
HC—OH   + CoA-S-CO-C$_n$H$_{2n+1}$
H$_2$C—O-P—OH   (or ACP-S-CO-C$_n$H$_{2n+1}$)
        OH

$\longrightarrow$

H$_2$C—O-C—C$_n$H$_{2n+1}$
           O
HC—O-C—C$_n$H$_{2n+1}$ + CoA-SH
H$_2$C—O-P—OH   (or ACP-SH)
        OH

Fig. 2.7. Schematic pathway of phosphatidic acid biosynthesis

in case of other phospho- and glycolipids; however, this has not yet been observed in bacteria.

2.1.5.3.4 Biosynthesis of ACL.

The antigen carrier lipid (ACL) plays an important part in the biosyntheses of both lipopolysaccharides and peptidoglycan; it is described in more detail there (see Sects. 2.2.5.3 and 3.2.2.2). Its biosynthesis proceeds distinctly differently from that of the lipids described so far; however, acetyl-CoA represents here as well the starting compound (Fig. 2.11.). It reacts with the already mentioned acetoacetyl-CoA to β-hydroxy-β-methylglutaryl-CoA. By means of NADH, the carboxyl group bound to coenzyme A is reduced in two stages, yielding mevalonic acid, which then again in two stages is transformed to 5-pyrophosphomevalonic acid, consuming two molecules of ATP. A third molecule of ATP is required to furnish 3-phospho-5-pyrophosphomevalonic acid. This compound is very unstable; it splits off phosphoric acid and is subsequently decarboxylated to 3-isopentenyl-pyrophosphoric acid. By allyl shift (migration of the double bond) dimethylallyl pyrophosphate is then formed. This compound

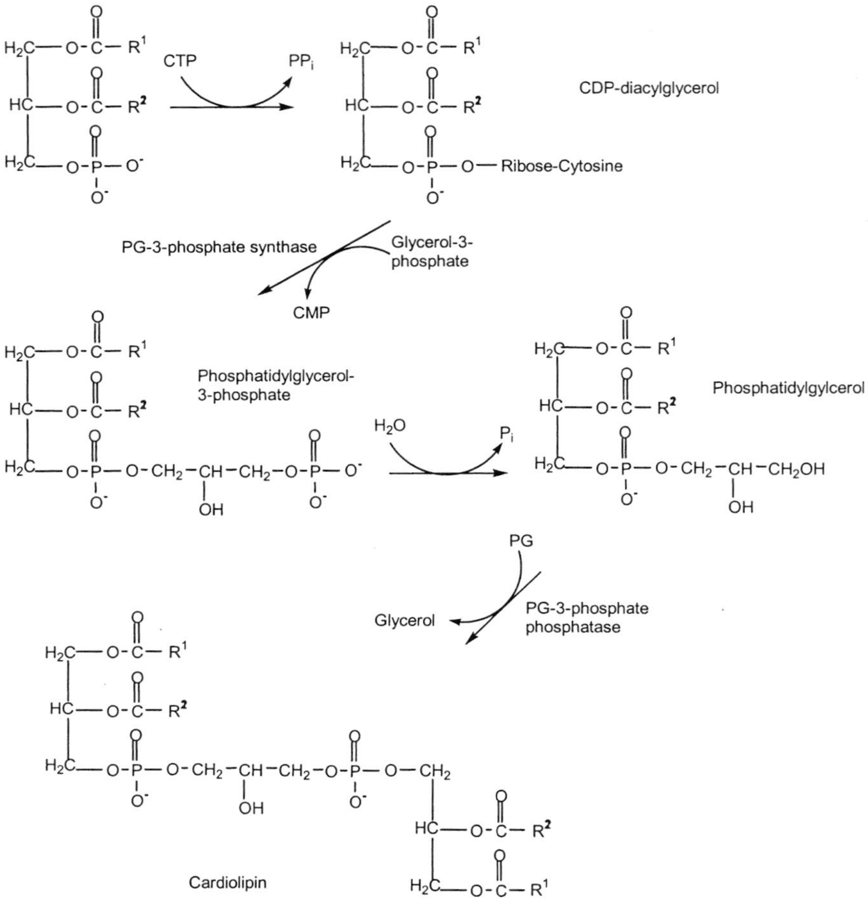

Fig. 2.8. Schematic pathways of cardiolipin as an example for phospholipid biosynthesis in bacteria

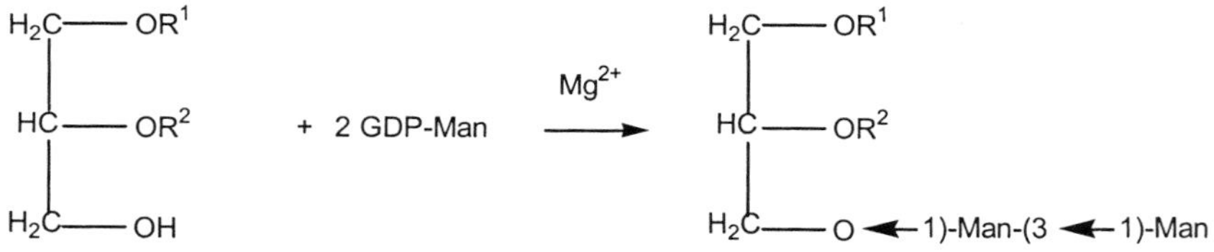

Fig. 2.9. Biosynthesis of a glycolipid

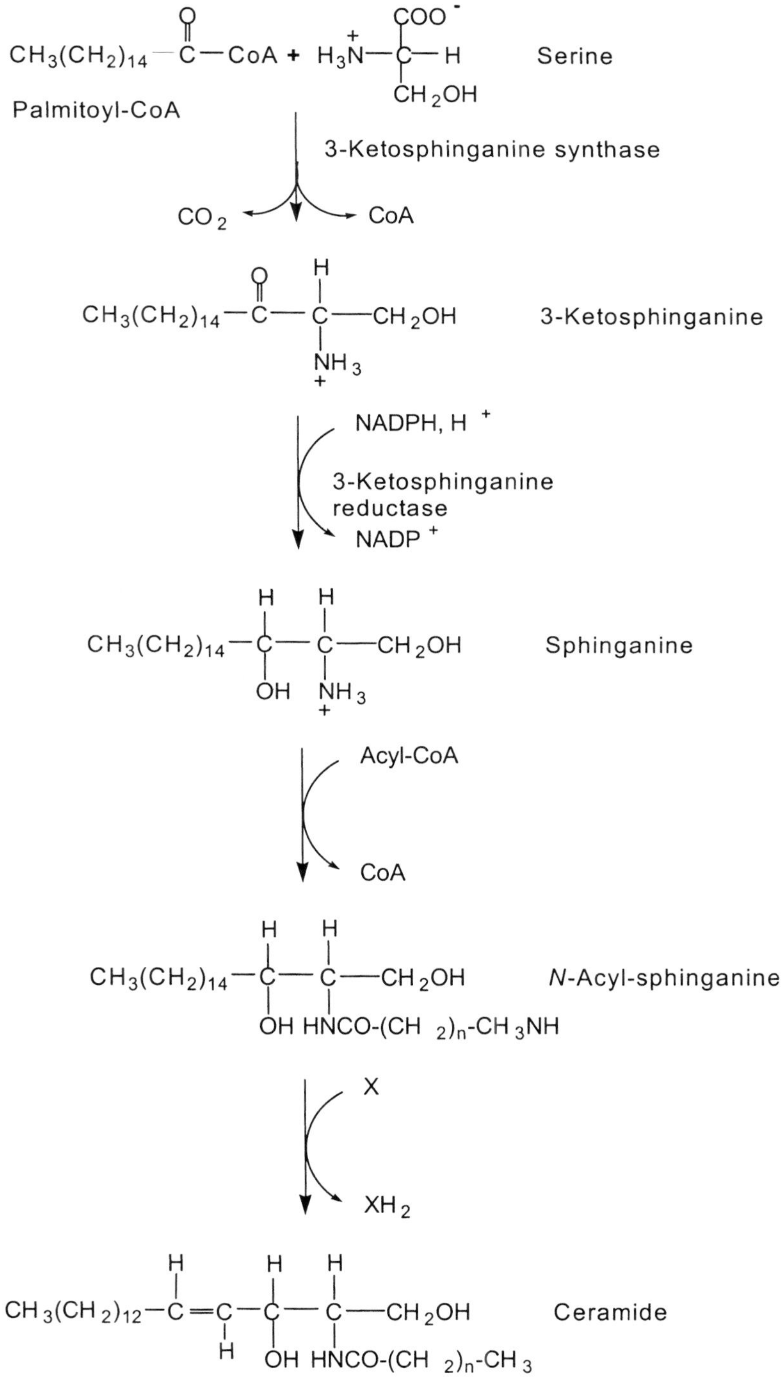

$$CH_3(CH_2)_{14}-\overset{O}{\underset{}{C}}-CoA + H_3\overset{+}{N}-\underset{CH_2OH}{\overset{COO^-}{\underset{|}{C}}}-H \qquad \text{Serine}$$

Palmitoyl-CoA

3-Ketosphinganine synthase

$CO_2 \longleftarrow \qquad \longrightarrow CoA$

$$CH_3(CH_2)_{14}-\overset{O}{\underset{}{C}}-\underset{\overset{NH_3}{+}}{\overset{H}{\underset{|}{C}}}-CH_2OH \qquad \text{3-Ketosphinganine}$$

NADPH, H$^+$

3-Ketosphinganine reductase

NADP$^+$

$$CH_3(CH_2)_{14}-\underset{OH}{\overset{H}{\underset{|}{C}}}-\underset{\overset{NH_3}{+}}{\overset{H}{\underset{|}{C}}}-CH_2OH \qquad \text{Sphinganine}$$

Acyl-CoA

CoA

$$CH_3(CH_2)_{14}-\underset{OH}{\overset{H}{\underset{|}{C}}}-\underset{HNCO-(CH_2)_n-CH_3}{\overset{H}{\underset{|}{C}}}-CH_2OH \qquad \textit{N}\text{-Acyl-sphinganine}$$

X

XH$_2$

$$CH_3(CH_2)_{12}-\underset{H}{\overset{H}{\underset{|}{C}}}=C-\underset{OH}{\overset{H}{\underset{|}{C}}}-\underset{HNCO-(CH_2)_n-CH_3}{\overset{H}{\underset{|}{C}}}-CH_2OH \qquad \text{Ceramide}$$

Fig. 2.10. Biosynthesis of ceramides

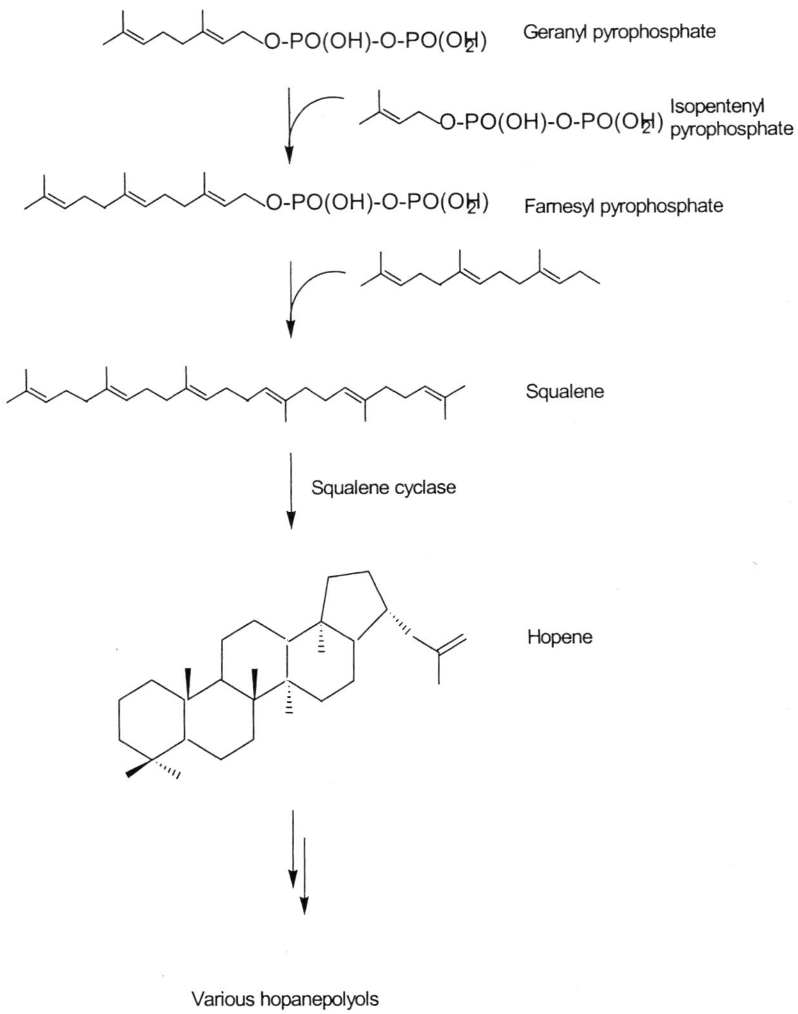

Fig. 2.13. Biosynthesis of hopanoids. For the biosynthesis of geranyl pyrophosphate see Fig. 2.11

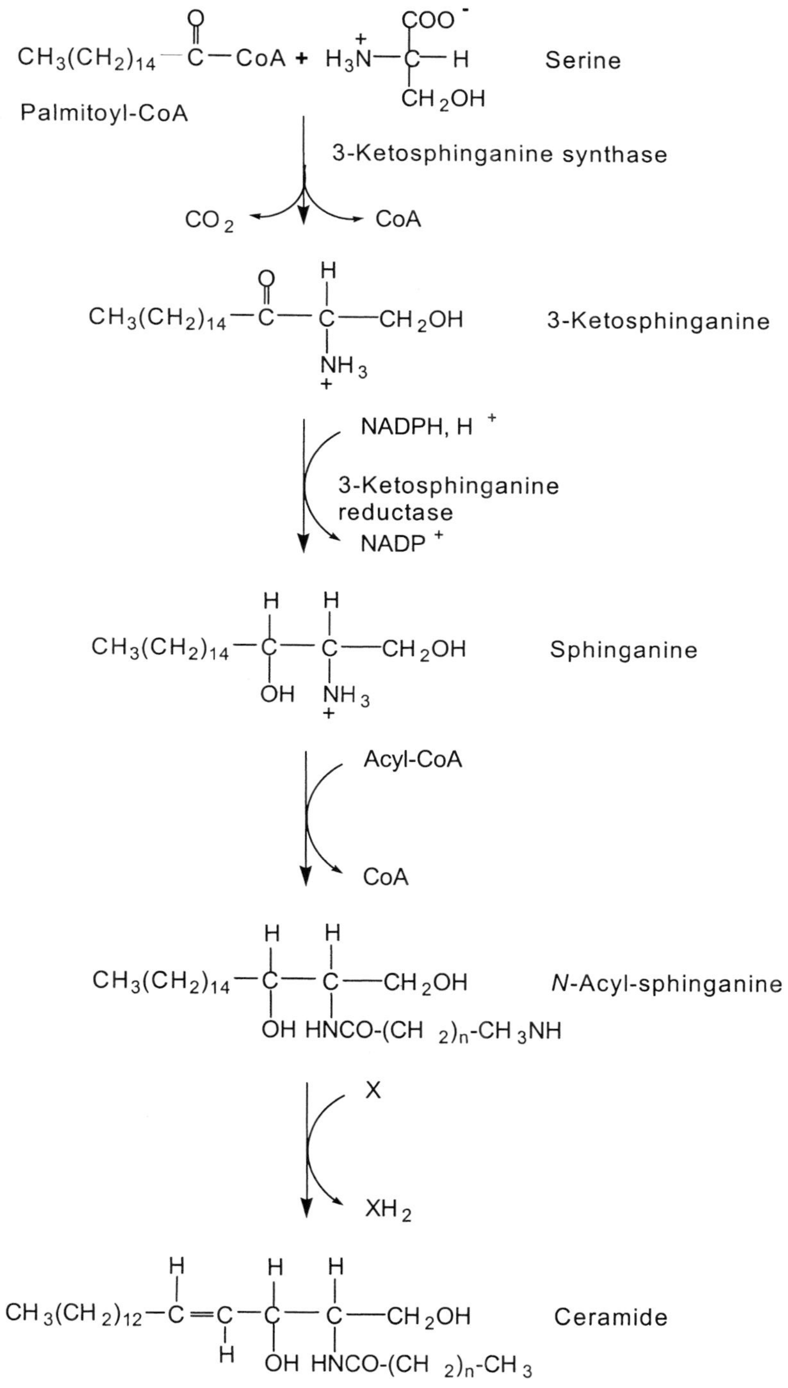

Fig. 2.10. Biosynthesis of ceramides

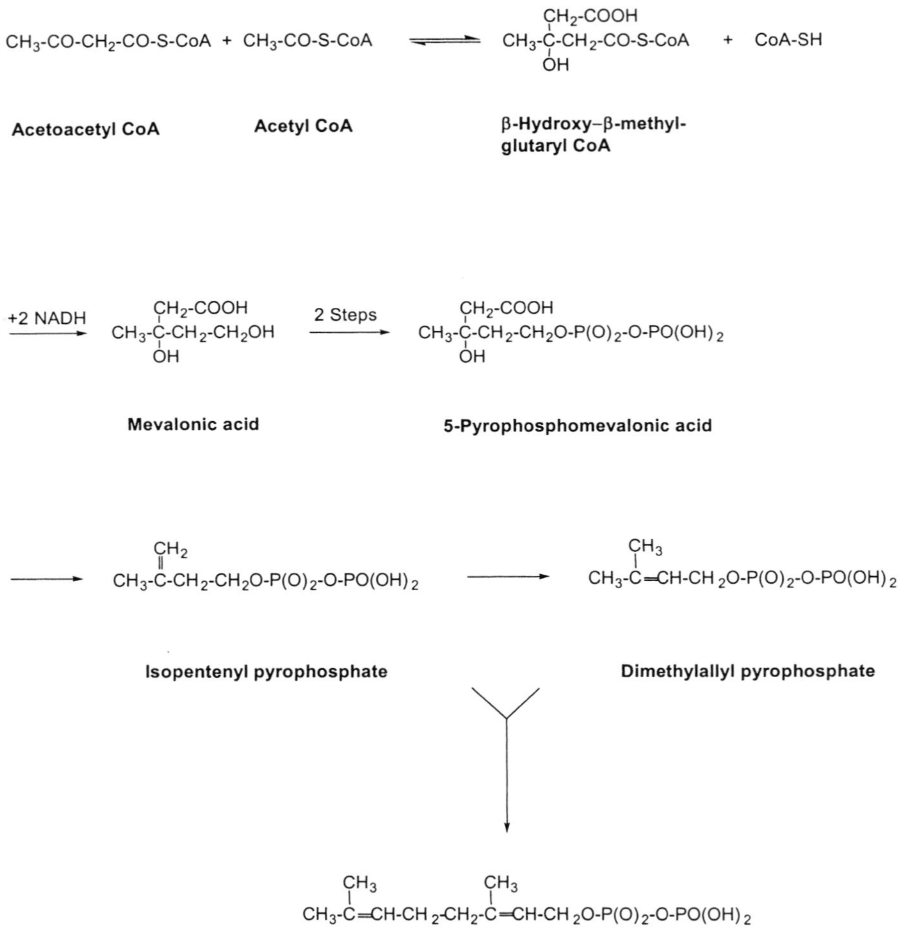

$$CH_3\text{-}CO\text{-}CH_2\text{-}CO\text{-}S\text{-}CoA \ + \ CH_3\text{-}CO\text{-}S\text{-}CoA \ \rightleftharpoons \ CH_3\text{-}\underset{\underset{OH}{|}}{\overset{\overset{CH_2\text{-}COOH}{|}}{C}}\text{-}CH_2\text{-}CO\text{-}S\text{-}CoA \ + \ CoA\text{-}SH$$

Acetoacetyl CoA   **Acetyl CoA**   **β-Hydroxy–β-methyl-glutaryl CoA**

$$\xrightarrow{\text{+2 NADH}} CH_3\text{-}\underset{\underset{OH}{|}}{\overset{\overset{CH_2\text{-}COOH}{|}}{C}}\text{-}CH_2\text{-}CH_2OH \xrightarrow{\text{2 Steps}} CH_3\text{-}\underset{\underset{OH}{|}}{\overset{\overset{CH_2\text{-}COOH}{|}}{C}}\text{-}CH_2\text{-}CH_2O\text{-}P(O)_2\text{-}O\text{-}PO(OH)_2$$

Mevalonic acid   **5-Pyrophosphomevalonic acid**

$$\longrightarrow CH_3\text{-}\underset{}{\overset{\overset{CH_2}{\|}}{C}}\text{-}CH_2\text{-}CH_2O\text{-}P(O)_2\text{-}O\text{-}PO(OH)_2 \longrightarrow CH_3\text{-}\underset{}{\overset{\overset{CH_3}{|}}{C}}\text{=}CH\text{-}CH_2O\text{-}P(O)_2\text{-}O\text{-}PO(OH)_2$$

Isopentenyl pyrophosphate   **Dimethylallyl pyrophosphate**

$$CH_3\text{-}\underset{}{\overset{\overset{CH_3}{|}}{C}}\text{=}CH\text{-}CH_2\text{-}CH_2\text{-}\underset{}{\overset{\overset{CH_3}{|}}{C}}\text{=}CH\text{-}CH_2O\text{-}P(O)_2\text{-}O\text{-}PO(OH)_2$$

Fig. 2.11. First steps of ACL biosynthesis

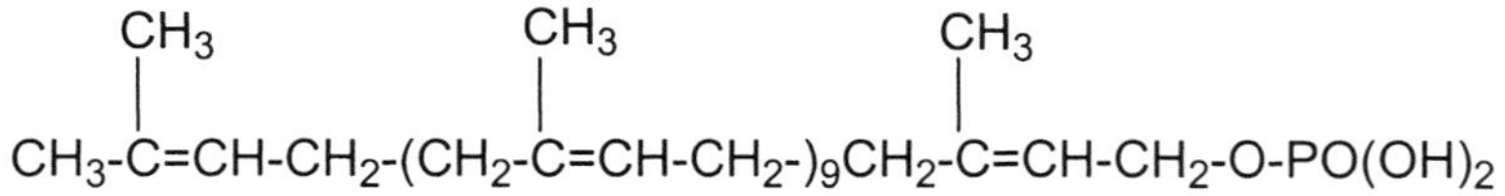

$$CH_3\text{-}\underset{}{\overset{\overset{CH_3}{|}}{C}}\text{=}CH\text{-}CH_2\text{-}(CH_2\text{-}\underset{}{\overset{\overset{CH_3}{|}}{C}}\text{=}CH\text{-}CH_2\text{-})_9CH_2\text{-}\underset{}{\overset{\overset{CH_3}{|}}{C}}\text{=}CH\text{-}CH_2\text{-}O\text{-}PO(OH)_2$$

Fig. 2.12. Undecaprenyl phosphate. Both double bonds adjacent to the phosphate residue have the *trans*-configuration, all the other the *cis*-configuration

easily splits off pyrophosphate and the remaining carbonium cation reacts with a further molecule 3-isopentenyl-pyrophosphoric acid to geranyl pyrophosphate.

The following stages are not demonstrated in Fig. 2.11, i.e. the analogous reaction of geranyl pyrophosphate with a further carbonium cation to farnesyl pyrophosphate and the subsequent attachment of further C_5 building blocks until the desired chain length (mostly C_{55} = undecaprenyl pyrophosphate; see Fig. 2.12) is achieved.

2.1.5.3.5 Biosynthesis of Hopanoids. Hopanoids are synthesised from isopentenyl units that are furnished in a new biosynthetic route. Six of these units form squalene, which represents the immediate precursor of hopanoid biosynthesis (Fig. 2.13). The hopane skeleton is furnished from squalene by the squalene-hopene cyclase, an enzyme with rather unusual properties that forms hopene, to which then a polar side chain is attached, resulting in a hopanoid.

2.1.5.4 Incorporation into the Outer Membrane

The biosynthesis of the first stages of membrane lipids takes place in the cytoplasma, that of the final stages in the cytoplasmic membrane. Between this and the outer membrane, in which the (phospho-)lipids are finally localised, a quick interchange of lipids takes place, presumably by diffusion via the fusion points mentioned above. The half-life ($t_{1/2}$) of the transport from the cytoplasmic membrane to the outer membrane mounts up to 30 s for anionic phospholipids and to 2.8 min for phosphoryl ethanolamine. The transport is bidirectional and rather unspecific with respect to the nature of phospholipid.

Bibliography

Huijbregts RPH, de Kroon AIPM, de Kruiff B (2000). Topology and transport of membrane lipids in bacteria. Biochim Biophys Acta 1469:43-61

IUPAC-IUB-JCBN (1998) Eur J Biochem 257:298

Kannenberg EL, Poralla K (1999) Hopanoid biosynthesis and function in bacteria. Naturwissenschaften 86:168-176

Keweloh H, Heipieper HJ (1995) *Trans*-ungesättigte Fettsäuren bei Bakterien — Vorkommen, Synthese und Bedeutung. Biospectrum 2:18-25

Magnuson K, Jackowski S, Rock CO, Cronan JE jJr (1993) Regulation of fatty acid biosynthesis in *Escherichia coli*. Microbiol Rev 57:522-542Raetz CHR (1986) Molecular genetics of membrane phospholipid synthesis. Annu Rev Genet 20:253-295

Sahm H, Rohmer M, Bringer-Meyer S, Sprenger GA, Welle R (1993) Biochemistry and physiology of hopanoids in bacteria. Adv Microbial Physiol 35:247-273

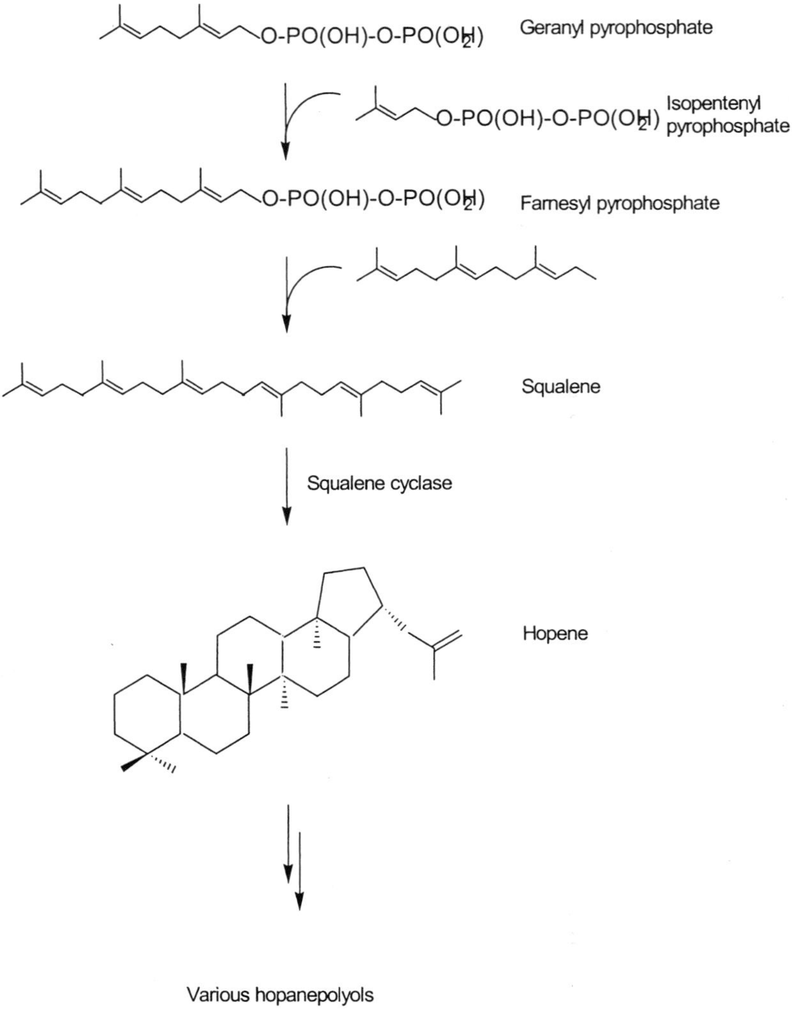

Geranyl pyrophosphate

Isopentenyl pyrophosphate

Farnesyl pyrophosphate

Squalene

Squalene cyclase

Hopene

Various hopanepolyols

Fig. 2.13. Biosynthesis of hopanoids. For the biosynthesis of geranyl pyrophosphate see Fig. 2.11

2.2 Lipopolysaccharides

2.2.1 General Remarks

Polysaccharides are hydrophilic macromolecules consisting of relatively small building units, the monosaccharides. Monosaccharides are linear polyhydroxy-aldehydes or -ketones being able to form six-membered (pyranosides; in formula sign p) or five-membered (furanosides; f) rings by reaction with one of their hydroxyl groups at C-4 or C-5 under acetal or ketal formation. The equilibrium between linear and the cyclic forms (α, β) is strongly shifted to the latter (Fig. 2.14).

Besides polyhydroxy-aldehydes or -ketones also some of their derivatives are found as monosaccharides, such as e.g. uronic acids formed by oxidation of the primary alcohol group, saccharic acids (on-acids) formed by oxidation of the aldehyde group, deoxy sugars formed by reduction of OH groups or amino sugars formed by substitution of OH by NH_2 groups.

In polysaccharides the monosaccharides exist nearly exclusively in their cyclic form. The linkage of the rings to each other in most cases proceeds via O-bonds directed from the acetalic or ketalic OH group of one monosaccharide to an OH group of an other. These bonds are called glycosidic (Fig. 2.15).

Six- and five-membered rings represent rigid structures and this equips polysaccharides in particular for signal storing. Besides this, in the case of polysaccharides, a given number of building blocks is able to store a much larger quantity of information units than in the case of proteins. This is connected with the fact that two different amino acids can form two different dipeptides (A-B and B-A), two monosaccharides containing each five asymmetric C-atoms, however, 5120 ($= 5 \times 2^{10}$) different disaccharides. Additionally, chain branchings are possible in the case of oligosaccharides up from four monosaccharidic building blocks, whereas peptides appear only in linear structures. Thus, the number of isomers rises considerably. For this reason, polysaccharides are frequently used in Nature for information storage and transmission.

Theoretically, monosaccharide rings in an oligo- or polysaccharide can rotate freely around the glycosidic bonds. Only few rotation angles, however, between two rings are energetically favoured and, additionally, particular conformations between the rings are stabilised via hydrogen bonds. Therefore polysaccharides in general show a rather stable spatial structure.

In the Gram-negative cell wall, polysaccharides, both unattached to other binding classes and in combination with peptides and lipids, can be found, especially as peptidoglycans (see Sect. 3.2) and as lipopolysaccharides (LPS).

Lipopolysaccharides are regarded as characteristic and essential for Gram-negative bacteria, including the more Gram-variable cyanobacteria. However, Gram-negative bacteria have been found where LPS are completely replaced by other molecules, e.g. by a glycosphingolipid in *Sphingomonas paucimobilis* and by a glycolipid in the oral spirochete *Treponema denticola*.

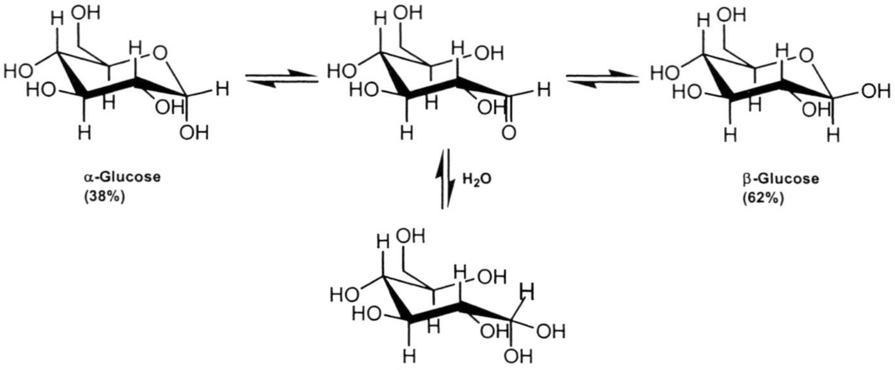

Fig. 2.14. Mutarotation of α-D-glucopyranose

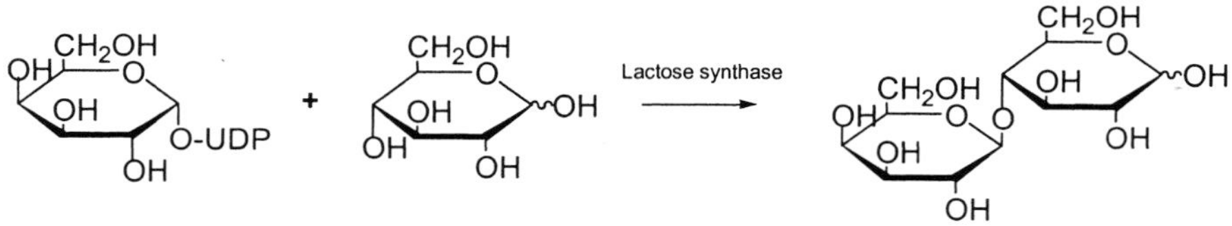

Fig. 2.15. Formation of a disaccharide (lactose)

Because of two characteristics, LPS had already been the object of scientific investigations at a time when their preparation as a pure substance was not even imaginable and nothing was known about their chemical nature. One of these characteristics is the toxicity, which gave the synonym endotoxins for LPS, endo due to the fact that, contrary to exotoxins, these toxins are not actively released by growing bacteria. Already in the 19th century, LPS-containing bacterial extracts were investigated. Besides toxicity, pyrogenicity was detected, which is extraordinarily characteristic for numerous endotoxins, as well as their tumornecroticity. Already very early, people tried to use both attributes in therapy. However, not all LPS represent toxic and/or pyrogenic molecules.

The second biological property of LPS was discovered during the investigation of immunogenicity[1] and antigenicity[1] of Gram-negative bacteria. Namely O-

---

[1] Immunogenicity means the ability to induce the production of specific antibodies in a macroorganism. On the contrary, antigenicity is the ability to react with these antibodies in an externally perceivable reaction such as agglutination or precipitation.

antigens[2], important for the serological classification of many Gram-negative bacteria, are localised in the polysaccharide moiety of the LPS.

The first attempts to prepare pure LPS were undertaken more than 60 years ago. The crucial breakthrough, however, was achieved about 1950 when Westphal and Lüderitz created the possibility of a relatively simple and fast extraction and purification of LPS from numerous enterobacteria by modification of the phenol/water procedure.

The biological characteristics of LPS have been examined using the purified substances. Table 2.3. gives an impression of their extraordinary diversity. Since not all characteristics are determined by the same molecular region, The attempt has been made to separate the individual properties by corresponding modifications of the molecule. This has been carried out by both chemical (synthesis of LPS with modified structures) and biological methods (utilisation of bacterial species producing "unusual" LPS; see above).

2.2.2 General Structure

Like many components of the cell wall, LPS represent amphiphilic substances, i.e. they consist of a hydrophilic and a hydrophobic part. Size and structure, especially of the hydrophilic part, may vary within a wide range according to the respective LPS. Roughly speaking, it is the hydrophilic part of the LPS which is responsible for their serological (antigenic) qualities, and the hydrophobic part for the complex of characteristics referred to with the term endotoxicity.

The hydrophilic part elongates into the surrounding medium away from the bacterial surface. It represents an often long-chain polysaccharide consisting of two different regions, which are called the O-specific polysaccharide and the core region. The first can be either a homo- or a hetero-polysaccharide. It is present only in the so-called smooth forms (S-forms) of the bacteria which frequently represent their wild types. It is by far the largest part of LPS and, in general, is composed of 20-40 repeating identical building blocks (repeating units). One repeating unit may contain between two and eight monosaccharide units. The number of repeating units and therefore the chain length of the polysaccharide may vary in a given LPS within a wide range and often differs from batch to batch even at constant cultivation conditions. S-Form LPS isolates contain also a fraction that is devoid of the O-specific polysacchaide. All this can very well be detected by sodium dodecylsulfate-polyacrylamide gel electrophoresis (SDS-PAGE) of the LPS, as will be demonstrated below (Fig. 2.17).

The great variability in the O-specific polysaccharide becomes evident also by the fact that up to 60 different kinds of monosaccharides have been found and that further structural modifications may occur, e.g. by acetalisation, acylation (also with amino acids), alkylation, amidation and phosphorylation. A selection of monosaccharide species found in the total LPS is listed in Table 2.4.

---

[2] Unfortunately, this term is not unequivocal. In Salmonellae e. g. we talk of antigen factors, in *Shigella flexneri* of antigens, although in both cases the same is meant and chemically comparable structures are present.

Table 2.3. Selection of biological properties of LPS

Pyrogenicity	Hageman factor activation
Hypothermia in mice	Induction of plasminogen activator
Lethal toxicity in mice	Induction of unspecific resistance to infections
Leukocytosis	Induction of tolerance to endotoxin
Local Shwartzman reaction	*Limulus* lysate gelation
Bone marrow necrosis	Adjuvance activity
Complement activation	Mitogenic activity for cells
Depression of blood pressure	Macrophage activation
Platelet aggregation	Induction of prostaglandin synthesis
Tumor necrotic activity	Induction of interferon production

The core is very much smaller, representing an oligosaccharide that consists of up to 15 monosaccharide units which can be substituted by monophosphate groups, 2-aminoethanol phosphate and diphosphate, and other molecules. Rough (R-) forms bacteria are mutants lacking the ability either to synthesise the O-specific polysaccharide or to incorporate it. Such bacteria can be recognised without difficulty by their altered colony morphology and an essentially reduced suspendibility in diluted salt and similar solutions. R-form bacteria possess R-form LPS, consisting only of lipid A and the core region. A number of bacteria, e.g. certain Neisseriaceae and Vibrionaceae, produce R-type LPS as wild-type LPS.

For historic reasons, the hydrophobic moiety is called lipid A. Its structure is relatively invariable and, in numerous kinds of bacteria, consists of glucosaminyl-glucosamine units, the OH groups of which are substituted by phosphate residues and long-chain fatty acids, and the NH_2 groups by long-chain fatty acids. Further substituents may occur. The fatty acids are necessary for anchoring the LPS in the cell wall.

Both core and lipid A show molecular heterogeneity such as the O-specific polysaccharide, though to a significantly reduced extent.

The chemical structure of the O-specific polysaccharide differs from serotype to serotype. The core structure is far less variable. Thus, most Salmonella serovars possess the same core region, and in *E. coli* only five different core types have so far been discovered. Lipid A represents the least variable part of the LPS. It is supposed that early bacteria synthesised only lipid A and that, in the course of evolution, at first the inner core, then the outer core, and finally O-specific polysaccharide were created to produce an always growing multitude of forms. The schematic assembly of LPS is depicted in Fig. 2.16.

2.2.3 Isolation and Analysis

2.2.3.1 Isolation

As will be demonstrated below, the cell wall of Gram-negative bacteria represents a complex structure with a very heterogeneous composition. The individual components are linked to each other by a great diversity of chemical bonds

Table 2.4. Selection of monosaccharides found in bacteria. (Knirel and Kochetkov 1994)

Substance class	Name
Pentoses	D-Arabinose, D-ribose, D-xylose, L-xylose
Pentuloses	D-Xylulose
Hexoses	D-Glucose, D-mannose, D-galactose
4-Deoxyhexoses	4-Deoxy-D-arabinose
6-Deoxyhexoses	D-Rhamnose, L-rhamnose, D-fucose, L-fucose, 6-deoxy-D-talose, 6-deoxy-D-altrose, 6-deoxy-L-gulose
3,6-Dideoxyhexoses	3,6-Dideoxy-D-*arabino*-hexose (tyvelose), 3,6-dideoxy-L-*arabino*-hexose (ascarylose), 3,6-dideoxy-D-*ribo*-hexose (paratose), 3,6-dideoxy-D-*xylo*-hexose (abequose), 3,6-dideoxy-L-*xylo*-hexose (colitose)
Hexuloses	D-Fructose
Heptoses	D-*Glycero*-D-*manno*-heptose, L-*glycero*-D-*manno*-heptose, D-*glycero*-D-*altro*-heptose
6-Desoxyheptoses	6-Deoxy-D-*manno*-heptose, 6-deoxy-D-*altro*-heptose
4-Amino-4-deoxy-pentoses	4-Amino-4-deoxy-L-arabinose
2-Amino-2-deoxyhexoses	D-Glucosamine, L-glucosamine, D-mannosamine, D-galactosamine
2-Amino-2,6-dideoxyhexoses	2-Amino-2,6-dideoxy-D-*gluco*-hexose (D-quinovosamine), L-quinovosamine, D-fucosamine, L-fucosamine, L-rhamnosamine
3-Amino-3,6-dideoxyhexoses	3-Amino-3-deoxy-D-quinovose, 3-amino-3-deoxy-L-quinovose, 3-amino-3-deoxy-D-fucose
4-Amino-4,6-dideoxyhexoses	4-Amino-4,6-dideoxy-D-*gluco*-hexose (viosamine), 4-amino-4,6-dideoxy-D-*manno*-hexose (perosamine), 4-amino-4,6-dideoxy-D-*galacto*-hexose (tomosamine)
2,3-Diamino-2,3-dideoxyhexoses	3-Amino-3-deoxy-D-glucosamine
2,3-Diamino-2,3,6-trideoxyhexoses	3-Amino-3-deoxy-D-rhamnosamine
2,4-Diamino-2,4,6-trideoxyhexoses	2,4-Diamino-2,4,6-trideoxy-D-*gluco*-hexose (bacillosamine), 4-amino-4-dideoxy-D-fucosamine
Hexuronic acids	D-Glucuronic acid, D-galacturonic acid
2-Amino-2-deoxy-hexuronic acids	D-Glucosaminuronic acid, D-mannosaminuronic acid, D-galactosaminuronic acid, L-galactosaminuronic acid, L-gulosaminuronic acid, L-altrosaminuronic acid

Table 2.4, continued

2,3-Diamino-2,3-dideoxy-hexuronic acids	2,3-Diamino-2,3-dideoxy-D-glucuronic acid, 2,3-diamino-2,3-dideoxy-D-mannuronic acid, 2,3-diamino-2,3-dideoxy-D-galacturonic acid, 2,3-diamino-2,3-dideoxy-D-guluronic acid
Hexulosonic acids	3-Deoxy-D-*threo*-hexulosonic acid
3-Deoxyheptulosaric acids	3-Deoxy-*lyxo*-heptulosaric acid
Octulosonic acids	D-*Glycero*-D-*talo*-oct-2-ulosonic acid
3-Deoxy-octulosonic acids	3-Deoxy-D-*manno*-oct-2-ulosonic acid,
5-Amino-3,5-dideoxy-non-2-ulosonic acids	5-Amino-3,5-dideoxy-D-*glycero*-L-*galacto*-non-2-ulosonic acid (neuraminic acid)
5,7-Diamino-3,5,7,9-tetradeoxy-nonulonic acids	5,7-Diamino-3,5,7,9-tetradeoxy-L-*glycero*-L-*galacto*-non-2-ulosonic acid (pseudaminic acid), 5,7-diamino-3,5,7,9-tetradeoxy-D-g*lycero*-L-*galacto*-non-2-ulosonic acid
4-C-(1-Hydroxyethyl)-3,6-dideoxyhexoses	4-C-(1-Hydroxyethyl)-3,6-dideoxy-(*R*)-D-*xylo*-hexose (yersiniose A), 4-C-(1-hydroxyethyl)-3,6-dideoxy-(*S*)-D-*xylo*-hexose (yersiniose B)

(covalent, H-bridges, chelate-like, ionic, hydrophobic). Therefore it very much depends on the extraction medium, in which form and together with which other compounds the LPS are extracted from the cell wall.

In the course of the years, numerous extraction and purification procedures have been described. Only some classical methods discussed in this connection as they are still, in part frequently, used today.

Extraction with 0.25 M trichloroacetic acid (Boivin procedure) and subsequent precipitation with ethanol yields a protein-containing LPS preparation. At least part of the protein is bound to the LPS. This preparation represents a considerably more efficient immunogenic than the free LPS.

Extraction of the bacteria with a 1:1 mixture of phenol and water (method after Westphal and Lüderitz) at 65-68 ^{0}C (i.e. at a temperature at which phenol and water are just completely miscible with each other), subsequent cooling to temperatures near 0 ^{0}C, separation of the water phase and precipitation from this phase with ethanol or acetone in the presence of Mg^{2+} result in a preparation that contains LPS besides other polysaccharides (e.g. capsular polysaccharides) and ribonucleic acid. In this preparation proteins are present at most in low quantities. Purification of the LPS occurs either from the water phase before the ethanol precipitation or from the aqueous solution of the ethanolic precipitate by repeated ultracentrifugation at 100000 g; accompanying polysaccharides and ribonucleic acid remain in the supernatant, the LPS are pelleted. The method is relatively drastic and at least partial destructions cannot be excluded. Besides this, there are a number of phenol-soluble LPS. This method still represents one of the most frequently applied.

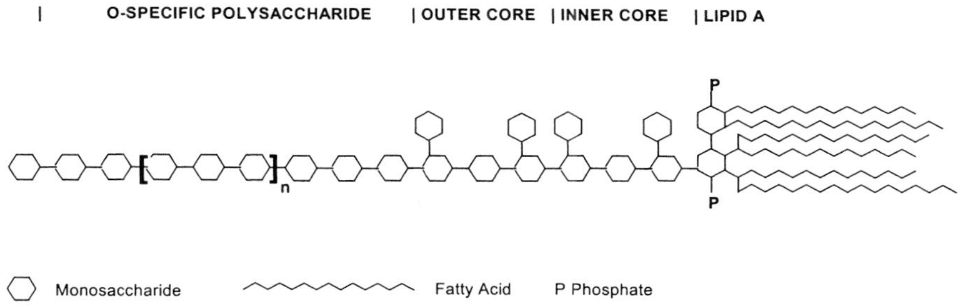

| | O-SPECIFIC POLYSACCHARIDE | | OUTER CORE | INNER CORE | | LIPID A |

○ Monosaccharide   〜〜〜〜〜〜 Fatty Acid   P Phosphate

Fig. 2.16. Schematic structure of a smooth form (S-form) LPS

The method has been modified (Galanos procedure) for the extraction of rather hydrophobic LPS, i.e. R-form LPS. Dried bacteria are extracted at 10-20 °C with a mixture of phenol, chloroform and light petroleum, low-boiling components are removed, and the LPS are precipitated from the remaining phenol phase by dropwise addition of water. This method is more protective than the phenol/water method; it fails, however, in the case of more hydrophilic LPS.

If bacteria are treated with ethylenediaminetetraacetate (EDTA), nearly half the LPS go into solution. This indicates that on binding this part in the cell wall, divalent cations, especially magnesium and calcium, play a role. LPS isolation procedures using EDTA rely on this effect.

2.2.3.2 Analysis

The assembly of LPS in three structurally and functionally different parts made the investigation of the complete structure very difficult. It has to be regarded as a matter of lucky chance that already rather early (1952) a successful separation of the lipid and the polysaccharide moiety by mild hydrolysis was achieved without significant modifications or destruction. The investigation of the polysaccharide moiety was facilitated by the fact that in wild-type strains the O-specific polysaccharide in general comprises more than 90%. Analytical errors caused by the presence of the core region are in most cases so small that they hardly give rise to misinterpretations of the results. A further facilitation results from the fact that the monosaccharide distribution in the polysaccharide chains of the O-specific polysaccharide is not irregular, but that they are composed of the already mentioned repeating units.

It was possible to elucidate the composition and structure of the core region because mutants had been found which did not contain the O-specific polysaccharide. These are the already mentioned R-forms. During investigation of the S→R-mutation, it became evident that not only R-forms with a complete core exist; thus, mutants with a more and more incomplete core region were isolated.

A number of factors considerably complicate the analyses. The stability of a glycosidic linkage is not identical for individual monosaccharides and the diversity of the monosaccharides given in Table 2.4 gives insight into a whole range of possibilities. In addition, LPS molecules are not composed only of carbohydrate constituents, but contain further substituents which influence the stability of the glycosidic linkages. The analyses are also complicated by microheterogeneities which are characterised by different chain lengths or incomplete substitutions and occur in all three molecular regions. They are mainly based on the fact that the biosynthesis of LPS, unlike that of proteins, is not controlled directly by RNA but via the substrate specificity of the enzymes, and this is not in all cases absolute (see LPS biosynthesis, below). Thus, compositional and structural analysis of such complicated and heterogeneous molecules as LPS can by no means be carried out according to a scheme which is identical for all kinds of bacteria. Therefore only some methodical possibilities are described in the following, on whose respective use the experimenter has to decide. It is recommendable to combine several independent procedures in order to eliminate potential objective misleadings of a given method.

As a first step it is advisable to bring the purified LPS into solution by boiling in SDS buffer and then separating it using SDS-PAGE. The bands are made visible by staining with alkaline silver nitrate solution after periodate oxidation (see Fig. 2.17). This method gives a good impression of the molecular mass distribution. At a gel concentration of 11-12% in the pherogram of many smooth-type LPS, the distinctly ladder-shaped band structure is visible. The interval of the individual "rungs" depends on the molecular mass of the repeating units. The intense bands present in the lowest part of the gel are caused by core molecules not substituted with repeating units. By increasing the gel concentration or by application of deoxycholate (DOC) instead of SDS in PAGE, this region can better separated.

Thus, it can be shown that S-form LPS may contain incompletely synthesised core portions. An enlargement of the potential of this method can be achieved by a combination with immunoblotting using anti-LPS antisera of different specificities.

The LPS are then submitted to compositional analysis. After acidic hydrolysis, neutral sugars are reduced with boron hydride and subsequently acetylated, and the obtained alditol acetates are then characterised by a combination of gas-liquid chromatography and mass spectrometry (GLC-MS analysis). From these results one can draw conclusions on the composition the O-specific polysaccharide and the core region, since monosaccharides appearing in large quantities derive from the first. The determination of fatty acids takes place as described for lipids, and the phosphate content is determined using a photometric assay.

In a following step, LPS can be cleaved into lipid A and the polysaccharide portion by mild acidic hydrolysis, and both can be fractionated. In doing so, partial decompositions may occur due to the diverse binding stabilities mentioned above.

In the case of the polysaccharide, fractionation can be carried out by gel chromatography or, after de-*N*-acylation, by ion-exchange chromatography. Lipid A can be fractionated by thin-layer chromatography on hydrophobic matrices, such as silica gel, using neutral solvents.

Another, recently developed, methodology leads to the isolation of pure oligosaccharide phosphates from R-form LPS which can be readily analysed by nuclear magnetic resonance (NMR) spectroscopy. Briefly, the LPS is completely

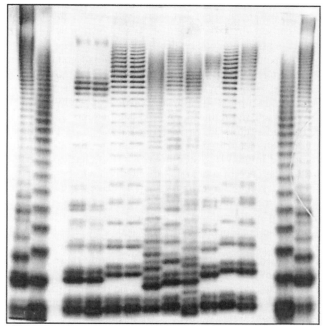

N o.	S t r a i n n o.	S e r o v a r
1	S . T y p h i m u r i u m 6 1 8	
2	E . c o l i 6 4	
3	4 2 1 / 9 9	4 5 : z 1 0 : -
4	9 5 9 / 9 3	1 , 4 5 : z 1 0 : -
5	5 6 0 / 9 3	3 8 : z 1 0 : -
6	7 0 7 / 9 3	3 8 : z 1 0 : -
7	1 0 1 3 / 9 3	3 , 1 5 : m , t : -
8	1 1 2 2 / 9 3	1 3 , 2 3 : k : -
9	1 1 5 1 / 9 3	1 8 : r : 1 , 5
1 0	1 1 8 6 / 9 3	4 3 : b : -
1 1	1 2 0 5 / 9 3	3 8 : z : -
1 2	1 2 4 3 / 9 3	8 , 2 0 : h v : -
1 3	E . c o l i 6 4	
1 4	S . T y p h i m u r i u m 6 1 8	

Fig. 2.17. SDS-PAGE-pherogram of LPS. (Courtesy W. Voigt)

deacylated using hydrazinolysis and treatment with hot KOH. Oligosaccharide phosphates are isolated from such samples by high-performance anion-exchange chromatography. Here, the linkage between lipid A and the core region is not cleaved, thus, the carbohydrate backbone of both LPS moieties, their linkage point and their phosphate substitution can be determined.

First essential conclusions on the structure of polysaccharides can be drawn from methylation analysis. By treatment with NaH in dimethyl sulfoxide and subsequent addition of methyl iodide, all free hydroxyl groups in the polysaccharide are transformed into the acidoresistant methyl ethers. Afterwards, the sample is hydrolysed and reduced, and the components of the obtained mixture of partially methylated alditols are characterised after their acetylation by GLC-MS analysis. The non-methylated hydroxyl groups in the liberated methyl sugars are those that participated in the glycosidic linkages. If the hydroxyl groups at C-2, C-3, C-4 and C-6 (in the case of a hexose) are methylated, the sugar was terminally linked. One further acetylated hydroxyl group indicates a sugar bound linearly in the chain, and two of these a chain branching.

More recently, a growing number of physical procedures have been introduced which may shorten the analytics of polysaccharides and lipid A significantly, and which also create new analytical possibilities in general. However, they have not yet completely replaced all classical procedures. Among these methods are various kinds of mass spectrometry and NMR spectroscopy.

In mass spectrometry, several procedures can be distinguished with regard to the ionisation of molecules, i.e. bombardment with electrons (electron impact, EI), with atoms or ions (fast atom bombardment [FAB]-ionisation) or with photons (laser-desorptions/ionisation, LDI). Finally, a dissolved sample may also be sprayed into an electric field (electrospray ionisation, ESI). Essential indications, especially with regard to the molecular mass distribution in molecule mixtures in the range of up to several 10 kDa is given by matrix-assisted laser desorption ionisation mass spectrometry (MALDI-MS). A high-energy laser beam ionises the sample molecules, and the ions are caught by an electric field and analysed by ion mass spectrometry.

Of high analytical value are the proton (^{1}H-), carbon (^{13}C-) and also phosphorus (^{31}P-)NMR spectra, which are recorded both in one- and two-dimensional homo- and hetero-nuclear correlated experiments.

Finally, in addition to physicochemical analyses of LPS, the use of monoclonal antibodies in the elucidation of structures has been applied very successfully.

2.2.4 Composition and Structure

2.2.4.1 The O-Specific Polysaccharide

The portions of the O-specific polysaccharides in the whole LPS amount to about 90% in most cases. They represent polysaccharides in which a diversity of monosaccharide species has been detected that up to now is unique. The by far prevailing part of monosaccharides depicted in Table 2.4 were identified in the O-specific polysaccharide and quite a number of them had been unknown before LPS was investigated. This is true for the 3,6-dideoxy-hexoses, the names of which derive from the bacteria in which they were found (e.g. tyvelose from Salmonella Typhi, colitose from *E. coli*). As mentioned above, the O-specific polysaccharides are composed of oligosaccharidic building blocks, the repeating units.

No differences are found when comparing the qualitative composition of LPS in various strains of one serotype (serovar). All strains of one serotype belong to the same chemotype. A chemotype, however, may comprise strains of several serotypes, which in this case differ in the chemical structure of the repeating units with the same monosaccharide composition. The serological specificities of the individual serotypes are localised in the repeating units; thus the chemical structure of O-specific polysaccharide determines the serological behaviour of the bacteria.

In the course of the years, the structures of several hundred O-specific polysaccharides from various bacteria have been elucidated (for a listing see Knirel and Kochetkov 1994; Jansson 1999).

The construction of the O-specific polysaccharides of several LPS furnished each from identical repeating building blocks may result in a relative uniformity of their spatial structure. For LPS from *Salmonella enterica*, containing repeating units of the elementary structure Gal-Man-Rha plus a 3,6-dideoxyhexose bound to mannose (O-group A, B, or D1), the three-dimensional structure has been calculated and was shown to represent a helix with about three repeating units per turn. The model indicates that certain regions of the helix are especially exposed. In host organisms, these regions (partial structures called antigenic determinants) are able to induce the production of specific antibodies directed against them. Usually, anti-O-specific polysaccharide antisera are mixtures of antibodies directed against different O-specific polysaccharide determinants. In some cases (e.g. in Salmonellae or in *Shigella flexneri*) the complex serological specificities could be successfully differentiated into the partial antigens (antigen factors) governed by the individual antigenic determinants. The experimental assignment of the individual antigen factors to defined chemical partial structures mostly occurs by determination of the inhibitory effect of defined oligosaccharides obtained from the repeating units in the serological system LPS/homologous antiserum. In this respect, monoclonal antisera turned out to be especially significant. Their specificity may be high enough to distinguish between different anomeric configurations of one glycosyl residue of the polysaccharide chain.

The extraordinary diversity of serological specificities of bacteria is caused by a corresponding diversity of chemical structures of the repeating units. In most cases, the repeating units represent heteropolysaccharides. There are repeating units, however, composed of only a single monosaccharide, but in different linkage or ring size (pyranosidic/furanosidic), such as:

$$\rightarrow\!3)\text{-}\alpha\text{-D-Gal}p\text{-}(1\!\rightarrow\!3)\text{-}\beta\text{-D-Gal}f\text{-}(1\!\rightarrow$$

in *Serratia marcescens* O20. There are also LPS in which the O-specific polysaccharide consists of one monosaccharide in one linkage and ring size, thus representing a real homopolymer, such as:

$$\rightarrow\!6)\text{-}\beta\text{-D-Gal}p\text{-}(1\!\rightarrow$$

in *Actinobacillus pleuropneumoniae* O5.

In many cases, biologically related serotypes (e.g. those of *Shigella dysenteriae*) are characterised by quite different compositions and structures of their O-specific polysaccharides and are in this regard not related to each other.

On the other hand, there are cases in which O-specific polysaccharides of individual serotypes of a (sub)genus are derived from the same basal structure by relatively minor changes, such as e.g. those from *Shigella flexneri* (Table 2.5.). All serotypes have the same basal structure, which represents the structure of the so-called variant Y. The individual serotypes differ from each other only in the substitution site of the chain monosaccharides with glucose or/and O-acetyl. The identification of the substitution sites and the respective linkages leads to conclusions with regard to the chemical structure of the individual *S. flexneri* O-antigen factors (see Table 2.6).

O-Specific polysaccharides preponderantly represent neutral polysaccharides. However, also acidic polysaccharides have been found. Strictly speaking, acidic polysaccharides basically are the domain of capsule polysaccharides. There are cases in which the O-specific polysaccharide and the capsule polysaccharide are structurally identical (e.g. in *E. coli* O141:K85).

2.2.4.2 The Core Region

This region is distinctly smaller than the O-specific polysaccharide and does not contain repeating units. Composition and structure vary much less depending on the investigated serotype; thus, e.g. most LPS from Salmonella studied so far have the same core region. It must be noticed, however, that microheterogeneities as mentioned for the O-specific polysaccharides have also been described for this LPS region. Compared to the O-specific polysaccharide, distinct differences exist with regard to its genetic determination and biosynthesis. There are a number of wild-type LPS that consist only of lipid A and the core region (R-form LPS). In many LPS, the core region possesses two different partial regions, namely the outer core (hexose region) and the inner core (heptose-Kdo-region) adjacent to lipid A.

Inner and outer core differ in composition and structure from each other (Figs. 2.18. and 2.19.). In the inner core, two monosaccharides, i.e. L-*glycero*-D-*manno*-heptose (LD-Hep) and 3-deoxy-D-*manno*-oct-2-ulosonic acid (formerly called 2-keto-3-desoxy-D-*manno*-octonic acid, Kdo) are bound. In some cases, the 3-hydroxy derivative of the latter, D-*glycero*-D-*talo*-octulosonic acid (Ko) has additionally been identified, e.g. in LPS of *Acinetobacter* and *Burkholderia*. Kdo is present in all LPS investigated so far, whereas heptose-free LPS have been described.

The inner core shows a more conservative composition and structure than the outer; it is supposed that most enterobacterial LPS contain the basal structure Hep_3-Kdo_1 shown in Fig. 2.18, though with diverse substitutions. The high degree of phosphorylation together with the carboxyl groups of the Kdo-molecule create an important preponderance of negative charges in this region being important for the function of LPS as components of the outer membrane. This is also true for core regions devoid of phosphate groups which possess uronic acid residues instead (e.g. in LPS of *Klebsiella pneumoniae*).

Several core types may be found in LPS of one species, e.g. five different types have been identified in LPS of *E. coli* which, with the exception of that one derived from strain K-12, differ only slightly from each other. Other monosaccharides like

Table 2.5. Interdependence of structure and serological specificity in *S. flexneri*

Serotype	Antigenic factors	Structure
Variant Y	-:3.4	→3)-GlcNAc[a]-(1→2)-Rha-(1→2)-Rha-(1→3)-Rha-(1→
1a	I:4	Glc ↓1,4 →3)-GlcNAc-(1→2)-Rha-(1→2)-Rha-(1→3)-Rha-(1→
1b	I:6	Glc O-Ac ↓1,4 |2 →3)-GlcNAc-(1→2)-Rha-(1→2)-Rha-(1→3)-Rha-(1→
2a	II:3.4	Glc ↓1,4 →3)-GlcNAc-(1→2)-Rha-(1→2)-Rha-(1→3)-Rha-(1→
2b	II:7.8	Glc Glc ↓1,3 ↓1,4 →3)-GlcNAc-(1→2)-Rha-(1→2)-Rha-(1→3)-Rha-(1→
3a	III:6,7.8	Glc O-Ac ↓1,3 |2 →3)-GlcNAc-(1→2)-Rha-(1→2)-Rha-(1→3)-Rha-(1→
3c	III:6	O-Ac |2 →3)-GlcNAc-(1→2)-Rha-(1→2)-Rha-(1→3)-Rha-(1→
4a	IV:3.4	Glc ↓1,6 →3)-GlcNAc-(1→2)-Rha-(1→2)-Rha-(1→3)-Rha-(1→
4b	IV:6	Glc O-Ac ↓1,6 |2 →3)-GlcNAc-(1→2)-Rha-(1→2)-Rha-(1→3)-Rha-(1→
5a	V:3.4	Glc ↓1,3 →3)-GlcNAc-(1→2)-Rha-(1→2)-Rha-(1→3)-Rha-(1→
5b	V:7.8	Glc Glc ↓1,3 ↓1,3 →3)-GlcNAc-(1→2)-Rha-(1→2)-Rha-(1→3)-Rha-(1→
Variant X	-:7.8	Glc ↓1,3 →3)-GlcNAc-(1→2)-Rha-(1→2)-Rha-(1→3)-Rha-(1→

[a] The GlcNAc residue possresses the β-D- and the Glc residue the α-D-configuration.
The rhamnose residues are α-L-configured

Table 2.6. Chemical structure of the *S. flexneri* O-antigenic factors

Factor	Structure
I	α-Glc-(1→4)-β-GlcNAc-
II	α-Glc-(1→4)-α-Rha-
III	2-O-Ac-α-Rha-(1→3)-β-GlcNAc-
IV	α-Glc-(1→6)-β-GlcNAc-
V	α-Glc-(1→3)-α-Rha-(1→3)-Rha-
6	2-O-Ac-α-Rha-
7.8	α-Glc-(1→3)-Rha-(1→2)-GlcNAc-

Antigen 3.4 has not yet been definitively localised. It has to be suspected that the determinant group is located in the basal chain; however, a still unknown substituent cannot be excluded

rhamnose, fucose, sedoheptulose, chinovosamine or neuraminic acid can be found in non-enterobacterial LPS core regions. Several LPS of different bacteria contain significant amounts of D-*glycero*-D-*manno*-heptose (DD-Hep). In the outer core region of LPS from *Klebsiella pneumoniae* strain R20 it forms an unbranched (1→2)-linked tetrasaccharide that is bound α-(1→6) to *N*-acetylglucosamine.

As already mentioned, S→R-mutation need not necessarily result in only one particular rough form. In Salmonella Minnesota, which was the first to be systematically examined, five mutants of the chemotypes designated Ra, Rb, Rc, Rd and Re have been discovered, the LPS of which in this order contain a decreasing number of monosaccharide residues (Table 2.7). The R-types formed by other Salmonella serovars are identical with the ones described. Other Enterobacteriaceae produce comparable, but not identical R-form types.

The presumption that rough forms in this order show a more and more incompletely constructed core (increasing "deeper" rough forms) could be confirmed by investigation of their structures (Table 2.8). As can be seen, most of these R-type LPS can be further subdivided according to their chemical structure.

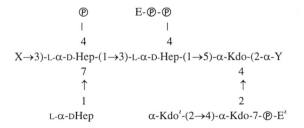

Fig. 2.18. Structure of the inner core of Salmonella LPS. *X* Binding site of the outer core; *Y*, linkage to lipid A; E, ethanolamine; L,D-Hep, L-*glycero*-D-*manno*-heptopyranose; Kdo, 3-deoxy-D-*manno*-oct-2-ulopyranosonic acid; ᶻ non-stoichiometrically substituted

```
        R^z                          α-D-Gal
         1                              1
         ↓                              ↓
         2                              6
X→4)-α-D-Glc-(1→2)-α-D-Gal-(1→3)-α-D-Glc-(1→Y
```

Fig. 2.19. Structure of the outer core of Salmonella LPS. *X* Binding site of the O-specific polysaccharide; *Y*, linkage to the inner core; *R*, α-D-Glc*p*NAc in LPS of Salmonella Minnesota and Salmonella Typhimurium, and α-D-Glc*p* in LPS of Salmonella Arizonae; *z* non-stoichiometrically substituted

As far as has been investigated, the core region of S-type LPS is generally rather complete. Namely, in the case of Salmonella, the first repeating unit of the O-specific polysaccharide can be bound either to the complete Ra or to the Rb core types. The linkage, however, does not occur at the terminal *N*-acetylglucosamine, but at the second distal glucose. This fact can be observed particularly well in analyses of the so-called SR-mutants, i.e. mutants in which only a single repeating unit is bound to the core region. Recently, the presence of a fourth LPS region (interlinking unit) interlinking the O-specific polysaccharide and the core region has been proposed. It may either comprise a monosaccharide (e.g. sedoheptulose in *Vibrio cholerae* strain H11) or an oligosaccharide [e.g. a heptoglycan built up from α-(1→3)-linked D,D-Hep in *Helicobacter pylori*]. However, biosynthesis and biological function(s) of this region are not known.

In summary, the number of different core types discovered so far in LPS of Gram-negative bacteria is smaller than that of the O-specific polysaccharide types. An adequate listing of structures can be found in Holst (1999).

2.2.4.3 The Lipid A

Lipid A represents in most, but not all, Gram-negative bacteria the endotoxic centre of LPS as well as the hydrophobic anchor for its incorporation into the outer membrane. Like the O-specific polysaccharide and the core region, lipid A represents a structural principle. Its structure is in most cases rather similar; however, considerable variations exist. The amphoteric as well as the amphiphilic properties of the lipid A molecule are of great importance for their biological properties.

In Nature, lipid A does not occur in the free form, but has to be liberated from the LPS by mild hydrolysis. During this procedure it precipitates. It became a matter of doubt whether a lipid A liberated by this procedure is identical to one present in the native LPS. It could be demonstrated that all procedures leading to a cleavage of lipid A from LPS cause at least partial modifications in composition and structure of lipid A. The resulting inner heterogeneities complicate the analyses and their interpretations. The errors can be minimised by comparative analysis of an Re-mutant LPS and the lipid A prepared therefrom.

The structures of enterobacterial lipid A molecules elucidated by classical procedures have been confirmed by modern analytical methods (see Sect. 2.2.3.2)

Table 2.7. Monosaccharide composition of LPS from Salmonella R forms

	Kdo	L,D-Hep	Glc	Gal	GlcN
Ra	+	+	+	+	+
Rb	+	+	+	+	-
Rc	+	+	+	-	-
Rd	+	+	-	-	-
Re	+	-	-	-	-

as well as by chemical synthesis. These lipids A consist of a β-D-glucosaminyl-(1→6)-D-glucosamine disaccharide which is substituted in positions 2, 3, 2′ and 3′ with long-chain 3-hydroxy fatty acids [mostly (R)-3-hydroxymyristic acid, 3-OH-14:0] and in positions 1 and 4′ with phosphate. Position 6′ represents the binding site for the first Kdo residue of the core region and position 4 is not substituted.

This structure is widespread in Nature, thus representing a basal structure. In *E. coli* the hydroxyl group of 3-hydroxymyristic acid in position 2′ of the glucosamine disaccharide is substituted with tetradecanoic acid (14:0) and in position 3′ with dodecanoic acid (12:0; Fig. 2.20.). In lipid A of LPS from Salmonella Minnesota, hexadecanoic acid (16:0) was identified to be non-stoichiometrically linked to amide-bound 3-OH-14:0 of GlcN I. The pattern of fatty acids bound in lipid A varies according to the type of bacteria. The number of fatty acids so far detected is considerable. Saturated fatty acids have been found which are unsubstituted straight-chained, hydroxylated, methoxylated, branched-chained or containing a cyclopropane ring; unsaturated fatty acids, on the contrary, are rare.
The phosphate residues in positions 1 and 4′ may carry further substituents. In *E. coli*, the phosphate in position 1 carries a second phosphate residue in non-stoichiometric amounts. In Salmonella Minnesota it is non-stoichiometrically substituted with phosphoryl ethanolamine, and 4-amino-4-deoxy-L-arabinose (LAra*p*4N) is non-stoichiometrically bound to the 4′-phosphate via its glycosidic OH-group. The latter substitution renders the bacteria resistant to the effect of the polycationic antibiotic polymyxin B (see Sect. 8.1.1).

Variations like those described so far are not regarded as type-modifying. More serious, however, are the following modifications:
- Positions 4 or 4′ are substituted by sugars instead of phosphate.
- Instead of amide bound 3-hydroxyfatty acids, 3- or 4-oxofatty acids are present.
- Positions 4 or 4′ of the disaccharide are not substituted at all.
- Glucosamine is replaced by 2,3-diamino-2,3-dideoxy-D-glucopyranose (DAG), which results in disaccharides of either two DAG or one DAG and one GlcN.

For lipid A, there is a quite unequivocal structure-function relationship. A comparison of numerous natural and synthetic lipid A revealed that the structure depicted in Fig. 2.20 with six fatty acids of a chain length of 12 to 14 C-atoms in the molecule is most appropriate for their endotoxic activity. Modifications of this structure lead to alterations of the properties listed in Table 2.3, though not in the

Table 2.8. Schematic structures of the core regions from LPS of Salmonella R-forms

Form	Structure
Re	Kdo- | Kdo
Rd$_2$	L,D-Hep-Kdo- | Kdo
Rd$_{1P}^-$	L,D-Hep-L,D-Hep-Kdo- | Kdo
Rd$_{1P}^+$	L,D-Hep-L,D-Hep-Kdo- | | E-P-P Kdo
Rc$_P^+$	L,D-Hep | Glc-L,D-Hep-L,D-Hep-Kdo- | | | P E-P-P Kdo
Rb$_4$	L,D-Hep | Gal-Glc-L,D-Hep-L,D-Hep-Kdo- | | | P E-P-P Kdo
Rb$_3$	L,D-Hep | Glc-L,D-Hep-L,D-Hep-Kdo- | | | | Gal P E-P-P Kdo
Rb$_2$	L,D-Hep | Gal-Glc-L,D-Hep-L,D-Hep-Kdo- | | | | Gal P E-P-P Kdo
Rb$_1$	L,D-Hep | Glc-Gal-Glc-L,D-Hep-L,D-Hep-Kdo- | | | | Gal P E-P-P Kdo
Ra	L,D-Hep | X-Glc-Gal-Glc-L,D-Hep-L,D-Hep-Kdo- | | | | Gal P E-P-P Kdo

E, ethanolamine, P, phosphate, X, GlcN in the R1S core type and Glc in R2S core type.

; degree and also not always into the same direction. Many lipid A that differ from the structure depicted in Fig. 2.20 are less toxic or not at all. The hope is to come across structures possessing the desired useful but not the harmful properties. This could be achieved in the case of pyrogenicity. As for tumornecroticity, the situation is unfortunately much more problematic. Though there are encouraging attempts now (Reisser et al. 1999 in the case of a 3,3′-deacylated *E. coli* lipid A), a crucial breakthrough has obviously not yet been achieved.

Recently, a genetically constructed Salmonella mutant has been described lacking the myristic acid (C14:0) bound to the β-hydroxymyristic acid that has the same growth abilities as the wild type, but a distinctly reduced lethality (toxicity). This is the first direct proof that death caused by a Salmonella infection is directly induced by lipid A toxicity.

As is described below, LPS are well suited for embedding in membranes due to their spatial structure. The chain length of their fatty acids corresponds to the thickness of a membrane monolayer. The higher number of fatty acids per molecule, compared with the phospolipids, provides the membrane layer with additional stability. The negative charges in lipid A are situated directly on the membrane surface and bind considerable amounts of cations there (K^+, Ca^{2+}, Mg^{2+}).

2.2.4.4 Physical Properties and Spatial Structure

LPS represent a large family of anionic amphiphilic substances more or less similar to each other. The physical properties, especially the solubility of the individual kinds of LPS, depend very much on their structure and on the size relationship of the hydrophilic to hydrophobic part, as well as on the kind of the positive counterions.

As can be seen in the spatial models in Fig. 2.20, all fatty acids in lipid A point in the same direction. This is one prerequisite for the incorporation of lipid A into the membrane. Generally, the chain length of a fatty acids comprises a range corresponding to the depth of a monolayer of the outer membrane. One exception to this rule is the lipid A from LPS of *Rhizobium trifolii*, which contains a 27-OH-28:0 fatty acid that penetrates both monolayers. The phosphate residues in lipid A are arranged in such a way that their charges are directly situated on the membrane surface. They bind considerable quantities of cations there and thus have a membrane-stabilising effect (especially via Mg^{2+} and Ca^{2+} bridges).

In aqueous media, both the complete LPS and the lipid A form aggregates. Since lipid A represents the endotoxic component of the LPS, correlations have been soughtbetween the supramolecular structure of its aggregates (as an expression of its molecular structure) and its endotoxicity. Under physiological conditions (37 °C, pH 7, >90% water content, presence of Mg^{2+}) the aggregates of non-active lipid A are lamellar. On the contrary, endotoxically active lipid A form either cubic (Salmonella Minnesota) or hexagonal (*Rhodocyclus gelatinosus*) structures (Fig. 2.21). These structures are predetermined by the molecular geometry of the respective lipid A, mainly by the tilt angle between the diglucosamine backbone and the membrane surface which, in turn, is dependent on the number and distribution of the acyl chains of lipid A. It is emphasised, however, that it is not

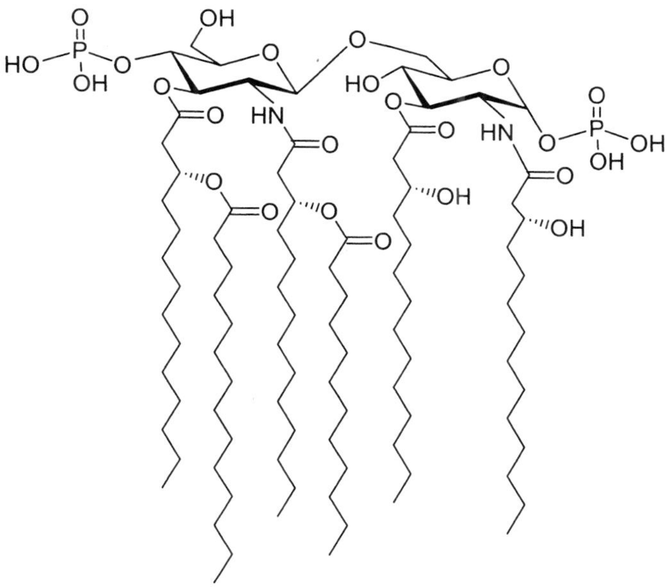

Fig. 2.20. Structure of lipid A from *E. coli*. The right ring is designated GlcN I, the left one GlcN II

the aggregate but the monomeric molecule that represents the endotoxically active form of LPS or lipid A (see also Sect. 8.3.6.2). Therefore, it is supposed that active lipid A molecules possess a conical endotoxic conformation which is expressed in the described aggregate structures. Molecules showing a cylindrical molecular structure express an antagonistic activity, a property that can be utilised in antisepsis treatment. It was found that the tilt angle between the sugar backbone of lipid A and the membrane surface represents an important parameter for understanding the molecular prerequisites for endotoxic activity. This angle depends on the number and distribution of the hydrocarbon chains. It is high in endotoxic active molecules of conical shape and low in inactive, antagonistic ones of cylindrical shape.

Like lipids, LPS also appear in two thermal phases: the ordered gel-like β-phase in which the fatty acid chains are all-*trans* arranged, and the more fluid liquid-crystalline α-phase, in which *gauche* conformations appear (see also Sect. 2.1.3.3.4). The transition temperature between the two phases depends on a number of structural parameters.

The polysaccharide chains of cell wall-bound LPS do not possess a distinct spatial orientation: they are flexing back and forth.

2.2.5 Biosynthesis

The biosynthesis of LPS takes place mainly at the cytoplasmic membrane. The molecule is not constructed in a single process, but the lipid A/core region on one hand and O-specific polysaccharide on the other are synthesised on different pathways and then linked to each other. The genetic determination of the process is not directly available from the nucleic acids as in the case of proteins, but via the substrate specificity of the enzymes necessary for the individual stages of the synthesis. Different loci of the bacterial chromosome encode these enzymes, i.e. the *lpx* genes for lipid A biosynthesis, the *rfa* gene cluster the core region biosynthesis, and the *rfb* region the O-specific polysaccharide biosynthesis. Recently, a new nomenclature system for bacterial polysaccharide genes has been proposed (Reeves et al. 1996) in which genes encoding glycosyltransferases are identified by the prefix waa, and genes involved in precursor formation are named after the precursor.

As in the case of the elucidation of the LPS structure, for investigation of the pathways of their biosynthesis mutants (both natural ones and genetically produced) play an essential role.

2.2.5.1 Biosynthesis of the Precursors

Many bacteria are able to grow by means of glucose as the sole C source and to synthesise from it the diversity of sugars bound in their LPS. In general, the metabolic pathways are known for *S. enterica* and *E. coli*, of which Fig. 2.22 gives a small choice. The mutual transformations are not carried out with the monosaccharides themselves but with their much more energy-rich nucleotide diphosphate derivatives which are also used for the incorporation of sugars into the growing polysaccharide. Some of them (UDP-Glc, UDP-Gal, UDP-GlcNAc, GDP-Man) are common intermediates of normal metabolism. They can both be incorporated directly into the growing LPS or be transformed to other monosaccharidic components characteristic of LPS.

Besides glucose, other monosaccharides may serve as precursors. Galactose, for example, can be transformed via galactose-1-phosphate into UDP-galactose, which stands in equilibrium with UDP-glucose. Also mannose can be taken in directly and then be phosphorylated. The biosynthesis of activated mannose from glucose is, in fact, distinctly more complicated and proceeds via Glc-6-P, Fru-6-P, Man-6-P and Man-1-P to GDP-Man.

CDP-glucose (see Fig. 2.23) serves as a precursor for the biosynthesis of all 3,6-dideoxyhexoses. It is metabolised into CDP-6-deoxy-L-*threo*-D-*glycero*-4-hexulose (CDP-4-keto-6-deoxyglucose in Fig. 2.22) and subsequently via two intermediate stages into CDP-3,6-dideoxy-D-*glycero*-D-*glycero*-4-hexulose. From this key substance ascarylose (epimerisation at C-5, then stereospecific reduction at C-4), abequose and paratose (each by stereospecific reduction at C-4) and tyvelose (epimerisation at C-2) are produced as CDP derivatives.

Not depicted in Fig. 2.22 are L-*glycero*-D-*manno*-heptose, Kdo and the uronic acids.

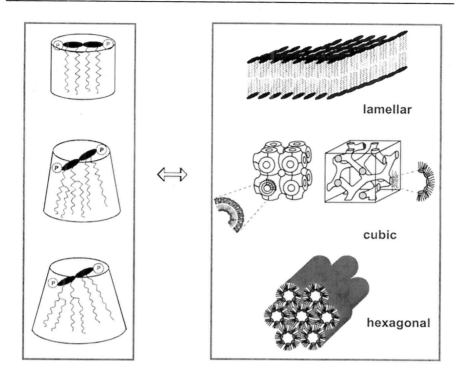

Fig. 2.21. Schematic conformations of lipid A (left) in relation to their supramolecular structures (right). (Seydel et al. 1996, courtesy U. Seydel)

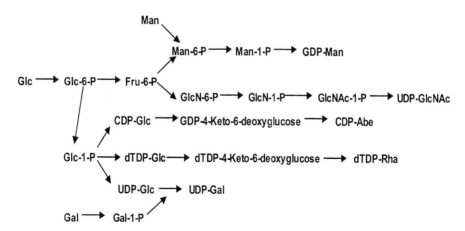

Fig. 2.22. Metabolic pathways of some monosaccharides present in the LPS of Salmonella Typhimurium

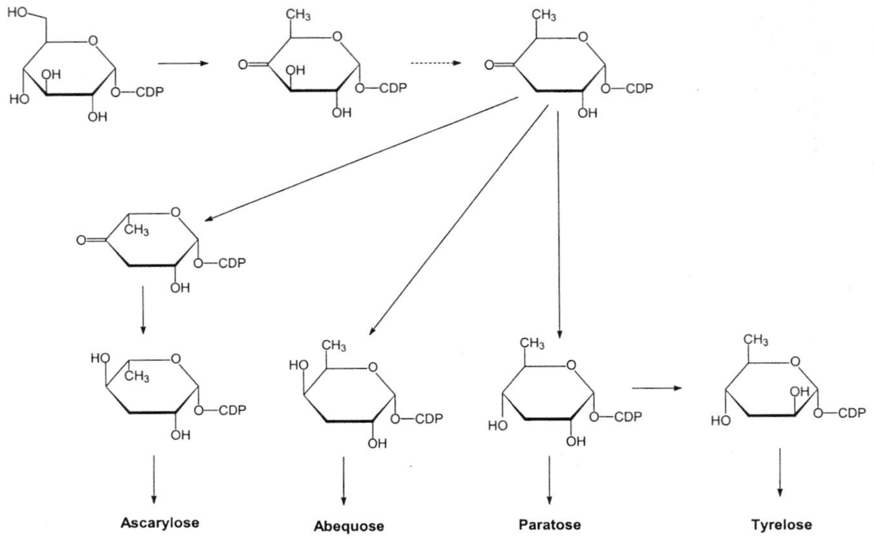

Fig. 2.23. Biosynthetic pathways of the 3,6-dideoxyhexoses

Activated L-*glycero*-D-*manno*-heptose derives from sedoheptulose-7-phosphate, an intermediate of the pentosephosphate pathway, via D-*glycero*-D-*manno*-heptose-7-phosphat, D-*glycero*-D-*manno*-heptose-1-phosphat and ADP-D-*glycero*-D-*manno*-heptose by epimerisation to ADP-L-*glycero*-D-*manno*-heptose. Recently, it was shown that ADP-L-*glycero*-β-D-*manno*-heptopyranose is the physiological sugar donor for heptosyltransferases I and II as precursor.

Kdo is synthesised from arabinose-5-phosphate and phosphoenolpyruvate by aldolcondensation. The Kdo-8-phosphate formed thus is transformed via the free Kdo by CTP into CMP-Kdo which in turn can be used in LPS biosynthesis.

Uronic acids mostly originate from the corresponding neutral sugars by oxidation with NAD^+ (both as nucleoside diphosphates), but mutual transformations (epimerisations) of the individual uronic acids in form of their nucleoside diphosphates are also known.

Hydroxylated fatty acids stem from the bacterial metabolism. The D forms represent probably intermediates of the fatty acid biosynthesis, the L forms those of fatty acid degradation.

2.2.5.2 Biosynthesis of Lipid A and the Core Region

The biosynthetic pathways of both lipid A and inner core are genetically as well as biochemically connected.

The starting substance UDP-*N*-acetyl-D-glucosamine (Fig. 2.24) is esterified with (*R*)-3-hydroxymyristic acid (reaction 1 in Fig. 2.24) at the 3-hydroxy group by enzymatic transfer from (*R*)-3-hydroxymyristoyl-acyl carrier protein (-ACP, see Sect. 2.1.5.1). The *N*-acetyl group is then cleaved and the free amino group is also substituted with (R)-3-hydroxymyristic acid (reactions 2 and 3). From one part of the formed UDP-2,3-diacyl-glucosamine, UMP is cleaved by a pyrophosphatase and thus transformed into 2,3-diacyl-glucosamine-1-phosphate, the so-called lipid X (reaction 4). UDP-2,3-diacyl-glucosamine and 2,3-diacyl-glucosamine-1-phosphate react with each other forming the tetra-acylated β-(1→6)-glucosamine disaccharide-1-phosphate (reaction 5) representing the major intermediate of lipid A biosynthesis.

The sites within the cell where these processes take place are not completely known. The first steps possibly take place in the cytoplasm and, with increasing hydrophobicity of the intermediates, presumably at the inner surface of the cytoplasmatic membrane.

By means of a membrane-bound 4´-kinase the glucosamine disaccharide-1-phosphate is phosphorylated to form β-(1→6)-glucosamine disaccharide 1,4´-biphosphate, the so-called precursor Ia (or precursor IVa, reaction 6). Starting from CMP-Kdo, at first one Kdo in a 2→6´-linkage to GlcN and then a second in an α-(2→4)-linkage to the first Kdo are bound stepwise to this precursor. These reactions are catalysed by a bifunctional Kdo transferase.

The sequence of the following steps is not exactly known yet; however, it appears that in further acylation of lipid A of *E. coli* at first 12:0 in 2´- and then 14:0 in 3´-position are introduced to the 3-hydroxy groups of 3-OH-14:0 to form the acyloxyacyl substituents. These reactions are followed by an α-(1→5)-substitution of the terminal Kdo with L-*glycero*-D-*manno*-heptose (L,D-Hep I) by means of heptosyltransferase I. The fact that viable bacteria exist whose LPS consist only of lipid A and Kdo (Re mutants) shows that acylation occurs without a preceding heptose transfer.

Subsequently, L,D-Hep II is bound to L,D-Hep I. Then, the first Glc residue of the outer core region is linked to L,D-Hep II, and after this, L,D-Hep III is transferred to L,D-Hep II. Thus, the biosynthesis of the inner core is finished, followed by the completion of the outer core region via transfer of further hexose residues.

The structural unit built up in this way is identical for many bacteria. It may, however, be modified by variable and partly non-stoichiometric substitutions with Kdo, L-Ara*p*4N, rhamnose, phosphate and/or phosphoryl ethanolamine employing species- or even strain-specific pathways. These mechanisms are still to a great extent unknown, and it is possible that besides genetic determinations, also adaptations to the respective (cultivation) medium plays a role.

The biosynthesis of the outer core occurs by stepwise incorporation of the individual hexoses starting from their energy-rich nucleotide phosphates. In *Salmonella enterica*, the sequence is as follows: attachment of GlcI (nomenclature of hexoses see Fig. 2.19) from UDP-Glc to Hep II followed by attachment of the side-chain galactose (Gal II) to Glc I. The following stages are the attachment of Gal I to Glc I, of Glc II to Gal I and of GlcNAc or Glc III to Glc II.

Except for one case (so-called Rc$_p$ mutant), phosphorylation of the inner core (Hep I and Hep III) is a prerequisite for this process. The biosynthesis may be stopped at every intermediate stage, and the resulting mutants have been isolated and investigated.

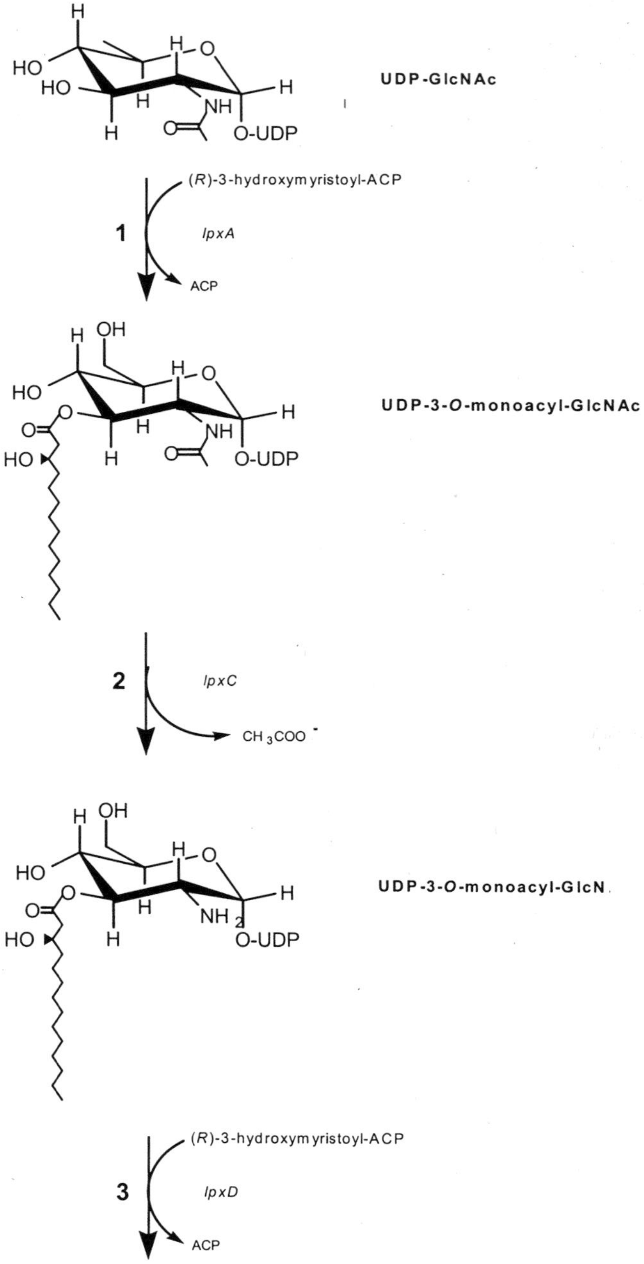
UDP-GlcNAc
O-UDP
(R)-3-hydroxymyristoyl-ACP
1
lpxA
ACP
UDP-3-O-monoacyl-GlcNAc
2
lpxC
CH 3COO -
UDP-3-O-monoacyl-GlcN
(R)-3-hydroxymyristoyl-ACP
3
lpxD
ACP

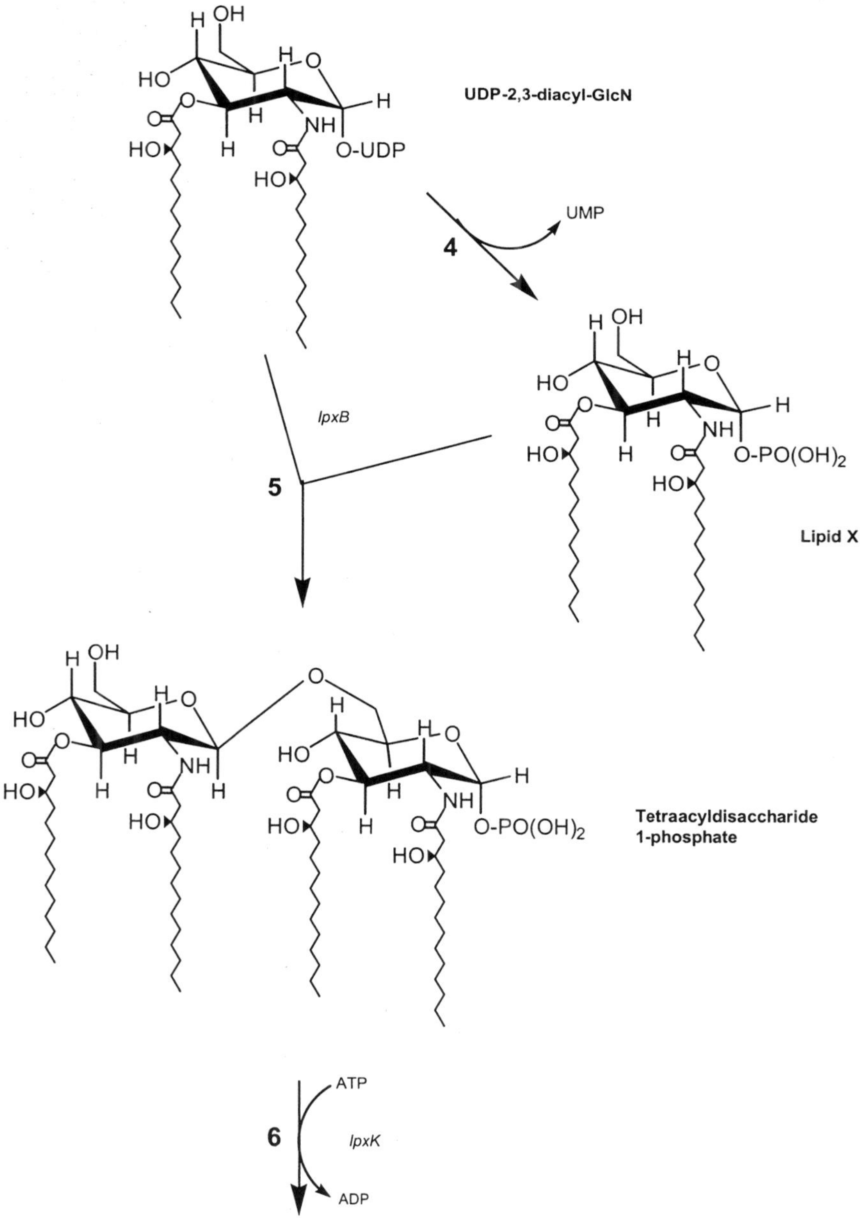

UDP-2,3-diacyl-GlcN

UMP

4

lpxB

5

Lipid X

O-PO(OH)$_2$

Tetraacyldisaccharide
1-phosphate

O-PO(OH)$_2$

ATP

6   *lpxK*

ADP

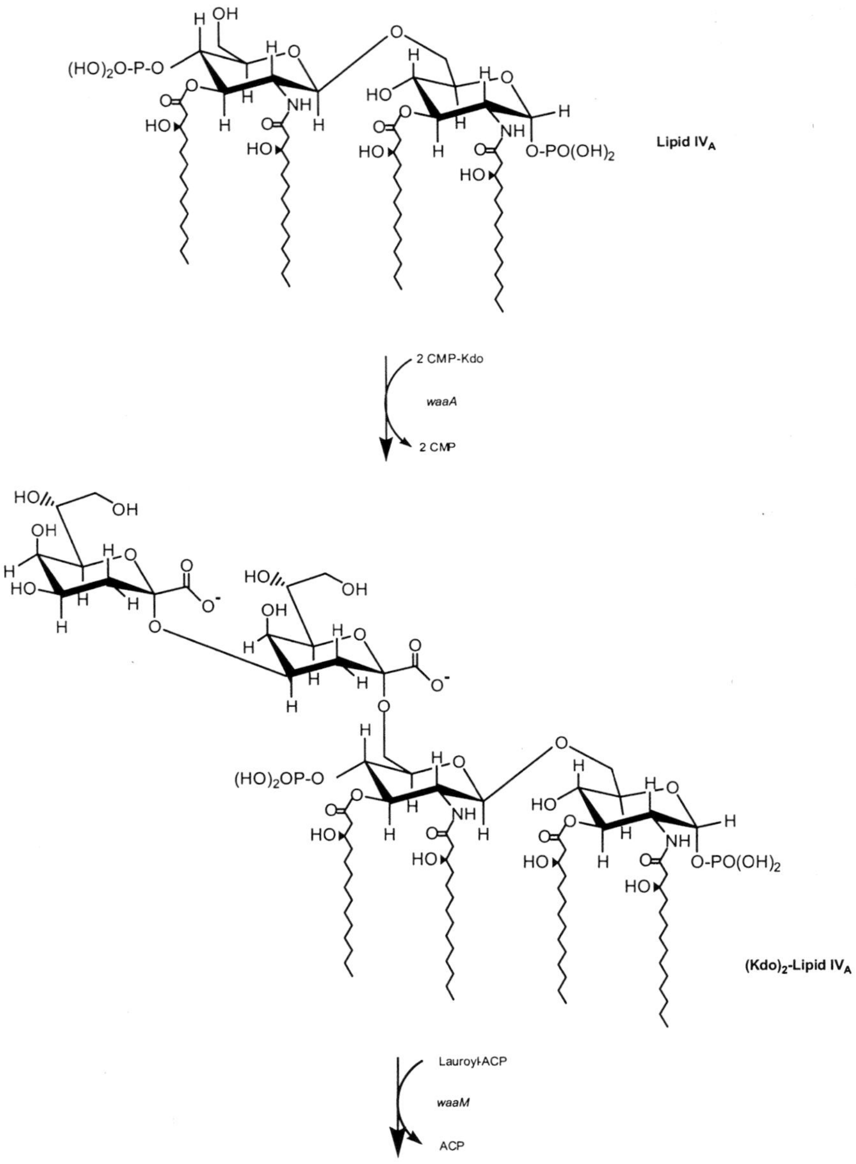

Lipid IV$_A$

2 CMP-Kdo

waaA

2 CMP

(Kdo)$_2$-Lipid IV$_A$

Lauroyl-ACP

waaM

ACP

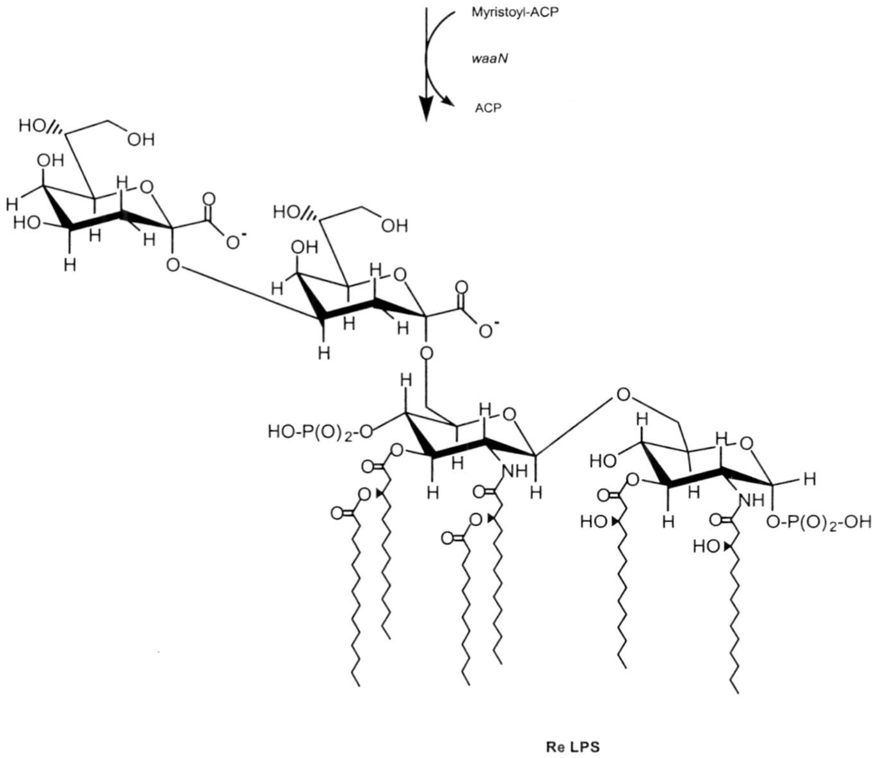

Fig. 2.24. Biosynthesis of Re chemotype LPS. For genetic control see Table 2.9

The synthesis of the core occurs completely at the inner surface of the cytoplasmic membrane. The transport across the membrane to its periplasmic surface is then carried out in a "flipping"-process needing a protonmotive force.

2.2.5.3 Biosynthesis of O-Specific Polysaccharides

Like the major part of the core region biosynthesis, that of the O-specific polysaccharides also proceeds at the cytoplasmic membrane. The mechanisms, however, are quite different. In general, two main pathways for O-specific polysaccharide biosynthesis have been described, i.e. the Wzy-dependent one and the ABC-transporter (ATP-binding cassette, see Sect. 8.1.2.2)-dependent. In the first, the monosaccharides are not successively incorporated into the growing chain, but repeating units consisting of several monosaccharides are formed as large-scale building blocks, and only these are united to the complete chain. This biosynthesis depends on the presence of ACL (see Sect. 2.1.5.3.3). In Salmonellae of group E, 10^5 ACL molecules have been found per cell. This ACL is basically identical with that appearing during the biosynthesis of the peptidoglycan (Sect. 3.2.2.2).

Nevertheless, both seem to derive from different pools that do not affect each other. However, the function of ACL is in both cases the same. With its hydrophobic terminus it is anchored in the lipid bilayer of the membrane and "passes on" the growing chain from enzyme to enzyme. The membrane binding of the individual transferases and their spatial arrangement as a multienzyme system are prerequisites for an effective course of the reactions.

All reaction stages described in the following have been demonstrated in cell-free extracts, i.e. in in vitro experiments, using radioactively labelled monosaccharide nucleotides. During LPS biosynthesis in Salmonellae of groups B and E, which have in the O-specific polysaccharide repeating units the common basal structure Man→Rha→Gal, galactose is added to ACL in an initial reaction starting from UDP-Gal (Fig. 2.25). The reaction product could be isolated. Its analysis revealed that the binding of galactose to ACL occurs via pyrophosphate bridge. By using ^{32}P-labelled UDP-Gal it could be demonstrated that the galactose-bound phosphate derives from UDP-Gal, i.e. galactose had been transferred as phosphate.

Subsequently, rhamnose (from TDP-Rha) and mannose (from GDP-Man) are added to Gal-P-P-ACL as unsubstituted sugars. Both intermediates could be isolated using model reactions. Thus the basal chain of the repeating unit is completed and then can be "polymerised" to the complete O-specific polysaccharide in those cases in which no further monosaccharides are present in the chain.

If the repeating units contain further components, the following reactions depend very much on their nature. In the case of monosaccharides such as abequose or tyvelose, first their incorporation takes place (starting from the CDP derivative). Only after that can the polymerization to the entire polysaccharide occur. In contrast, in the case of glucose or O-acetyl, the polymerization is independent of their incorporation. This is due to a completely different incorporation mechanism, especially for glucose.

Because of the water solubility of the nucleoside diphosphates, all previous reactions resulting in repeating units proceed in the cytoplasm or at the inner surface of the cytoplasmic membrane. Since the completed O-specific polysaccharide chains are formed at its periplasmic surface, transport of the repeating units has to be carried out across the membrane. This is achieved by a protein Wzx ("flippase"); however, the mechanism of this transport is still unclear.

Now the linking of the repeating units to the O-specific polysaccharide chain proceeds, which in literature is not quite correctly called polymerisation. The polymerase Wzy necessary for this process represents a very hydrophobic transmembrane protein, the functional domains of which are presumably located in the periplasm. It has been proposed that Wzy has two binding sites specified R (receiving) and D (donating). One ACL-bound repeating unit is linked to the R region. Then the growing polysaccharide chain located in the D region is bound to the new repeating unit in such a manner that it is localised at the reducing end of the chain (scheme see Fig. 2.26).

At this reaction stage, a modification of the polysaccharide may also occur via incorporation of O-acetyl (from acetyl-coenzyme A) and glucose (from ACL-P-

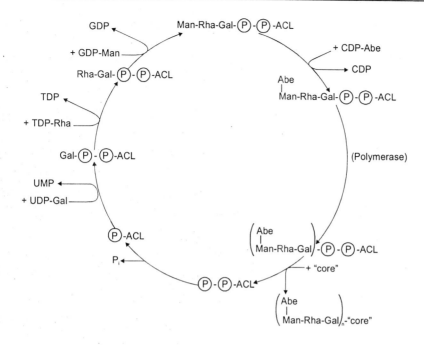

Fig. 2.25. Main pathway of the biosynthesis of the O-specific polysaccharide of LPS from Salmonella Typhimurium (without glucose and acetyl; see text)

Glc, formed by UDP-Glc and ACL-P). In both cases, the incorporation may take place during or after polymerization. Thus, only some of the repeating units are substituted with glucose or *O*-acetyl, whereas the incorporation of galactose, mannose, rhamnose and abequose can take place only at the preceding stages owing to the substrate specificity of the individual enzymes, and the polymerisation occurs only after a completed incorporation of abequose, the incorporation of glucose or *O*-acetyl represent no prerequisite for other, following reactions. The enzymes catalysing the incorporation of the latter two components may thus "overlook" unsubstituted sites, and the resulting microheterogeneity may be quite desirable.

After having reached a certain chain length, the incorporation of further repeating units is stopped. As can be shown by means of SDS-PAGE (see Fig. 2.17), this is in most cases valid not only for one single molecule size. Besides bands probably resulting from R-type LPS fractions (not substituted with repeating units), one can find several bands originating from LPS-containing 15-20 repeating units (bimodal distribution). In some cases, additionally a weak occurrence of shorter and longer chains can be found. A mathematical model has shown that this bimodal distribution cannot be established by simple competition between the polymerization and ligation processes but is produced by a regulator protein Wzz, which might interact as a molecular chaperone (for definition see Sect. 2.3.1.1). The lack of this protein leads to electrophoretic patterns showing bands of equal

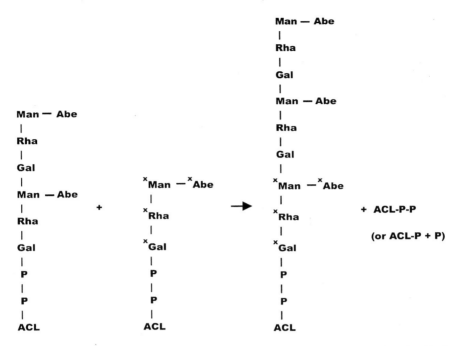

Fig. 2.26. Elongation of the O-specific polysaccharide during biosynthesis of LPS from Salmonella Typhimurium by incorporation of new repeating units

intensity in the range of 1-30 and more repeating units. In the case of *Shigella flexneri*, for instance, it is maintenance of a modal LPS chain-length distribution that is essential for full virulence.

At the last stage of LPS biosynthesis, the O-specific polysaccharide is transferred to the core, a reaction being catalysed by a ligase. This enzyme is able to transfer polysaccharides of every chain length. The mechanism of this biosynthesis stage is as yet completely unclear. In addition, there is still no generally accepted model explaining the transport of the complete LPS across the outer membrane and its fixation in its outer monolayer.

As already described (see Fig. 2.19), the O-specific polysaccharide is in LPS of *Salmonella enterica* not bound to the terminal *N*-acetylglucosamine or glucose of the core, but to the second distal glucose. During this reaction ACL-P-P is liberated. This molecule and those released during polymerization are split into inorganic phosphate and ACL-P, and the latter restarts the reaction cycle as a Gal-P acceptor. This cleavage, as well as the comparable one taking place during the biosynthesis of peptidoglycan, is inhibited by the antibiotic bacitracin, thus inducing enrichment of ACL-P-P. The target of bacitracins is the pyrophosphatase catalysing this cleavage.

Besides the described Wzy-dependent pathway, two further mechanisms for the biosynthesis of the O-specific polysaccharide have become known, i.e. one ABC-transporter-dependent and, rather recently, a synthase-dependent one.

The ABC-transporter-dependent pathway is used, e.g. for the biosynthesis of the O-specific polysaccharide of *E. coli* O9. It represents a homopolymer of the structure

[→3)-α-D-Manp-(1→3)-α-D-Manp-(1→2)-α-D-Manp-(1→2)-α-D-Manp-(1→2)-α-D-Manp-(1→]$_n$.

Its biosynthesis basically differs from the Wzy-dependent one. The starting reaction takes place at the cytoplasmic side of the cytoplasmic membrane:

ACL-P + UDP-GlcNAc → ACL-P-P-GlcNAc + UMP.

Subsequently, deviating from the scheme in Fig. 2.26 the attachment of mannose takes place at the non-reducing end in single steps, i.e. not as repeating unit, according to

ACL-P-P-GlcNAc + GDP-Man → ACL-P-P-GlcNAc-Man + GDP.

The incorporation of (1→2)- or (1→3)-bound mannose molecules is governed by different genes. In this pathway, a polymerase is not necessary.

The completed chain is transported across the cytoplasmic membrane and the periplasm by means of an ABC-transporter system, namely one ABC and two auxiliary proteins; however, the transport across the outer membrane is still unknown to a great extent. It has been hypothesised that the fusion zones between the cytoplasmic and the outer membrane are essential for this transport; the completed LPS could be pushed through it from the inner surface of the cytoplasmic membrane directly to the outer surface of the outer membrane; thus, a passage across both membranes and the cytoplasm would not be necessary.

In addition, a synthase-dependent pathway has been proposed in which the nascent O-specific polysaccharide is simultaneously elongated and extended across the plasma membrane. This pathway has so far been characterised only for plasmid-encoded LPS biosynthesis in *Salmonella enterica* serovar Borreze (O-group 54). Plasmids represent self-replicating genetic elements which consist of circular DNA of an essentially lower molecular mass than chromosomal DNA. In the bacteria they mostly exist separately from the chromosome, but may also be integrated by it. In this case they are called episomes.

2.2.5.4 Genetic Determination of the LPS Biosynthesis

As already described, LPS biosynthesis is not governed directly by the nucleic acids but via enzymes. Only a few of these have a transmembrane character; most of them are bound at the inner surface of the cytoplasmic membrane or are even water-soluble. They cannot be found in the periplasm or the outer membrane. The events are described in the following for Salmonella Typhimurium, but they are comparable, e.g. for *E. coli* K-12.

The biosynthesis of the individual LPS blocks is determined by different gene cluster (loci) of the bacterial chromosome. Besides this, there are also other single genes involved (see Fig. 2.27).

As shown in Table 2.9, the first biosynthesis steps for the LPS are governed from the *lpx* region which is located at 4 min on the *E. coli* K-12 gene map. The *waaA* gene responsible for the incorporation of both Kdo residues is located near the right end of the *waa* cluster (see below) and determines the biosynthesis of a bifunctional enzyme [α-(2→6´)- and α-(2→4)-linkage]. Acylation with 12:0 is controlled by the gene *waaM* and that with 14:0 by *waaN*; however, for both loci other functions are discussed. The *waa* cluster is responsible for the biosynthesis of the heptose/hexose core region (Table 2.10). However, it contains not only genes for glycosyl transferases, but also genetic determinants for core modifications, attachment of the O-specific polysaccharide as well as for Kdo and heptose transfer. It consists of three operons of which the central, largest one is responsible for the biosynthesis of the outer core.

The biosynthesis of the repeating units is determined by the *wba* gene cluster. Table 2.11 shows the individual genes in case of the biosynthesis of Salmonella Typhimurium LPS.

Posttranslational modifications of the polysaccharide chain are not controlled by *wba* cluster, but by the *oafA* locus in case of O-acetylation and by the *oafR-oafE* locus in case of glucosylation.

The chainlength is controlled by the *wzz* gene which is also responsible for the bimodal distribution. The regulation proceeds via a 36.5-kDa protein which is anchored in the cytoplasmic membrane, but mainly located in the periplasmic space.

Genes responsible for LPS biosynthesis are not in all cases exclusively located on the chromosome. Plasmids and phages may also serve as carriers for such genes. One example is *Shigella sonnei*, where a plasmid of 120 MDa is responsible for the formation of smooth-type LPS. Transfer of this plasmid, e.g. to a *Shigella flexneri* serotype 2a strain, induces the formation of an LPS possessing the O-specificities of both "parents".

The role of phages is discussed in detail in connection with the lysogenic conversion (Sect. 8.5).

The gene loci responsible for the above-described ABC-transporter-dependent pathway for biosynthesis of the O-specific polysaccharide are listed in Table 2.12.

The effect of a loss of individual genes on the bacteria is variable. In the case of those listed in Table 2.9, the bacteria are generally no longer viable. In contrast, the loss of the individual *waa* genes is not lethal. The resulting loss of enzymes responsible for the core biosynthesis leads to the formation of incomplete cores and therefore to different deep rough forms (chemotypes Ra, Rb etc.; Table 2.8). In this case it is possible, however, that the O-specific polysaccharide is still produced. Because of the lack of the appropriate attachment sites, however, despite the presence of a translocase they are not incorporated, but excreted. The same may occur in the presence of the appropriate attachment sites but absence of the translocase.

Rough forms also develop if the bacteria lack the enzymes necessary for the biosynthesis of the repeating units in the broadest sense; i.e. not only those encoded by the genes listed in Table 2.11, but also those responsible for the biosynthesis of monosaccharides or their nucleoside diphosphates. In mutants lacking the polymerase, transfer of only one single repeating unit occurs (so-called SR mutants). In contrast, if the ligase is missing, the resulting rough strain accumulates the long O-specific polysaccharides bound to ACL.

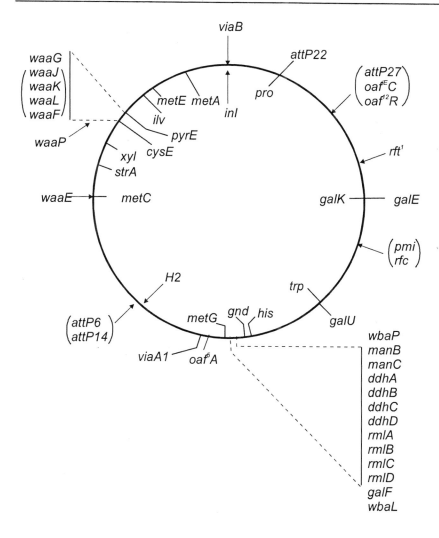

Fig. 2.27. Localisation of some genes involved in the biosynthesis of enterobacterial LPS. The genes *manB* and *manC* are involved in the GDP-mannose pathway, *ddhA*, *ddhB*, *ddhC*, and *ddhD* in the CDP-abequose pathway, and *rmlA*, *rmlB*, *rmlC*, and *rmlD* in the dTDP-rhamnose pathway.

In the case of capsule-forming strains, a lack of O-specific polysaccharides can be "covered", the bacteria appear to be smooth. The capsular polysaccharide takes over the function of the O-specific polysaccharide. This is the case, e.g. in *E. coli* O14.

Table 2.9. Genetic control of the reactions listed in Fig. 2.24

Reaction	Type of reaction	Gene
1	Acylation with 3-OH-14:0	*lpxA*
2	*N*-Deacetylation	*lpxC*
3	Acylation with 3-OH-14:0	*lpxD*
4	UMP-Liberation	
5	Condensation	*lpxB*
6	Phosphorylation	*lpxK*
7	Incorporation of 2→6′-Kdo	*waaA*
8	Incorporation of 2→4-Kdo	*waaA*
9	Acylation with 12:0	*waaM*
10	Acylation with 14:0	*waaN*

Table 2.10. Genetic control of the biosynthesis of the heptose/hexose region of the core in S. Typhimurium

Type of reaction	Gene
Binding of Hep I to Kdo I	*waaC*
Binding of Hep II to Hep I	*waaF*
Partial binding of Hep III to Hep II	*waaQ*
Binding of GlcI to Hep II	*waaG*
Binding of phosphate to Hep I	*waaP*
Binding of Gal II to Glc I	*waaB*
Binding of Gal I to GlcI	*waaO*
Binding of Glc II to Gal I	*waaJ*
Binding of GlcN to Glc II	*waaK*

Designation of the monosaccharide see Fig. 2.19.

Table 2.11. Genetic control of the biosynthesis of the repeating units and of their transport across the cytoplasmatic membrane

Reaction	Gene
Binding of Gal to ACL	*wbaP*
Binding of Rha to Gal	*wbaN*
Binding of Man to Rha	*wbaU*
Binding of Abe to Rha	*wbaV*
Binding of *O*-Acetyl	*oafA*
Binding of Glc	*oafR/oafE*
Flippase	*wbaX*
Polymerase	*wzy*
Control of the chain length	*wzz*

Kind of reactions see Fig. 2.25.

An interesting question is whether it is possible to graft the O-specific polysaccharide of the strain from one genus onto the core region of another one.

Table 2.12. Genetic control of the biosynthesis using the ABC-transporter-dependent pathway

Reaction	Gene
Binding of GlcNAc to ACL	wecA
Binding of the first Man to GlcNAc	wbdC
Binding of 1→2-Man	wbdA
Binding of 1→3-Man	wbdB
Chain transport through the membrane	wzt
Chain transport through the membrane	wzm

Such a possibility would be of importance for the development of vaccines. An *E. coli* strain with the O-specificity of Salmonella Typhi would not cause typhoid fever in humans but induce the production of antibodies directed against Salmonella Typhi and thus represent an efficient living vaccine. Successful attempts in this direction have been repeatedly described in the literature.

2.2.6 The Enterobacterial Common Antigen (ECA)

ECA is found in all enterobacteria and obviously only there. Though it does not belong to the LPS, there are, however, considerable correlations between them, because:
- it is linked to LPS in the outer membrane, partly via hydrophobic, partly via covalent bonds,
- it represents a polysaccharide with an amphiphilic character,
- its biosynthesis is partially determined by gene loci which are also essential for LPS biosynthesis.

ECA is present in bacteria in the haptenic and immunogenic forms. In the first, the polysaccharide moiety of ECA is covalently bound to the 1-phosphate of an L-glycerolphosphatide carrying in position 2 a C16:0 fatty acid and in position 3 mostly a C16:1 one, sometimes a C16:0 or a C18:0 one. The hydrophobic bonds, possible via the glycerolphosphatide, obviously suppress the immunogenicity of ECA. This is proven by the fact that purified "haptenic" ECA is immunogenic in rabbits, however, this immunogenicity can be suppressed by addition of LPS or (phospho-)lipid. In its immunogenic form ECA is covalently linked to the core of LPS carrying no O-specific chains. This bond does not alter the serological specificity of ECA. This ECA modification is limited to bacterial rough forms.

In showing these characteristics, ECA recalls certain capsular polysaccharides and there are indications that ECA forms a kind of microcapsule.

Finally, a cyclic form has been described for *Shigella sonnei*.

The polysaccharide moiety of all ECA forms consists of repeating units of the structure

→3)-α-Fuc*p*4NAc-(1→4)-β-ManN*p*AcA-(1→4)-α-Glc*p*NAc-(1→,

the GlcNAc-molecules of which are partially *O*-acetylated at C-6.

The *N*-acetylmannosaminuronic acid residue is serologically dominant, the *N*-acetylglucosamine residue of the most inner repeating unit mediates the linkage to the lipid moiety.

In SDS-PAGE, haptenic ECA shows a similar ladder pattern as S-type LPS with bands (in Salmonella Typhimurium) in the range of 12-35 kDa.

Besides the already mentioned LPS-specific gene loci, also one locus designated *rff* takes part in ECA biosynthesis; it is responsible for biosynthesis and transfer of mannosaminuronic acid.

Bibliography

Heinrichs DE, Whitfield C, Valvano MA (1999) Biosynthesis and genetics of lipopolysaccharide core. In: Brade H, Morrison DC, Opal S, Vogel S (eds) Endotoxin in health and disease. Marcel Dekker, New York, pp 305-330

Holst O (1999) Chemical structure of the core region of lipopolysaccharides. In: Brade H, Morrison DC, Opal S, Vogel S (eds) Endotoxin in health and disease. Marcel Dekker, New York, pp 115-154

Holst O (2000) Deacylation of lipopolysaccharides and isolation of oligosaccharide phosphates. In: Holst O (ed) Methods in molecular biology, vol 145, Bacterial toxins: Methods and protocols. Humana Press, Totowa, New Jersey, pp 345-353

Jansson PE (1999) The chemistry of O-polysaccharide chains in bacterial lipopolysaccharides. In: Brade H, Morrison DC, Opal S, Vogel S (eds) Endotoxin in health and disease. Marcel Dekker, New York, pp 155-178

Keenleyside WJ, Whitfield C (1999) Genetics and biosynthesis of lipopolysaccharide O-antigens. In: Brade H, Morrison DC, Opal S, Vogel S (eds) Endotoxin in health and disease. Marcel Dekker, New York, pp 331-358

Knirel YuA, Kochetkov NK (1994) The structure of lipopolysaccharides of Gram-negative bacteria. 3. The structure of O-antigens (a review). Biochemistry (Moscow) 59:1325-1383

Kuhn H-M, Meier-Dieter U, Mayer H (1988) ECA, the enterobacterial common antigen. FEMS Microbiol Rev 54:195-222

Lottspeich F, Zorbas H (eds) (1998) Bioanalytik. Spektrum Akademischer Verlag, Heidelberg

Mamat U, Seydel U, Grimmecke D, Holst O, Rietschel ET (1999) Lipopolysaccharides. In: Barton D, Nakanishi K, Meth-Cohn O, Pinto BM (eds) Comprehensive natural products chemistry, vol 3: Carbohydrates and their derivatives including tannins, cellulose, and related lignins. Elsevier, Oxford, pp 179-239

Rauen HM (1964) Biochemisches Taschenbuch Bd. II. 2. Auflage, Springer-Verlag Berlin-Heidelberg-New York

Reeves P (1994) Biosynthesis and assembly of lipopolysaccharide. In: Ghuysen M, Hakenbeck R (eds) Bacterial cell wall. New comprehensive biochemistry, vol 27. Elsevier; Amsterdam, pp 281-317

Reeves PR, Hobbs M, Valvano MA, Skurnik M, Withfield C, Coplin D, Kido N, Klena J, Maskell D, Raetz CRH, Rick PD (1996) Trends Micobiol 4:495-503

Reisser D, Arnould I, Maynadie M, Belichard C, Coudert B, Jeannin JF (1999) Lipid A OM-174 increases the natural killer activity of peripheral blood cells from breast cancer patients. J Endotoxin Res 5:189-195

Rick PD, Raetz CRH (1999) Microbial pathways of lipid A biosynthesis. In: Brade H, Morrison DC, Opal S, Vogel S (eds) Endotoxin in health and disease. Marcel Dekker, New York, pp 283-304

Schnaitman CA, Klena JD (1993) Genetics of lipopolysaccharide biosynthesis in enteric bacteria. Microbiol Rev 57:655-682

Seydel U, Wiese A, Schromm AB, Lindner B, Brandenburg K (1996) Function and activity of endotoxins: a biophysical view. Jahresbericht, Research Center Borstel, Germany, pp 7-52

Stryer L (1991) Biochemie. Spektrum Akademischer Verlag Heidelberg

Wilkinson SG (1996) Bacterial lipopolysaccharides - themes and variations. Prog Lipid Res 35:283-343

Zähringer U, Lindner B, Rietschel ET (1994) Molecular structure of lipid A, the endotoxic center of bacterial lipopolysaccharides. Adv Carbohydr Chem Biochem 59:211-276

Zähringer U, Lindner B, Rietschel ET (1999) Chemical structure of lipid A: recent advances in structureal analysis of biologically active molecules. In: Brade H, Morrison DC, Opal S, Vogel S (eds) Endotoxin in health and disease. Marcel Dekker, New York, pp 93-114

2.3 Proteins of the Outer Membrane of Gram-Negative Bacteria

2.3.1 General Remarks

Proteins represent higher- and high-molecular compounds which are completely or mainly composed of amino acids peptidically linked with each other. The respective names reflect the components that are additionally present in the protein molecule: glycoproteins possess a sugar component, lipoproteins fatty acids, and phosphoproteins phosphate. They have all been found in the Gram-negative cell wall, glycoproteins predominantly in S-layers, flagellae and fimbriae.

2.3.1.1 Composition and Structure

Most proteins are composed of up to 20 different amino acids (essential amino acids). This small number of the basic constituents of proteins is sufficient to create the extraordinary diversity of their structures and functions. The drastic differences between, for example, structural proteins, transport proteins, receptor proteins, enzymes and toxins are possible because each of the 20 amino acids is unique in its structure and function (spatial requirements, charge as well as hydrophilicity/hydrophobicity of the side chain). Thus, each amino acid provides a clearly defined site in the peptide chain with clearly defined properties. Therefore the amino acid sequence in a protein is not accidental, but determined by the qualities of the protein that have been developed in the course a selection lasting millions of years. Moreover, it is often not only one property that renders an amino acid suitable for its incorporation at a defined point of a protein chain, but a combination of several characteristics. This is proved, e.g. by the fact that norleucine does not occur as a constituent of proteins. Methionine, incorporated in ist stead, develops the same spatial parameters, but contains in addition a lone electron pair at the sulfur atom with which it is able to furnish metal-ligand bonds. Thus, methionine is superior to norleucine and the use of both would be "uneconomical" for metabolism.

The essential amino acids can be subdivided into unpolar hydrophobic, polar, but uncharged, and charged amino acids. The first group is of great importance for the formation of hydrophobic regions in the membrane proteins. The other two groups are essential for the formation of intra- or inter-molecular H-bridges or ionic bonds as well as for the formation of defined charge distributions in the proteins.

For the description of the complete protein structure one has to distinguish between the linear primary structure and the three spatial ones, secondary, tertiary, and quarternary structure.

The primary structure stands for the sequence of the amino acids in the proteins, from which the other three structures result. Up to now, the relations between primary structure, tertiary structure and functions are still incompletely understood. It is expected, however, that it may be possible in the future to model the relations in such a way that the spatial structure of the proteins may be calculated from a given amino acid sequence. The primary structure is schematically outlined in Fig. 2.28.

Except for a very few cases, the peptide bond serving as linking region for the amino acids exclusively extends from the carboxylic group of the first amino acid to the amino group at the α-carbon atom of the second. Thus, chain branchings are impossible; the peptide chains in the proteins are linear without exception.

The peptide bond represents a mesomeric system which can be approximatively reflected by the limiting structures depicted in Fig. 2.29. The free electron pair of the nitrogen can be used for the formation of a mesomeric system with a double bond between carbon and nitrogen atom as a possible limiting form. This system is plane, it prevents free rotability, thus reducing the number of possible conformations, and therefore represents an important stabilising factor of the spatial structure of the peptide chains.

The expression secondary structure stands for the conformation of the basal scaffold of the peptide chains. The most important types for cell wall proteins are the α-helix and the β-sheet structure.

In the case of the α-helix, the peptide chains appear to be screw-shaped, coiled around a cylinder (Fig. 2.30). The stabilization of this structure occurs by means of hydrogen bridges between NH groups of one winding and CO groups of the respective next winding. Each turn contains 3.7 amino acids, the height of one helix reaches 5.44 Å. Some amino acids interfere with the formation of an α-helix for steric reasons (e.g. proline or leucine) or because of their charge (e.g. glutamic acid or lysine), especially in those cases when they repeatedly appear adjacent to each other. In space models of proteins, α-helical regions are represented as spirals (see e.g. Fig. 2.33).

The β-sheet structure is created in such a manner that H bridges are formed between parallel-running peptide chains (Fig. 2.31). Because of the binding angles at the C- and N-atoms, the resulting structure resembles a manifold folded sheet (sheet structure). The side chains of the amino acids stretch more or less perpendicularly out of the plane. Depending on whether two adjacent peptide chains run in codirectional or opposite direction, the sheet is named parallel or antiparallel, respectively. In structural protein models, β-sheet regions are represented as flat arrows (see e.g. Fig. 2.32).

The three-dimensional arrangement of the polypeptide chains is called the tertiary structure. Low-energy linkages as well as mainly hydrophobic linkages and hydrogen bridges, but also ionic and van der Waals bonds, play an important role in the stabilisation of this structure. Essentially more stable is a fixation by S-S bridges between two cysteine residues. For energetic reasons, the peptide chains in the proteins are generally arranged in such a way that the hydrophobic regions point inwards. However, membrane proteins, especially the integral ones, are exceptional since they are located in a hydrophobic environment. In many proteins there are densely packed regions of 40-400 amino acids that can be regarded as structural entities within the tertiary structure; they are called domains.

In general, the natural tertiary structure represents the thermodynamically most favourable three-dimensional arrangement. α-Helical regions may coexist in this

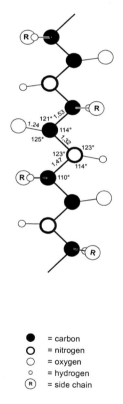

●	= carbon
○	= nitrogen
◯	= oxygen
○	= hydrogen
Ⓡ	= side chain

Fig. 2.28. Primary structure of proteins: part of a peptide chain. (H. Rauen, Biochemisches Taschenbuch Bd. II, 1964, Springer Verlag, Heidelberg)

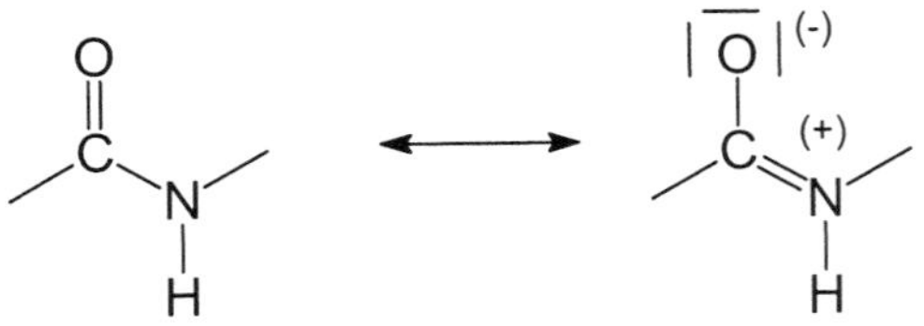

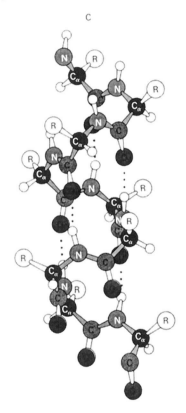

Fig. 2.29. Mesomeric limiting structures of the peptide bond

Fig. 2.30. Model of an α-helix. *C, H, N* Element symbols; C_α α-carbon in amino acids; *R* specific residues in amino acids; *dotted* hydrogen-bonds between NH- and CO-groups (Stryer, Biochemistry, 1997, W.H. Freeman Publ., New York, with permission)

structure with β-sheets (so-called secondary structural regions), loops and even disordered areas. Interactions of secondary structural regions with each other may take place, frequently under formation of so-called super-secondary structuralregions. As an example, the β-barrel (Fig. 2.32) is given, which often

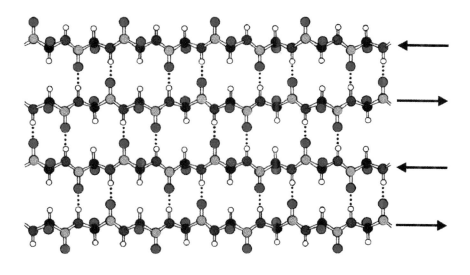

⬤	= C
●	= C$_\alpha$
◐	= O
◕	= N
o	= H
◕	= side chains
•••	= H-bridges

Fig. 2.31. Model of a antiparallel β-sheet (Stryer, Biochemistry, 1991, W.H. Freeman Publ., New York, with permission)

represents a combination of β-sheet regions with smaller α-helical ones and which plays an important role in the proteins of the outer membrane of Gram-negative bacteria. Such arrangements are called motifs. β-Barrel structures are typical for outer membrane proteins.

Because of the mostly low energy of its stabilising linkages, the tertiary structure is a rather labile arrangement. Heating, high energy radiation, changes in pH, organic solvents, certain precipitating agents or detergents may cleave the bonds and cause disruption of the tertiary structure (denaturation). However, in accordance with the fact that the tertiary structure is determined by the primary structure and mostly represents a structure of lowest energy, it could be proven that a complete renaturation of a denatured protein is possible under suitable conditions.

Quarternary structure means the association of several peptide chains to a defined complex molecule. The reason why large proteins are composed of subunits is explained by their biosynthesis. It is quite obviously disadvantageous to synthesise polypeptide strands exceeding a particular size as a whole. It is better to

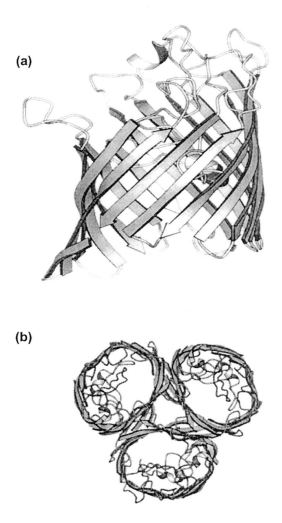

Fig. 2.32 a and b. Ribbon diagram of an OmpF porin. (a) View of a monomer within the membrane plane towards the face of the barrel; (b) View of a trimer looking down the threefold axis (Watanabe M,   et al. (1997): Computer simulation of the OmpF porin from the outer membrane of *Escherichia coli*. Biophys. J. *72*, 2094-2102, with permission)

synthesise parts of them that contain the information aligned in such a way that the fulfilment of the functions of the whole protein will be guaranteed and, in this manner, form the macromolecule.

Even if the naturally occurring three-dimensional structure of a protein represents the thermodynamically most favourable one, it is not in every case the only possible one. In pursuance of their biological functions (e.g. opening and closing of pores in the outer membrane of Gram-negative bacteria), alterations in the three-dimensional structure may occur.

The development of the complete spatial protein structure during biosynthesis is a process not yet completely understood. It is possible that chaperones play a role; these are substances turning the folding process into defined directions. In general, chaperones represent proteins, but chaperone functions have also been proved for phosphatidyl ethanolamine (a lipid). In addition, the LPS, in particular their core regions, are necessary for an appropriate folding of (mainly porin) proteins.

An important characteristic for membrane proteins is the hydropathy index. It defines the ratio of hydrophobic to hydrophilic amino acids in a given region of the primary structure.

2.3.1.2 Isolation and Analysis

The isolation of proteins from bacteria depends very much on their nature and on that of their accompanying molecules. The bacteria usually have to be destroyed by mechanical disintegration (sonication, X-press) or autolysis, and afterwards a crude solutions. Enzymatic extractions, also, play a role, however mainly with Gram-positive bacteria. Finally, bacteria constructed by genetic manipulation or by cultivation in the presence of antibiotics (e.g., penicillin) lack the stabilising peptidoglycan, and therefore can often be disintegrated already by osmotic shock.

The purification is mostly a multistage process. In a first stage the protein is often precipitated by means of ammonium sulfate, alcohol or trichloroacetic acid, followed by ion-exchange chromatography, gel chromatography, hydrophobic chromatography and/or preparative electrophoresis. Immunochromatography is especially efficient, using as solid phase carrier substances that are loaded with specific antibodies; the proteins adsorbed at neutral pH are eluted, e.g. by urea or guanidine solutions. In this case, the procedure used depends also on the nature of the protein and of the accompanying molecules. The more purified the proteins are, the more easily they undergo denaturation.

Estimation of the purity of a protein occurs mostly by means of electrophoretic methods (often by using antibodies: "blotting"), but also by means of chromatography or ultracentrifugation. It is recommended to carry out the determination of purity using at least two methods that are based on different methodical principles. Problems may still occur due to the microheterogeneity of the proteins, i.e. small deviations in the molecules of a particular protein in composition, charge and/or molecular mass which during analysis may lead to band broadening or splitting.

Prior to analysis of the primary structure of a protein, its molecular mass is determined. Polyacrylamide gel electrophoresis, gel chromatography, ultracentrifugation or light scattering serve well for these purposes.

The determination of the primary structure of a polypeptide or a protein starts with a mostly acidic total hydrolysis and the quantitative determination of the amino acids (compositional analysis). To identify the amino acid sequence (sequencing) according to Edman (1967), the N-terminal amino acid is transformed into the phenylhydantoin derivative by reaction with phenylisothiocyanate and then selectively cleaved. The resulting peptide shortened by one amino acid is submitted to the same procedure. These steps are repeated until the total sequence of the peptide is determined. This method has been automated.

The described method is appropriate to determine sequences of peptides consisting of up to 50 amino acids. In order to analyse larger molecules, these have to be selectively cleaved into defined fragments. Enzymes like trypsin (cleavage of the chain C-terminally at arginine or lysine) or chymotrypsin (cleavage C-terminally at phenylalanine or tyrosine), but also bromcyan (cleavage C-terminally at methionine) are used for this purpose. Before such reagents can be applied, the disulfide bridges of the protein have to be reductively cleaved. All resulting peptides are submitted to Edman sequencing. By combining the sequences of the peptides created by the individual cleaving reagents, the sequence of the whole protein can be concluded.

Recently, it has become possible to deduce the amino acid sequence from the DANN base sequence of the gene region responsible for the encoding of the biosynthesis of the respective protein.

The three-dimensional structure of the proteins is mainly determined by means of X-ray structure analysis of protein crystals. The crystallization of high molecular substances like proteins demands great experimental skill and has not been successful in all cases. X-ray structure analysis frequently starts with low resolution (5-6 Å) as this simplifies the interpretation. In this case, the degree of resolution is not sufficient to exactly localise the individual atoms, but the manner of folding of the peptide chain can be detected. To gain detailed information, resolutions of 1.5-2 Å are necessary. Much less expensive, but also much less informative than X-ray structure analysis are electron microscopy or the recording of vibrational spectra.

To determine the quarternary structure, in principle, the same methods are used. In addition, however, it is necessary to investigate the possibilities of cleaving the molecules into subunits. For this purpose, mild procedures are used, e.g. the application of detergents or urea.

2.3.2 Proteins of the Outer Membrane

For the following descriptions, *E. coli* mostly serves as an example. It is emphasised, however, that comparable proteins appear also in other Gram-negative bacteria.

2.3.2.1 Classification

Up to 50% of the total mass of the outer membrane consists of proteins (outer membrane proteins, OMP). The predominant part of this mass, however, is distributed among only a few protein species (major proteins). In addition, the membrane contains quite a number of proteins in low concentrations (minor proteins). The distinction between major and minor proteins refers only to the quantity of the proteins, and not to their importance. The major proteins amount to $>10^5$, the minor proteins frequently to less than 10^3 copies per cell. There are cases, however, where minor proteins are produced in quantities corresponding to those of major proteins under certain growing conditions. Thus, the receptor protein for the *E. coli* phage λ apparently becomes a major protein during the growth of the bacteria on maltose as the sole C source.

Another kind of classification of the membrane proteins results from its spatial arrangement: integral proteins penetrate the membrane totally, peripheral only partially.

The major part of the OMPs plays a role in the uptake of nutrients and the stabilization of the membrane. In addition, some proteins exist that are important for secretory processes of macromolecules, and a few enzymes.

The nomenclature of OMPs is carried out in relation to the segment of the gene map responsible for its biosynthesis.

2.3.2.2 Detection and Isolation

Since the outer membrane is rather poor in major protein species, their detection and semi-quantitative determination is not very difficult. The bacteria are cultivated until mid-log phase, harvested and disrupted by sonication. The released cell envelopes are isolated by ultracentrifugation, from which the cytoplasmic membranes are removed by means of the detergent sarcosyl (*N*-lauroyl sarcosine), and the cell walls isolated by repeated ultracentrifugation. The OMPs are liberated by boiling with SDS buffer in the presence of mercaptoethanol. SDS-PAGE is used for the analytical separation of the proteins (see Sect. 2.2.3.2). The protein patterns obtained by staining with Coomassie blue or alkaline silver nitrate are very characteristic for a given strain under standardised cultivation and separation conditions. Therefore OMP pattern analysis can be used for the epidemiological characterisation of Gram-negative bacteria.

A number of procedures have been developed for the isolation and preparative purification of the individual OMPs. As an example, the method developed by Hindennach and Henning (1975) is mentioned, which provides a good yield of individual proteins. The cell envelope fraction obtained from *E. coli* cell paste by mechanical disintegration and centrifugation is pre-fractionated by extraction with SDS/MgCl$_2$ in Tris/HCl buffer and subsequent extraction with SDS/EDTA in Tris/HCl buffer and then separated into the individual proteins by gel-permeation chromatography. Purity control occurs by means of SDS-PAGE.

2.3.2.3 The Channel-Forming Proteins

The lipid bilayer of the outer membrane represents a barrier that can hardly be penetrated, especially by hydrophilic substances. To guarantee the supply of the bacteria with nutrients and the disposal of metabolic waste products, the outer membrane has to be rendered partially permeable. This is done by embedding channel-forming proteins into the membrane. Three kinds of such proteins are known: porin proteins, specific channel-forming proteins and high-affinity receptors.

2.3.2.3.1 Porin Proteins. Classical porins pass through the membrane vertically to its plane. In their interior, water-filled transmembrane channels permits, by simple diffusion, the penetration of hydrophilic molecules with a molecular mass of up to about 600 Da. The porin proteins consist of subunits in the range 30-50 kDa; they can be separated from each other by means of SDS-PAGE. In *E. coli* K-12 three porin proteins could be detected, two cation-selective ones (OmpC, OmpF) and one anion selective one (PhoE), the latter produced in the case of phosphate deficiency. These porin types show a high degree of sequence homology; further non-related channel-forming proteins show a different sequence, but nevertheless a similar tertiary structure.

Most porins possess a β-barrel structure formed by antiparallel β-strands, as well as by three short α-helices (Fig. 2.32). On the outside they are linked together by large loops, on the periplasmic side only by short turns. The α-helices are part of the loops. The β-strands in the β-barrel are mainly held together by hydrophobic linkages. The number of β-strands varies from porin to porin, in the case of OmpF and PhoE amounting to 16. In many porin proteins three β-barrels are associated to form a trimer; the linking of the barrels to each other is stabilised mainly by hydrophobic, but also by hydrophilic bonds. Thus, three channels are present directly adjacent to each other. The high degree of homology between OmpC, OmpF and PhoE is consistent with the observation that heterotrimers of all three porins can be formed (even with LamB). In case of OmpF, the β-strands are inclined 30-60° towards the trimeric axis. On the whole, this arrangement creates a relatively wide (about 23 Å) hydrophilic channel with an elliptic cross-section. One of the loops (L3, consisting of 43 amino acid residues) linking the β-strands is folded in such a way that it is directed towards the channel interior. Thus, in an only approximately 9-Å-wide zone the channels are narrowed into an eyelet with the dimensions 7×11 Å.

This arrangement, wide channel/narrow central eyelet, is of great advantage, since it allows a high diffusion rate without loss of the sieving function for molecules with a molecular mass of more than about 600 Da. An entirely narrow channel would distinctly reduce the diffusion rate due to the greatly increased interactions between the solutes and the channel wall. In the case of OmpC and OmpF, the eyelet in the interior of the channel is much more negatively charged than in the case of PhoE. Because of these charges, transport of hydrophobic substances is excluded on the one hand and, on the other, cations are preferred in the case of OmpC and OmpF and anions in the case of PhoE.

The walls of the porin channels predominantly contain hydrophilic amino acids, resulting in exclusion of transport of lipophilic substances which is reinforced by the construction of the loop forming the eyelet in the interior of the channel. On one side it contains several negatively charged amino acids, and on the other several positively charged. Due to the electrostatic forces caused by this arrangement, a very stable and well-defined spatial structure results with a constant inner diameter opposing the influx of lipophilic substances. PhoE contains in its channel many more lysine and arginine residues instead of neutral amino acid residues, as in the case of OmpF (glycine, valine, asparagine). The exchange of a glycine residue for lysine in the constriction zone seems to be of great importance.

This arrangement is very stable; this is also proved by the fact that the porins are amazingly resistant to proteases and detergents. A rather old isolation procedure is based on this fact. Upon dissolving intact cell envelopes in a 2% SDS solution at 60 ^{0}C, porins remain together with the peptidoglycan and Braun´s lipoprotein (see below) in the insoluble material. Then the porins can be extracted either by heating in the same SDS solution to 100 ^{0}C for 5 min or by incubation with SDS solution containing 0.5 M NaCl at 37 ^{0}C. In case of the first method, denaturation of the porins takes place, but not using the second one.

Because of the appropriate distribution of hydrophobic and hydrophilic amino acids in the trimers, their surface contains a hydrophobic ring which in its dimensions (about 25 Å) exactly corresponds to the non-polar (surface)parts of the outer membrane. The fixation of the trimer in the membrane plane caused by hydrophobic bonds is reinforced by the polarity of asparagic acid and glutamic acid residues in the porin region attached outwards at the hydrophobic zone. Their carboxylic groups bind divalent cations (Ca^{2+}, Mg^{2+}), through which bridges to the anionic groups of the LPS core are created. This part is about 20 Å large, the hydrophilic region directed into the periplasm only about 8 Å.

At both border lines between polar and non-polar zones of the porin, there are belts containing quite a number of aromatic amino acids; mainly tyrosine in the outer belt, phenylalanine on the non-polar side and tyrosine on the polar side of the inner belt. It is supposed that both belts cause a stabilization of the porin conformation, especially during mechanical stress in the membrane.

The eyelets of OmpC are somewhat smaller than those of OmpF. Under environmental conditions where the presence of higher concentrations of bile acids is probable (higher temperature and osmolarity, characteristic for the interior of the intestine), more OmpC and less OmpF is produced, thus opposing their influx into the cell. The processes in biosynthesis that control the ratio of these proteins are rather complicated and need sensor and regulator proteins. A further possibility to prevent the penetration of noxious substances into the cell consists in closing porin pores. This possibility has been described for several porins. It is supposed that this can be triggered by an electric potential or by changes to a more acidic pH. PhoE pores close at electric potentials of a polarity opposite to that of OmpF pores. It is discussed that the closing mechanisms set in after penetration of large quantities of charged substances (positively charged in the case of OmpF, negatively charged in the case of PhoE) into the cell and thus represent a protective mechanism. However, this opinion is not generally accepted.

Recently, for *E. coli*, but also for *Salmonella*, *Shigella* and *Pseudomonas*, an unspecific porin designated OmpG has been described, the pore diameter of which amounts to 20 Å and which thus is significantly larger than that of OmpC and OmpF. It represents a 16-stranded β-barrel lacking the L3 loop mentioned above. It obviously produces only monomers and is present in rather small concentrations (Fajardo et al. 1998).

Meanwhile, a considerable number of porins of diverse species has been investigated (e.g. from *Rhodobacter capsulatus*, *Haemophilus influenzae*, *Klebsiella pneumoniae*, *Neisseria meningitidis*, *N. gonorrhoeae*, *Bordetella pertussis*). Despite individual structural characteristics, the basal structural principle is present in all these species.

2.3.2.3.2 Proteins Forming Specific Channels. Some substrates necessary or favourable for growth and multiplication of Gram-negative bacteria are too large for a sufficiently fast diffusion through the porin channels. On the other hand, adequate-sized substrates are often present in such low concentrations that a sufficient supply of the bacteria by means of simple diffusion through the porin channels is not possible. In these cases, the transport is carried out through specific channels, of which the LamB-supported transport of maltose oligosaccharides has been best investigated.

In 1975 it was detected that *E. coli* mutants that lack the receptor protein for phage λ, possess a significantly reduced uptake velocity of maltose. Later, it was demonstrated that this LamB protein produces channels across the outer membrane which have a high specificity for maltose and malto-oligosaccharides (malto-porins), but also permit unspecific diffusion of amino acids and non-related monosaccharides. The unspecific diffusion can be inhibited by higher malto-dextrins. This indicates the existence of a specific binding site in the interior of the malto-porins. The concentration of the substrate at the cell wall is essentially increased by its binding to this site and therefore also the speed of its diffusion to the inside.

In normally growing bacteria, LamB is present in only small amounts, but its biosynthesis is inducible. Mutants lacking the LamB protein grow poorly at low concentrations of maltose as sole C source. *E. coli* wild-type strains loaded with anti-λ-receptor protein antibodies exhibit a highly reduced maltose transport. In contrast, the phage adsorption cannot be inhibited even by high maltose concentrations. This shows that there are different binding sites for the malto-dextrins and for phage λ.

The molecular mass of the LamB protein amounts to about 47 kDa. Like the porins, it is not excessively hydrophobic according to its amino acid composition. Moreover, there are many similarities between both kinds of proteins: they form stable trimers and are rich in β-transmembrane strands, forming an 18-stranded antiparallel β barrel. In the interior of the barrel, the pore-shaped translocation pathway is situated, in which six aromatic amino acid residues create a kind of greasy slide (Van Gelder et al. 1997) that is lined up by polar residues (ionic track). Saccharide molecules are in van der Waals´ contact with their greasy slide by their hydrophobic face, while hydrogen bonds are formed between their OH groups and the ionic track residues. The substrate moves through the channel by continuous disruption and formation of these hydrogen bonds.

Neither a significant homology nor a serological cross-reaction could be found between the LamB protein and the porin proteins.

In cell division of *E. coli*, newly synthesised λ-receptor protein is incorporated during septum formation exclusively at the septation site. Only then is it distributed over the whole cell surface, as could be proven by electron microscopy after induction with ferritin-labelled phage λ.

Besides LamB, Tsx (specific for nucleoside transport) and the plasmid-encoded ScrY (transport of sucrose) play a role in *E. coli*, in *Pseudomonas aeruginosa* OprB (transport of glucose), OprD (transport of basic amino acids), OprP (transport of phosphate) and OrpO (transport of pyrophosphate) have been described.

The Tsx protein represents the receptors for phage both T6 and colicine K. It is able to clearly distinguish between cytidine and thymidine and to transport deoxynucleosides better than comparable nucleosides. Free bases or nucleoside monophosphates are not transported. Tsx channels have only 1/100 of the conductivity of OmpF channels.

In addition, the TolC protein is mentioned here, although in most cases it does not contribute to the inward transport. It represents a multifunctional outer membrane channel which plays a role for OMP regulation and for the secretion of proteins (colicines, hemolysines; TolC = tolerance to colicins) from the bacterial cell.

2.3.2.3.3 High-Affinity Receptor Proteins.
Besides the two types of transport proteins in the outer membrane of Gram-negative bacteria described above, some transporters of high affinity for the uptake of substances like Fe^{3+} complexes or vitamin B_{12} exist, which do not form mere channels and are, additionally, dependent on energy supply from the cytoplasmic membrane.

Iron Transport Proteins. Iron ions play an essential role in cell metabolism. Therefore, the sufficient supply of the bacteria is of crucial importance. Because of the extremely sparing solubility of Fe^{3+} ions (ferric ions), which represent the only stable ones in the presence of oxygen and in a neutral milieu, the uptake of sufficient quantities can take place only in the form of soluble chelates. The chelators involved in iron transport (see Sect. 8.1.2.1) are called siderophores. As the iron-loaded siderophores are too large for transport through the porin pores, the outer membrane contains specific receptor-/transport systems formed by transmembrane proteins. According to the respective siderophore system, these proteins show characteristic molecular masses in the range 74-83 kDa. Their biosynthesis is iron-regulated. Most bacterial strains develop more than one transport system.

Table 2.13 Iron transport proteins of the outer membrane of *E. coli*

Receptor	Molecular mass (kDa)	Uptake from	Receptor (colicin/phage)
FecA	80.5	Fe^{3+}-citrate	
FepA	81	Fe^{3+}-enterobactin	ColB
FhuA	78	Ferrichrome	ColM, phages T1, T5, UC-1,Ø80
FhuE	76	Fe^{3+}-coprogen	
IutA	74	Fe^{3+}-aerobactin	Cloacin DF13
Cir	74	Catecholate-mediated	ColI
FoxB	66+26	Ferrioxamine B	

The different receptor-/transport-proteins can be separated by SDS-PAGE; the patterns obtained are frequently used for taxonomic or epidemiological purposes. Table 2.13. presents a selection of some transport proteins.

As an example for the structural principles of the iron-uptake proteins, the enterobactin system (cyclic triester of 2,3-dihydroxy-N-benzoyl-L-serine, named also enterochelin) is described; other uptake systems do not differ in principle.

With regard to the spatial structure, there are basically no differences between the receptor/transport protein for enterobactin (FepA) and the porin proteins. The molecular mass of 79.9 kDa of the former is much higher, but, like the porins, it contains a great number (22) of amphiphilic β-sheet strands passing through the outer membrane and forming a pore by producing a huge β-barrel. Similarly to the porins, the proteins exist as trimers. One difference, however, consists in the fact that an N-terminal plug domain exists, which is located inside the barrel and thus obstructs the channel interior. On the top of the plug domain is a specific binding pocket for the iron-loaded siderophore that is mainly padded by aromatic residues of the plug domain. The removal of this plug by site-directed mutation transforms the iron transport protein into a variant forming an energy-independent unspecific channel with a larger diameter (20 Å) than the porin channels. Supply of energy to the intact siderophore, where the TonB protein (see below) plays a decisive role, causes a modification of the plug conformation, in this way opening the channel and leading to a specific transport into the periplasm of the chelate.

The receptor/transport proteins of the iron transport can be misused for the uptake of colicins and some antibiotics (Sect. 8.1.2.1), and as phage receptors.

Vitamin B$_{12}$ Receptor Protein. Vitamin B$_{12}$ is much too large to be transported through porin channels. Its uptake occurs by means of a protein which, in addition, was found to serve as a receptor for phage BF23 and the colicins E$_1$, E$_2$ and E$_3$, namely the *btu* gene encoded protein with a molecular mass of 60 kDa. The outer membrane contains about 200-300 vitamin B$_{12}$ receptors per cell. However, also mutants containing only one to two receptors per cell are still sensitive towards phage BF23 and are able to transport sufficient amounts of vitamin B$_{12}$. Binding of colicin E$_3$ leads to a blockade of vitamin B$_{12}$ uptake. Thus, the receptor protein carries out all three functions by making use of the same structural elements.

TonB. The above-mentioned high-molecular outer membrane proteins, which are involved in the active transport of diverse large substances like iron siderophores and vitamin B$_{12}$, but also for the penetration of group-B colicins or certain phages through the outer membrane, signal the presence of substrates from the cell surface into the cytoplasm together with a deeper-situated protein, named TonB (e.g. T1 = T-one → Ton). This represents a module transducing energy from the electrochemical potential of the cytoplasmic membrane.

The molecular mass of TonB (26 kDa in case of *E. coli*) is relatively low. In most cases, the TonB proteins of the individual bacterial species are similar but not identical to each other. In general, they are characterised by a high content of proline (17%).

The primary structure of TonB is known. It consists of three functional domains: an N-terminal hydrophobic one, a hydrophilic rigid central one, and (in *E. coli*, but not generally) a hydrophobic C-terminal one. TonB is anchored in the cytoplasmic membrane with its hydrophobic N terminus. In addition, it serves as a linking element for two proteins necessary for the development of full efficiency of TonB. These proteins are named ExbB and ExbD. ExbB (26.1 kDa) spans the cytoplasmic membrane threefold and forms a loop in the cytoplasm with a very large region; ExbD (15.5 kDa) spans the cytoplasmic membrane only once and is localised with its major components in the periplasmic space. The central domain of TonB contains two proline-rich regions of a rod-like structure spanning the periplasmic space. The C-terminus brings about the contact with the outer membrane. It is essential for the activity of TonB; the loss of 15 amino acids at its C-terminus eliminates this activity.

There are indications that the Ton system is responsible not only for the energy transport from the inside to the outside, but also for the signal transmission in the opposite direction.

2.3.2.4 The Structure Proteins

2.3.2.4.1 Braun´s Lipoprotein. When boiling EDTA-treated cell envelopes of log-phase grown E. coli in 4% SDS solution, all components (proteins, phospholipids, LPS) are removed and merely peptidoglycan is left behind containing a covalently bound lipoprotein, which can be cleaved by means of trypsin or lysozyme. Braun and Rehn in 1969 were the first to report on this lipoprotein. The molecular mass of Braun´s lipoprotein amounts to about 7.2 kDa, and its chemical structure is known (Fig. 2.33). It consists of 58 amino acid residues, among which the proteinogenic amino acids histidine, tryptophane, glycine, proline and phenylalanine are missing. The covalent linkage to the peptidoglycan occurs via the ε-amino group of the C-terminal lysine to the carboxyl group of every 10th-12th meso-diaminopimelic acid. The N-terminal part of the lipoprotein consists of S-glyceryl cysteine to which two fatty acids are ester-linked and one fatty acid is amide-linked. The ester-linked fatty acids are the same as found in the phospholipids, the amide-linked ones consist of up to 65% palmitate and of mainly mono-unsaturated fatty acids. The fatty acids are embedded in the inner leaflet of the outer membrane.

Besides the bound form of the lipoprotein, it is also present in a free form. The bound form migrates in SDS-PAGE a little more slowly than the free form because it still contains fragments of peptidoglycan emerging from lysozyme treatment. Both forms cannot be distinguished chemically. The lipoprotein is localised exclusively in the cell wall and occurs at about 2.5×10^5 molecules per cell in the bound form and about twice as much in the free form.

The amino acid sequence of the lipoprotein is remarkable, since a duplicated sequence of 15 residues is followed by four successive duplications of the 7 C-terminal residues of the original sequence. This points to a corresponding evolution of the structural gene. The N- and C-terminal sequences of the lipoproteins flank the repetitive motif and serve as attachment regions at the lipid (N-terminal) and the peptidoglycan (C-terminal). The three-dimensional arrangement of the amino acids in the lipoprotein is such that all charged and hydrophilic amino acids are

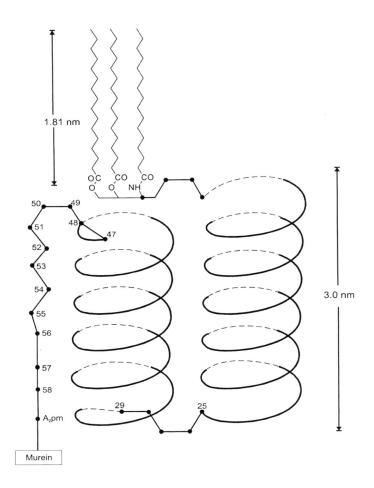

Fig. 2.33. Schematic structure of Braun´s lipoprotein (Braun V, et al. (1976): Schematic structure o f Braun's lipoprotein. Eur. J. Biochem. *70*, 601-610, Blackwell Science Ltd., with permission)

localised on one side of an α-helix, all hydrophobic on the opposite one. About 80% of the molecules are α-helically arranged.

Several spatial structures for the lipoprotein have been discussed, of which that proposed by Braun et al. (1976) seems to be the most probable. According to this, the molecule forms an α-helix interrupted by a β-insertion (amino acids 25-29). The amino acid chain thus forms a two- or three-stranded coiled coil. The three fatty acids protrude from the coil and can be incorporated into the inner monolayer of the outer membrane. As the polypeptide moiety is rather hydrophilic, it should not be embedded in the outer membrane, but rather serve to maintain the distance between outer membrane and peptidoglycan (distance 4,8 nm).

As revealed by cross-linking studies, the lipoprotein appears in the cell wall in trimeric form (one bound and two unbound units).

Braun´s lipoprotein is widespread among Gram-negative bacteria. It plays an important role in the stabilisation of the cell wall and is involved in vital processes like cell growth and multiplication. Even if mutants that lack the lipoprotein grow and divide normally, they show characteristic differences in their morphological and physiological properties. They are very sensitive to EDTA, cationic agents, detergents etc., and release considerable quantities of periplasmic enzymes. As could be shown by electron microscopy, all this occurs in connection with morphological and structural alterations. For example, the outer membrane shows a tendency to detach from the peptidoglycan layer. Such a detachment can be caused by cultivating these bacteria for some time in a minimal medium mainly containing small amounts of Mg^{2+}. In the same way, the removal of the stabilising Mg^{2+} bridges by means of EDTA leads, in the absence of lipoprotein, to damage of the outer membrane and then to cell lysis. This effect may be prevented by 0.5 M sucrose in the solution.

2.3.2.4.2 OmpA-Protein. This protein has been described for *E. coli* K-12. Comparable proteins (OmpA family) have been found especially in Enterobacteriaceae, but also in many other Gram-negative bacteria, such as OprF in Pseudomonas. OmpA is present in a quantity of up to 7×10^5 molecules per cell, its molecular mass has been determined to 35 kDa (325 amino acids).

Like the porin proteins, OmpA is embedded in the outer membrane as a transmembrane protein. Its N-terminal part with 171 amino acid residues passes through the outer membrane eight times in the form of antiparallel β-strands which are connected by three short periplasmic turns and four relatively large surface-exposed hydrophilic loops under formation of an amphiphilic β-barrel. Its C-terminal region with 155 amino acids protrudes into the periplasmic space; it has been proposed to interact specifically with the peptidoglycan layer. The interior of the β-barrel represents a water-filled cavity, a possible pore function is under discussion.

The exposed position of OmpA takes up an in the cell wall reflects its function as a receptor for certain phages and colicins (e.g. K and L). Its C-terminal part is not shielded against proteolysis in vitro.

OmpA carries out an important function during plasmid conjugation proceeding via F-pili. Strains lacking OmpA show a drastic decrease in plasmid transfer. If donor strains conjugating via F-pili are incubated with purified OmpA, a decrease in the transfer frequency hardly occurs. This takes place only after addition of LPS. However, OmpA does not, presumably, represent the real receptor for F-pili even in the presence of LPS. Rather, in its absence, the formation of the stable aggregates between donor and recipient strain indispensable for the conjugation, does not occur.

A characteristic feature of OmpA is its increased apparent molecular mass as determined by SDS-PAGE during heating to >50 °C (about 30 kDa of the non-heated versus 35 kDa of the heated). This is presumably connected with the binding of larger SDS quantities by the β-structure present at lower temperatures, which results in a higher negative charge and, thus, in faster migration in the gel, suggesting a lower molecular mass. OmpA denatured through heating under loss of the β-structure is able to refold in the presence of LPS into the native form and incorporate into vesicles again.

2.3.2.5 Further Proteins

Besides structure and transport proteins, the outer membrane contains a whole number of further (lipo-)proteins, in low quantities each, however. Their functions are only partly known. Some of them represent enzymes (phospholipase A_1, NADH oxidase, ATPase, nitrate reductase and esterases), others are involved in the export of high molecular substances (TonB-dependent receptors) or are important for virulence by neutralising host-defence mechanisms (OmpX). Some of them are summarised under the terminus OMA (outer membrane auxillary) proteins.

2.3.2.6 Biosynthesis of the Outer Membrane Proteins

In principle, the biosynthesis of the outer membrane proteins does not basically differ from that of other proteins. Some particularities are due to the fact that the polypeptide chain has to penetrate the cytoplasmic membrane from the inner to the outer side and afterwards has to be perfectly integrated into the outer membrane.

The proteins of both membranes and of the periplasmic space as well as the extracellular proteins are synthesised at polysomes, an arrangement of several ribosomes, bound at the inner monolayer of the cytoplasmic membrane and subsequently exported.

In the first step, the amino acids are activated by binding to t-RNA. The m-RNA controlling the amino acid sequence is located at the polysomes, rendering a fast translation sequence possible by means of a single m-RNA. According to the codon, the biosynthesis starts with that of a strongly hydrophobic signal sequence with a basic N-terminus important for the penetration through the cytoplasmic membrane of the growing peptide chain, followed by that of the complete primary structure in a form required for translocation. The positively charged N-terminus enables the signal sequence to interact with anionic headgroups of the membrane phospholipids, and the hydrophobic region enables it to span the membrane.

The mechanism of penetrating the cytoplasmic membrane has not yet been completely elucidated. The following model is possible:

The largely hydrophilic peptide chain passes the hydrophobic interior of the cytoplasmic membrane through specific channels being formed from one or several Sec proteins (from secretion). In the first step of this mechanism, the initially synthesised signal sequence is inserted as a loop into the inner monolayer of the cytoplasmic membrane. By means of diffusion, it finds the receptor of the translocase (SecY) of the Sec system, which represents a heterotrimeric integral me mbrane protein consisting of SecY, SecE and SecG. The respective channel (consisting of four heterotrimers) opens at the side to take up a short piece of the peptide chain to be transported which is located close to the signal sequence; at the same time, the signal sequence loop erects, and the signal sequence penetrates the whole cytoplasmic membrane (Fig. 2.34.). It should be mentioned that another mechanism is conceivable, in which the signal sequence is primarily also inserted into the channel by hydrophobic protein-protein interactions, which subsequently results in opening the channel towards the lipid bilayer through which the signal sequence finally moves. In each case, the peptide chain is then pushed unfolded through the channel into the periplasm in an energy-dependent process with the assistance of further Sec proteins (besides others SecA, an ATPase, which by cycling through different conformational states, at the expense of ATP, pushes the precursor forward through the channel); it is possible that translation and translocation are carried out in parallel.

Before being folded into their final shape, the outer membrane proteins represent rather hydrophilic molecules which are therefore able to exist in the periplasm in a soluble form, only after enzymatic cleavage of the signal sequence and folding into the β-barrel structure proceeding via several intermediate steps (unfolded → folded monomer → possibly dimer → metastable trimer → stable trimer in the case of porins; intermediate steps, but also others, have also been found in case of OmpA), the surface of the final molecular form (metastable or stable trimer) is sufficiently hydrophobic to render its incorporation into the outer membrane possible. Folding and trimerisation of the porins occur only in the presence of outer membrane fragments or of LPS bilayers; LPS during their biosynthesis are especially efficient (possibly coupling of some biosynthesis steps of both molecules). The early stages of both reactions take place on the outer surface of the cytoplasmic membrane or in the periplasmic space, the later ones either in the periplasmic space or during its incorporation into the outer membrane. Stimulation of the translocation of the

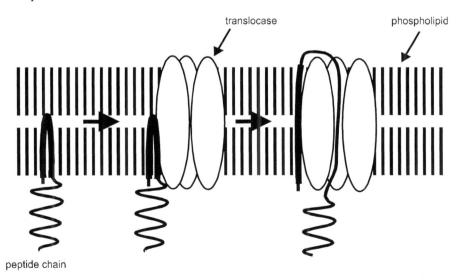

Fig. 2.34. Penetration of the cytoplasmic membrane during protein biosynthesis. Thick line: signal peptide (Pugsley AP (1993) Penetration of the cytoplasmic membrane during protein biosynthesis. Microbiol. Mol. Biol. Rev. *57*, 50-108, with permission)

unfolded early steps of the outer membrane protein biosynthesis by chaperones has been proved.

This mechanism explains the fact that membrane proteins after their incorporation into the membrane exist in a fixed state, whereas this incorporation can only be explained on assumption of a relatively great mobility.

Another problem is the targeting of the mature protein (protein targeting) to its destination site (cytoplasmic membrane, periplasm, outer membrane, secretion). This is partly solved via their hydrophobicity. The proteins remaining in the cytoplasmic membrane are already in their secondary structure (mostly α-helices) so hydrophobic that they cannot leave this membrane. The proteins destined for the periplasmic space remain hydrophilic even after folding into their final state (tertiary structure). On the contrary, the OMPs become hydrophobic by folding. For their incorporation into the membrane, sorting signals are additionally present (e.g. a nearly ubiquitous phenylalanine residue in the case of porins) that are identical in the diverse OMPs (porins, receptors, enzymes) and are located at the C-terminus. In Braun´s lipoprotein they are missing. Proteins of the PulD family situated in the outer membrane are involved in the secretion of extracellular proteins.

The biosynthesis of Braun´s lipoproteins (Fig. 2.35) offers some peculiarities. The prolipoprotein (peptide moiety plus N-terminal signal sequence) is synthesised and translocated through the cytoplasmic membrane like other outer membrane proteins. The lipid part, which is important for the incorporation into the inner monolayer of the outer membrane, is integrated in two steps: enzymatic transfer of (1.) glycerol from phosphatidylglycerol to the SH group of cysteine, and (2.) long chain fatty acids from the glycerolphosphatide pool to the free OH groups of the glycerol. From this diglyceride prolipoprotein, the signal sequence is split off and, subsequently, the free amino group of the cysteine is acylated, mainly with palmitic acid. Finally, a part of the lipoprotein is peptide-linked to every 10th-12th *meso*-diaminopimelic acid residue of the peptidoglycan by a still mainly unknown pathway via the ε-amino group of the C-terminal lysine.

The biosynthesis of the individual outer membrane proteins is to a variable extent inhibited by antibiotics, which indicates differences in their biosynthesis pathways. The biosyntheses of OmpA and of Braun´s lipoprotein are rather insensitive to puromycin, the latter being especially extreme with an inhibitory concentration of 300 µg/ml, an antibiotic that evokes a precocious interruption of the biosynthesis of the polypeptide chains at the ribosomes. Likewise, the m-RNA molecules that are specific to the individual major outer membrane proteins are differently stable against rifampicin.

2.3.3 Organisation and Function of the Outer Membrane Proteins

Electron microscopical examinations after freeze-fracture disruption (for the technique see Sect. 2.4.3), for example on S. Typhimurium, revealed the presence of many particles at the inner fracture planes of the outer membrane (OM_i) and, corresponding to this, scars at the outer fracture planes of this membrane (OM_a; Fig. 2.36). The particles consist to a great extent of outer membrane proteins, in most cases aggregated with LPS. In deep rough mutants with defective LPS

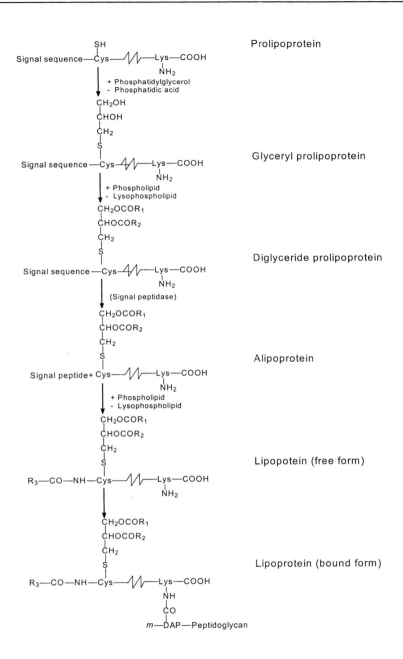

Fig. 2.35. Biosynthetic pathway of Braun's lipoprotein. R_1, R_2, R_3 Long-chain fatty acids

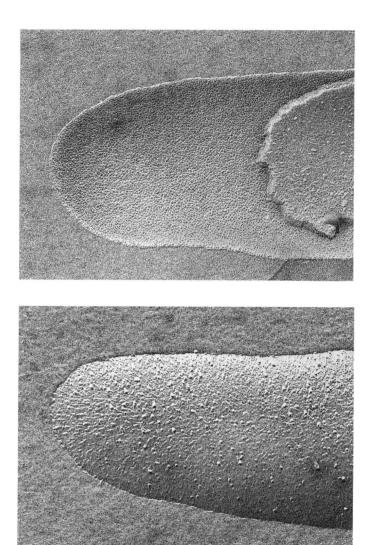

Fig. 2.36. Electron micrographs taken after freeze-fracture disruption of the outer membrane from Salmonella Typhimurium (courtesy K. Lounatmaa). x 75000. *Upper photo* OM$_a$, *lower photo* OM$_i$. In the *upper photo* parts of the cytoplasmatic membrane can be seen

structures, the amount of proteins in the outer membrane is reduced. This indicates specific interactions between proteins and LPS.

As described earlier, the proteins OmpC and OmpF are present as homotrimers, OmpA presumably as monomer and linked to Braun´s lipoprotein. An *E. coli* mutant lacking these three proteins still contained up to a maximum of 25% of the usual amount of OM$_a$ particles. In cells synthesising only one of the major proteins, the number of particles had risen again. This shows that independently of each other, each of the membrane proteins is suited for particle formation. All proteins mentioned above contain a β-barrel structure with a hydrophobic ring in the central region of the barrel surface, the width of which corresponds to that of the hydrophobic interior of the membrane. Thus, the structural prerequisites for the transmembrane localisation of the proteins and therefore for the particle formation are given. The fact that the proteins in intact bacteria are able to bind both phages and antibodies shows that their hydrophilic areas (loops) protrude relatively far outwards.

The nearly exclusive localization of the particles at the outer fracture plane indicates a stronger anchoring of the proteins in the outer monolayer. Its higher viscosity as well as hydrophilic interactions between the proteins on one hand and the core region of LPS on the other, may explain this fact.

In contrast to the proteins mentioned above, Braun´s lipoprotein cannot be detected at the bacterial surface. Due to its chemical structure, its hydrophobic part can only be incorporated into the inner monolayer of the outer membrane. Its hydrophilic part as a mechanically stable α-helix maintains the distance between outer membrane and peptidoglycan. On the other hand, the lipoprotein connects both cell wall components by anchoring them simultaneously in the outer membrane and in the peptidoglycan.

OmpA represents a structure protein providing the structural integrity of the outer membrane as well as the shape of the bacteria together with Braun´s lipoprotein as a stabilising element. Both proteins can in most cases be isolated as a complex. In addition, they can be chemically cross-linked in situ, which means that they are situated close to each other at least in the outer membrane. Strains lacking both proteins show a globular shape, need high concentrations of bivalent cations for growth and frequently show the blebbing phenomenon. They are essentially more sensitive to osmotic stress, EDTA, polycations or detergents. Their outer membrane is more permeable.

The channel-forming proteins play a crucial role in the nutrient supply of bacteria, the disposal of metabolic waste products, but also as protective components against the penetration of noxious substances. In the simplest case, the proteins forming the diffusion pores are well suited for the fulfilment of these tasks. In the case of *E. coli*, these are OmpC, OmpF and PhoE. They control the nature of the substances to be taken up. OmpC and OmpF manage the diffusion of low-molecular hydrophilic substances which are either not or positively charged. The OmpF pores (diameter 1.16 nm) are somewhat larger than those formed by OmpC (diameter 1.08 nm). This difference is negligible at molecular masses <200 Da, but sucrose (molecular mass 342 Da) diffuses about twice as fast through OmpF as through OmpC. In the case of even higher-molecular as well as of hydrophobic or anionic substances such as bile acids, penetration into the bacteria via OmpC pores is seriously reduced. Therefore, by elimination of the OmpF pores, a certain resistance can easily be created for which corresponding control mechanisms have been described. A further possibility of preventing penetration of harmful substances consists in closing the porin pores. It is assumed that this can be effected by means of an electric potential or changes of the pH values to the acidic range.

PhoE is anion-selective, it can be produced in larger quantities in the case of phosphate deficiency.

Recently, it has been proposed that OmpA forms diffusion channels through the outer membrane, though rather inefficiently. The real presence of such channels and their contribution to the total permeability of the outer membrane are still unclear. It could, however, be of importance in OmpC and OmpF deficiency.

Distinctly more selective and efficient than the diffusion through the porine pores is the transport by means of molecules that form specific channels, especially the high-affinity transport systems. In some cases (e.g. LamB), their biosynthesis is inducible. These systems are necessary to guarantee the supply of the bacteria with higher-molecular nutrients, e.g. substances too large for a passage through the diffusion pores, without leading to a general permeabilisation of the outer membrane towards higher-molecular substances, that are in most cases harmful.

A further group of proteins (e.g. the Tol complex) in the cell envelope is involved in the transport of higher-molecular substances (both outer membrane components such as LPS or porins and secreted substances like bacteriocins or toxins); their main function seems to maintain the outer membrane integrity. "Unwanted" functions of the Tol complex are the import of some colicins and uptake of filamentous phage DNA.

Bibliography

Braun V, Rehn K (1969) Chemical characterization, spatial distribution and function of a lipoprotein (murein-lipoprotein) of the E. coli cell wall. Eur J Biochem 10:426-438

Braun V, Wu HC (1994) Lipoproteins, structure, function, biosynthesis and model for protein export. In: Ghuysen JM, Hakenbeck R (eds) Bacterial cell wall. New Comprehensive Biochemistry, vol 27. Elsevier, Amsterdam, pp 319-341

Braun V, Rotering H, Ohms JP, Hagenmayer H (1976) Conformational studies on murein-lipoprotein from the outer membrane of E. coli. Eur J Biochem 70:601-610

Crichton RR, Charloteaux-Wauters M (1987) Iron transport and storage. Eur J Biochem 164:485-506

Earhardt CF (1996) Uptake and metabolism of iron and molybdenum. In : Neidhart FC, Curtiss R III, Ingraham IL, Lin ECC, Low KB, Magasanik B, Reznikoff WS, Riley M, Schaechter M, Umbarger ME (eds) Escherichia coli and Salmonella: cellular and molecular biology. 2nd ed. ASM Press, Washington, pp 1075-1090

Edman p (1967) A protein sequenator. Eur J Biochem 1:80-91

Fajardo DA, Cheung J, Ito C, Sugawara E, Nikaido H, Misra R (1998) Biochemistry and regulation of a novel Escherichia coli K-12 porin protein, OmpG, which produces unusually large channels. J Bacteriol 180:4452-4459

Hindennach I, Henning U (1975) The major proteins of the Escherichia coli outer cell envelope membrane. Eur J Biochem 59:207-213

Koebnik R, Locher KP, Van Gelder P (2000) Structure and function of bacterial outer membrane proteins: barrels in a nutshell. Mol Microbiol.37:239-253

Pugsley AP (1993) The complete general secretory pathway in Gram-negative bacteria. Microbiol Rev 57:50-108

Van Gelder P, de Cock H, Tomassen J (1997) Assembly of bacterial outer membrane proteins. In: von Heijne G (ed) Membrane protein assembly. R G Landes Comp, pp 63-82
Watanabe M, Rosenbusch J, Schirmer T, Karplus M (1997) Computer simulations of the OmpF porin from the outer membrane of *Escherichia coli*. Biophys J 72:2094-2102
Zubay GL (1998) Biochemistry, 4th edn. Wm C Brown Publishers, Dubuque, Iowa

2.4 The Total Membrane

2.4.1 General Structure of Biological Membranes

Biological membranes carry out a seemingly contradictory double function. On the one hand they draw a clear dividing line between cells or cell compartments and their environment, on the other, they have to manage the exchange of substances, energy and information between cell and environment. The exchange of substances does not only occur by unspecific diffusion processes, but the membrane is also able to preferentially transport substances that are necessary for the cell, even against a concentration gradient.

2.4.1.1 Composition, Structure, Function

The major membrane components are represented by polar lipids and proteins, but also polysaccharides, glycolipids, glycoproteins and lipoproteins occur in rather considerable quantities.

The kind and the amount of lipids contained in membranes correspond to the origin and the function of the respective membrane. They can amount between 30% (in mitochondria) and up to 70% (in myelin sheaths). Membranes which function mainly as a barrier possess a high lipid content, whereas the so-called functional membranes contain much protein. Phospholipids represent the major component of the membrane lipids; other lipids, even mostly amphiphilic ones, are more rare. Cholesterol, which is characteristic for animal membranes, is of no importance in bacteria. Instead, bacteria may use hopanoids as components that are analogous in structure and function. However, few bacterial species have been described that synthesise steroids.

Lipids may not be regarded only as building materials or structural elements, they also have important functional tasks. This is proved by the fact that the activity of many membrane-associated enzymes depends on the presence of quite particular lipids.

The membrane proteins have both structural and functional duties. They are divided into structural proteins, transport proteins and enzymes. They can partially or totally span the membrane (integral proteins) or be arranged on the membrane surface (peripheral proteins). Structural proteins often have the ability to form a two-dimensional lattice (2d proteins). They may be located both on and in the membrane and play an important role for the maintenance of structure, elasticity and asymmetry of the membrane. In contrast, transport proteins have to permeate the membrane; they are described in detail below.

In general, polysaccharides are located on the membrane, frequently anchored by a hydrophobic moiety. They are able to constitute a barrier layer especially against hydrophobic substances, but also to serve as information memory.

A number of models have been developed to explain the membrane properties. At present the fluid mosaic model (Singer 1977) is largely accepted. It proposes as basal structure a double layer formed by amphiphilic lipids in which proteins are embedded (Fig. 2.37). The hydrophobic tails of the lipids are oriented to each other in the interior of the bilayer, the hydrophilic heads are located on both surfaces of the membrane. Under physiological conditions, the interior of the membrane is in dependence of the temperature more or less thin-bodied (liquid; liquid-crystalline phase). Thus, the proteins are able to move quite rapidly in the membrane plane, both translations and rotations are possible. The immersion depth of the proteins is determined by the size of the hydrophobic areas on their surface.

The fatty acid composition is determined by the requirement that the lipid bilayer must be heterogeneously composed within its regions. Phospholipids that contain two identical fatty acid residues, e.g. mono-*cis*-unsaturated ones, tend to autaggregation and thus to the formation of quasicrystalline areas (gel-like phase) within the membrane. Such areas, which are also caused by lowering of the environmental temperature, distinctly disturb the fulfilment of the normal membrane functions. In general, however, quasicrystalline areas appear for only a short period in the absence of proteins and at normal temperatures.

The hydrophobic interactions between the immersing part of the protein and the hydrophobic lipid bilayer are very intense, and the assembly obtained is the energetically most favourable one. Each movement of a protein or lipid molecule perpendicular to the membrane plane is connected with energy consumption and is therefore not very probable under physiological conditions. Although energetic restrictions are not valid for movements in the membrane plane, they may not occur without any restrictions. Some of the proteins are fixed in stable arrangements, at least against each other, by interactions existing between them. In this case, so-called supersubunit structures occur, to which, for example, the mentioned 2d lattices belong. The sole task of some peripheral proteins is to fix integral proteins in such arrangements via electrostatic interactions or hydrogen bridges.

Since the proteins are able to strongly bind at least one layer of lipid molecules around their hydrophobic part, these are also spatially fixed. This provides the membrane with fixed structures in some areas, and the order existing there resembles that in liquid crystals.

2.4.1.2 The Asymmetry of Membranes

Membranes are asymmetrical in various respects.

First, both monolayers are differently composed. This kind of asymmetry is necessary because the environments separated by the membrane are in most cases differently composed. The asymmetry guarantees optimal transmembrane functions such as transport. It is especially well-developed in glycolipids, LPS and proteins, but also lipids may be asymmetrically distributed, though to a lesser extent. It can be maintained because an exchange of components between both membrane layers hardly occurs. All substances with a hydrophilic component have difficulties in migrating from one side of the membrane to the other because the immersion of this component into the hydrophobic interior of the membrane needs an extraordinarily high activation energy. The half-life period for a vertical penetration of the lipid bilayer amounts to magnitudes of weeks to months for phospholipids; flip-flop mechanisms are therefore not possible. However, there is a second, 10^3-fold more effective possibility, and, exclusively in growing membranes, a third, 10^5-fold more effective one for lipid exchange between both sides. In these cases, transmembrane proteins, so-called flippases, play a catalytic role, and the energetic restrictions cease to exist.

A second kind of asymmetry concerns the orientation of the membrane components. All molecules of a defined membrane protein take the same position relative to the membrane plane, i.e. they have the same immersion depth and spatial arrangement, and often also the same surrounding molecules. Unsubstituted polysaccharides cannot invade the lipid bilayer due to their lack of hydrophobic areas. Besides the distribution asymmetry, an orientation asymmetry may be found also that is established by binding at defined head groups on the membrane surface.

The third kind of asymmetry results from the domain structure of the membranes. This means that the components of the membrane are not homogeneously distributed, but in defined areas are at least significantly enriched. This becomes possible by the formation of supramolecular complexes or by the presence of matrix proteins. This kind of asymmetry is also necessary for an optimal function of the membrane.

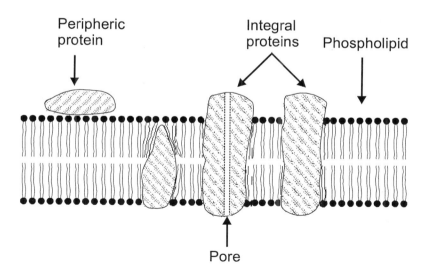

Fig. 2.37. Schematic presentation of a membrane cross-section

In the outer membrane of Gram-negative bacteria the lipopolysaccharides show an extreme asymmetry. Especially in the wild-type forms they are present exclusively in the outer monolayer, whereas the inner monolayer contains all the phospholipids. This asymmetry is caused by the long polysaccharide chains, which do not allow their incorporation into the inner monolayer.

2.4.2 Composition of the Outer Membrane

For the determination of its composition, the outer membrane must be separated from the other cell components.

Starting from intact bacteria, the cell wall is isolated by mechanical disrupture of the cells, then mild detergents are added to remove the cytoplasmic membrane, followed by centrifugation. The peptidoglycan which is still bound to the sedimented outer membrane disturbs the investigations in most cases only insignificantly; it has rather a stabilising effect on the composition and structure of the outer membrane. On the contrary, an incomplete removal of the cytoplasmic membrane may cause distinct analytical errors.

Starting from peptidoglycan-free spheroplasts (Sect. 3.2.1), the cells are mechanically disrupted, the membrane fraction is sedimented by centrifugation and separated into inner and outer membrane by means of differential centrifugation. This procedure is also not free from disadvantages. The lack of the stabilising peptidoglycan may lead to reorientations in the membrane. Besides this, an exchange of components between inner and outer membrane may occur, and parts of the outer membrane may be completely lost.

Another difficulty results from the limitations of the analytical techniques applied. Although there have been dramatic methodical improvements in the past 10 years, misinterpretations of data cannot be excluded. They can be caused both by objectively established experimental errors during analyses and from drawing false conclusions from obtained data. Thus, it is probable that present concepts on the construction of the outer membrane will have to be modified in the future.

Among Gram-negative bacteria, the qualitative composition of their outer membrane seems to be similar to a great extent. There are, however, distinct variations in the quantitative composition in dependence on both the nature of the bacteria as well as the environment, i.e. growth phase, composition of the culture medium and cultivation temperature. In general, the outer membrane contains 30-40% protein, 35-45% LPS and 25% lipid; the latter consists of approximately 75% phosphatidylethanolamine, 20% phosphatidylglycerol and 5% cardiolipin. If the biosynthesis of phosphatidylethanolamine is impossible, the compound can be replaced by its precursor phosphatidylserine.

Even within the individual substance classes variations occur:
- In LPS, differences in length distribution of the O-specific side chains, but also variations in the quantitative monosaccharide composition have been observed. The latter may influence the occurrence of an individual O-antigen factor.
  - Drastic variations may appear in the fatty acid composition of the lipids. The content of unsaturated fatty acids in *E. coli* phospholipids, which usually amounts to about 50%, may thus decrease to 15-20% without cessation of bacterial growth. In the range between 50 and 15%, the permeability of the membrane diminishes

according to the decreasing content of unsaturated fatty acid. In the case of even lower values, it rises drastically, the bacteria become leaky and finally lyse. The content of saturated fatty acids may also decrease to about 15% without serious consequences; at lower values, however, the membrane becomes more permeable.
- In proteins, both quantitative and qualitative variations have been observed. Lack of a single porin species may occur without a significant reduction in the vitality of the bacteria.

Although the quantitative proportions between individual phospholipids of the outer membrane are comparable to those of the cytoplasmic membrane, the total phospholipid content of the outer membrane is distinctly poorer. On the contrary, the cytoplasmic membrane contain no LPS. The protein composition of both membranes is also clearly different, especially with regard to enzymes. The outer membrane is markedly poor in enzymes (as well as in energy), whereas the cytoplasmic membrane contains a multitude of them.

2.4.3 Structure of the Outer Membrane

Like all biological membranes, the outer membrane of Gram-negative bacteria consists of a lipid bilayer. It is, however, asymmetric, especially with respect to the composition of its two monolayers.

The localisation of the individual components at the outer surface of the outer membrane occurs by the determination of their accessibility to non-penetrating molecules. In the case of LPS for example, ferritin-labelled anti-LPS antibodies are used. Thus, the polysaccharide chains of the O-specific polysaccharide can be presented as rows of black points in the electron microscope. For the localisation of the phospholipids, the cells are submitted to the action of different phospholipases. Subsequently, the non-hydrolysed portions of the phospholipids are analysed. Some phospholipids such as phosphatidylethanolamine can be labelled by means of CNBr-activated dextran.

The localisation of proteins may be deduced from the hydropathy index: areas of appropriate length (about 20 amino acids in the case of α-helical structures) with a high hydropathy index indicate a transmembrane segment. Regions with a low hydropathy index correspondingly suggest areas which are located outside the membrane bilayer and frequently link individual transmembrane segments with each other. The decision inside/outside can be made by monoclonal antibodies or phages.

In the smooth forms of many Enterobacteriaceae only the O-specific polysaccharide chains of LPS and some proteins (mainly porins) have been localised as surface component. In deep rough forms, especially in Rd and Re, additionally phospholipids are detectable on the surface. All investigations lead to the conclusion that in smooth forms, the outer monolayer of the outer membrane consists of LPS and (mainly transmembrane) proteins, the inner monolayer of phospholipids and proteins. The presence of phospholipids also occurs in the outer monolayer of deep rough forms, presumably due to its lower protein content. Based on the amounts of the different components, calculation of their required areas in the inner and the outer monolayer leads to similar data.

We owe essential information about the structure of the outer membrane to electron microscopy. Earlier investigations were carried out on ultrathin sections of Gram-negative bacteria in which the outer membrane shows the typical trilaminaric structure. Better insights are obtained by the freeze-fracture technique. In this procedure, the specimen is abruptly deep frozen in vacuo and subsequently cut with a microtome blade. During this process, fracture lines frequently occur in the membrane plane which can be made visible under the (scanning) electron microscope after application of the usual preparation techniques. This procedure thus allows direct observation of both cytoplasmic membrane planes. The convex planes derive from the inner and the concave from the outer monolayer.

If the interior of membranes consisted exclusively of hydrocarbon chains, only smooth fracture planes would appear. However, this is not the case (see Fig. 2.36). Instead, the already described (Sect. 2.3.3) particles and scars appear, the first of which presumably consist of membrane proteins or membrane protein-LPS aggregates. Lack of several major proteins causes a drastic decrease in the number of particles. The outer membranes of deep rough forms also show a smaller number of particles. This, also, is connected with the reduced amount of proteins in these mutants: deep rough LPS are not able to induce trimerisation of the porin proteins.

The determination of the environment of particular molecules is performed by means of cross-linking substances such as dialdehydes or diimido esters or by using fluorescence probes and subsequent recording of NMR spectra. Investigations revealed that phospholipids and LPS do not form mixed areas (domains) and that the matrix proteins create a two-dimensional lattice in the outer membrane.

The importance of the individual components for the function of the outer membrane is difficult to determine due to its extraordinary complexity. However, two procedures can be applied. The first consists of protective disassembling of the membrane into the individual components and a subsequent partial or complete reconstitution of the membrane from these. The second method uses defined mutants.

The protective disassembling of the outer membrane can be carried out by means of mild detergents such as deoxycholate (DOC). Nearly all LPS and phospholipids and at least 70% of the proteins become soluble in this way. After removal of the detergent and addition of Mg^{2+}, the reconstitution of the membranes takes place mostly in the form of vesicles. By leaving out one or two components, their importance for the complete membrane can be studied. In this connection, it must be considered that the components of a membrane do not simply sum up their properties on being added to each other, but partially multiply them so that new characteristics emerge. For example, some outer membrane proteins become resistant to proteolysis on addition of LPS.

A further advantageous method is the planar lipid bilayer technique. The device used for this purpose consists of a chamber separated into two compartments by a partition wall containing a hole of a diameter of less than 1 mm. On the surfaces of the solution located in both compartments beneath the hole, a lipid monolayer is produced by evaporating a solution of phospholipids in (for example) chloroform. By careful simultaneous raising of the solution in both compartments, a lipid bilayer is formed over the hole. By using differently composed monolayers, the bilayer can be arranged asymmetrically. Such bilayers can be used to examine, e.g.

the function of the individual membrane components, by the determination of the electric conductivity before and after their incorporation. The method is restricted to the investigation of deep rough LPS owing to the insolubility of most LPS in chloroform.

The use of mutants that lack one or several components of the outer membrane has become much more simple and expressive since they can be directly constructed by means of modern genetic procedures.

All investigations performed by these techniques indicate that the outer membrane forms a bilayer typical for all biological membranes. At least in the smooth and the higher rough forms, however, the outer monolayer contains all the LPS, and the inner monolayer all phospholipids (Fig. 7.1.6.). Embedded into the bilayer are the proteins.

For diverse reasons, such as a higher stability of the monolayer, a dense layer of polysaccharide chains and accumulation of negative charges on the surface, the presence of LPS results in a clearly reduced penetrability especially for hydrophobic substances. The bilayer is practically impermeable to amphiphilic substances, especially for negatively charged ones. Finally, the LPS render the outer membrane more stable towards membrane-active substances such as drugs, detergents or components of the immune system. This stability is distinctly lower in (especially deep) rough forms, which indicates the already-mentioned participation of the polysaccharide chains in resistances. Also, gentle extraction of a part of the LPS using substances like EDTA, not leading to an impairment of the vital functions of the bacteria, clearly renders the bacteria more sensitive.

By binding to LPS, the proteins both inside and outside the membrane become distinctly more resistant to denaturation and enzymatic degradation. For this reason, bacteria frequently excrete entero- and other toxins embedded in LPS vesicles which are formed from blebs of the outer membrane, i.e. the normal low curvature of this membrane is abruptly changed to the high curvature form of the vesicle. Such vesicles contain OMPs, LPS and phospholipids in their wall and periplasmic constituents in the interior, all located as they normally are in the bacterium. They possess diameters between 50 and 250 nm, are spherically shaped, and their wall represents a bilayered membranous structure. The vesicles can attack both Gram-negative and Gram-positive bacteria as well as eukaryotic cells and inject their "toxic" interior.

On the contrary, the accumulation of negative charges on the membrane surface due to the core region results in a weakness of the cell wall towards (poly-)cations such as aminoglycoside antibiotics or cationic peptides (defensins) of the serum.

The phospholipids contribute essentially to the fluidity of the membrane interior; the proteins are responsible for the membrane morphology and permeability. Phospholipid and porins together are already able to form membranes which resemble the natural ones in their transport abilities. However, the incorporation of these proteins into phospholipid bilayers clearly proceeds more slowly than into phospholipid-LPS bilayers. LPS obviously have also a recognition function, and they are necessary for the correct orientation of the proteins.

2.4.4 The Fluidity of the Outer Membrane

The liquid-crystalline state of the membrane interior is an essential prerequisite for the perfect course of all membrane processes. All membrane components are constantly in motion. They move at high speed laterally around the cell, and, at the same time, they rotate on their long axis. The degree of liquidity is regulated in the outer membrane mainly by the nature of phospholipids, whereas the contribution of LPS to the fine regulation is clearly less pronounced, due to its more constant fatty acid composition. The nature of the fatty acid, but also that of their head groups (size, charge, conformation, water-binding capacity) represents regulating factors in the phospholipids.

In order to investigate the importance of the individual fatty acids for the fluidity of the membrane, mainly mutants were used in which the biosynthesis of particular fatty acids is blocked. Another possibility rises from the cultivation of bacteria in the presence of cerulenin, an antibiotic inhibiting the complete fatty acid biosynthesis, and subsequent addition of the respective fatty acid (together with an emulsifier) into the cultivation medium. The catalytic hydration of the membranes after addition of homogeneous catalysts has also been described; it leads to the transformation of all unsaturated fatty acid into saturated ones.

The presence of saturated long-chain fatty acids increases the phase transition temperature T_t (see Sect. 2.1.4) of membranes, whereas that of unsaturated, branched-chain or cyclopropane fatty acid reduces it. The reason for this lies in the molecular geometry. As shown in Fig. 2.1, the hydrocarbon chains of the unsaturated, branched-chain or cyclopropane fatty acids are more bulky than those of the saturated ones, and therefore more difficult to be packed into ordered structures. In the solid (gel-like) phase, the hydrophobic tails of the fatty acids are mainly in the all-*trans* configuration, during transition into the liquid-crystalline state the percentage of the *gauche* conformation increases, and the packing density decreases simultaneously.

All kinds of the fatty acids described are equally necessary for the regulation of the membrane functions; however, their effect is different. For example, the *cis*-unsaturated and the cyclopropane fatty acid lower T_t drastically, the *trans*-unsaturated ones only moderately. Several metabolic parameters can be influenced by manipulating the content of unsaturated fatty acids. Investigations carried out in fatty acid auxotrophic mutants revealed that it is optimal for the outer membrane and their functions if the phospholipids contain a saturated fatty acid and an unsaturated one each. Depending on their extent, deviations from this norm lead to disturbances in growth and metabolic functions and finally induce lysis of the bacteria. Thus, a substantial number of physiological functions seem to be dependent on the properties of the membrane lipid phases.

As described in Section 2.1.3.3.4, the control of T_t (phase transition temperature from the gel-like to the liquid state of the membrane) and Δ_t (temperature range of this transition, in the case of the outer membrane up to 20 $^{\circ}$C) via the composition of phospholipids is more complicated in the outer membrane than in mere phospholipid bilayers. For example, at decreasing temperatures, not only a transition of the lipids from the unordered to the ordered state takes place, but also a partial separation of mere lipid regions and of areas enriched with proteins. In the former, the lipids are ordered and in the latter disordered. Both areas differ in their physical and chemical properties and therefore can be separated.

Since the membrane is only able to function perfectly if its interior is liquid, T_t can be found in the range between 0 °C and about 10 °C below the cultivation temperature of the bacteria, and Δ_t is rather wide. On lowering the cultivation temperature, the bacteria increasingly incorporate unsaturated fatty acids (homeoviscous adaptation). On increasing the cultivation temperature, the opposite effect appears. In both cases, T_t is again located distinctly below the cultivation temperature. In this way, a safety interval is probably created that guarantees the full activity of the bacteria even at a rapid drop in the environmental temperature. For this purpose, the bacteria have a further control system. *E. coli* cells cultivated at 40 °C have (seemingly paradoxically) an essentially more active "unsaturated fatty acid synthetase" than those cultivated at 10 °C. The regulation of the incorporation is undertaken by the acyl transferase system, it slows down the unsaturated fatty acid synthetase via a feedback mechanism (short-term control). By a rapid decrease of the environmental temperature from 40 to 10 °C, this slowing-down is interrupted, but also the degradation of the unsaturated fatty acid synthetase begins. This, however, demands a lag period (long-term control), and, thus, the enzyme can still provide a sufficient quantity of unsaturated fatty acids in order to dilute the saturated ones and thus increase T_t.

The conditions are reversed if bacteria cultivated at 10 °C are warmed up to 40 °C. At 10 °C the saturated fatty acid synthetase is the more active one, and at this temperature it is slowed down by the acyl transferase system. At 40 °C, on the contrary, this slowing-down is eliminated, and until the adaptation of the saturated fatty acid synthetase concentration to the new temperature, sufficient saturated fatty acids will be present for the dilution of the now harmful unsaturated ones.

The mechanisms described are only possible because the specificity of the acyl transferases is not absolute, but depends on the presence of sufficient quantities of the respective fatty acid needed. In addition, both transferases show different temperature-effect curves.

Another mechanism to regulate the liquidity of the outer membrane is the degradation of lipids. The degradation velocity is different for the individual kinds of lipids, e.g. in *E. coli*, phosphatidylglycerol has a half-life time of about 1 h, diphosphatidylglycerol of about 2 h with regard to the polar moiety and considerably longer with regard to the fatty acid moiety.

The 10-20-fold higher viscosity of the outer membrane compared to the inner one, in spite of the similar fatty acid composition of their phospholipids, is due to the presence of LPS. There are several reasons for this. First, each LPS molecule carries in most cases five to six saturated fatty acid residues in its lipid A, whereas phospholipids possess only two partially unsaturated ones. This results in a distinctly elevated packing density and increases the interactions between the molecules. In liquid-crystalline LPS bilayers, the area per hydrocarbon chains amounts to 2.6 nm, i.e. it is much tighter than in phospholipid membranes. Second, the core region with its accumulation of negative charges is located directly on the monolayer formed by lipid A molecules, causing stabilization via Me^{2+} bridges. Third, the long polysaccharide chains mutually influence each other. This is transmitted to the lipid A region and thus leads to a strengthening of the membrane.

Re Mutants with their rather saccharide-free LPS show a T_t of about 30 °C. Corresponding to an increasing chain length up to the S-type LPS, T_t gradually rises to 37 – 40 °C.

The fluidity of the outer membrane is still a matter of discussion. The outer monolayer containing the LPS is clearly less liquid (perhaps even quasicrystalline) than the phospholipid-containing inner one. Therefore, the fluid mosaic model can be applied only in a modified version for the description of the outer membrane. The above-mentioned strongly reduced penetrability of the outer membrane especially towards hydrophobic substances is attributed to the reduced liquidity of the outer monolayer. The degree of mutual influence of both monolayers (T_t, fluidity) is still uncertain. Likewise, the contribution of the outer membrane proteins to the liquidity of the membrane is also under discussion.

2.4.5 Assembly of the Outer Membrane

The outer membrane is highly asymmetric. A migration of components from one monolayer to the other is practically impossible in the case of polysaccharides and proteins, and possible to only a very low extent in the case of lipids. This means that the components of the outer membrane, which are mostly synthesised at or in the cytoplasmic membrane, have to penetrate both membranes and then must immediately reach their final place of destination. The mechanisms active in this process must be very selective because substances which may be very similar to each other have to be transported to very different areas of the membrane.

The incorporation of the phospholipids appears to be best elucidated. It occurs by means of their rapid exchange between both membranes proceeding in both directions, presumably via fusion points. The asymmetry of the membranes can be at least partially regulated by different transport velocities; phosphatidylglycerol, e.g. moves faster than phosphatidylethanolamine; but also the specific interactions between phospholipids and the respective membrane proteins might be responsible for it.

The mechanisms of LPS and protein transport from the periplasmic side of the cytoplasmic membrane to the outer side of the outer membrane is far less elucidated. Contrary to phospholipid transport, both are unidirectional. It was supposed earlier that the transport of LPS and other components of the outer membrane proceeds via fusion points (insertion points) between both membranes which were found by electron microscopic examination. These fusion points may be conceived as tubes formed by phospholipids (heads outwards, tails inwards) connecting both membranes with each other and forming hydrophobic pores. Hydrophobic substances could migrate from one membrane to the other through the interior of the tubes, amphiphilic ones by lateral diffusion via the surfaces of the tubes. Starting from the insertion points, the molecules could spread by diffusion over the whole outer membrane. This model is supported by the fact that newly incorporated porins are always located in immediate proximity of the fusion points. On the other hand, it presents decisive disadvantages (too low diffusion velocities of LPS and OMP to guarantee an even covering of the surface during maximal growth; the problem of how a molecule primarily located on the periplasmic side of the cytoplasmic membrane can reach the outside of the outer membrane), and thus the model is hardly accepted at present. A periplasmic route is preferred (see also Sect. 2.3.2.6), but detailed ideas about it are missing.

Ideas on the OMP biosynthesis in the cytoplasmic membrane and their transport across the periplasmic space into the outer membrane have already been discussed in Section 2.3.2.6. Although there are indications that the proteins of the outer membrane in the folded final state are by far more hydrophobic than those of the periplasmic space (see also Sect. 2.3.2.6) and therefore the localisation of both is predetermined; it seems, however, that already during transport of the unfolded chains through the cytoplasmic membrane a certain presorting takes place.

There are also differences in the incorporation of porins and OmpA. During folding into the final state, there are intermediate steps in both cases, i.e. in case of the porin proteins those described in Section 2.3.2.6 and in the case of OmpA at least one precursor form which is still not heat-modifiable. In contrast to the porins, however, folding into the final state and incorporation into the membrane of OmpA are independent of the presence of LPS. For incorporation in the outer membrane, the presence of the last transmembrane section of OmpA (amino acids about 160-170) is necessary, and the incorporation occurs in toto after terminal folding of the membrane localised part.

The porins are embedded in the outer membrane after folding into the hydrophobic final state as well as after trimerisation. In model experiments it was found that the transport of porins into a membrane consisting of an LPS and a phospholipid monolayer is only possible from the phospholipid side and proceeds more than tenfold faster than into a mere phospholipid bilayer. The incorporation takes place side-correct (inside/outside). The protein is obviously drawn into the membrane in its right position and, only in this case, by the help of LPS.

The transport of the newly synthesised Braun´s lipoprotein occurs on the same Sec-dependent pathway as that of the other OMPs until the signal sequence is cleaved. Its binding to the peptidoglycan has already been described, the incorporation of the hydrophobic part into the inner monolayer of the outer membrane proceeds analogously to the one of the phospholipids.

The specific interactions between the individual components of the membrane make those to be newly incorporated find their place of destination easily (self-assembly).

It is not anticipated that the biosyntheses of the individual membrane components occur completely isolated from each other or from the cellular requirements. The occurring regulatory mechanisms taking place differ in bacteria, but there are, in principle, similar characteristics. The biosynthesis of the major proteins is often mutually regulated. Thus, an overexpression of OmpC leads to a complete blocking of OmpA and LamB biosyntheses, overexpression of OmpA to the blocking of that of OmpC and OmpF, and lack of OmpA to an increased biosynthesis of OmpC and OmpF. On the contrary, there seems to be no influence on the OMP biosynthesis by lipids. Therefore, a failure of the lipid biosynthesis may result in the formation of protein-rich membranes. On the contrary, the biosynthesis of lipids can be regulated by that of the membrane proteins, a second control mechanism may occur via an absolute demand for molecules. In the latter case, the feedback mostly takes place already during the synthesis of fatty acids or a little later. This regulation is not absolute in all cases and can be partially bypassed.

Bibliography

Beveridge TJ (1999) Structures of Gram-negative cell walls and their derived membrane vesicles. J Bacteriol 181:4725-4733

Duong F, Eichler J, Price A, Leonard MR, Wickner W (1997) Biogenesis of the Gram-negative bacterial envelope. Cell 91:567-573

Hancock REW, Karunaratne DN, Bernegger-Egli C (1994) Molecular organization and structural role of outer membrane macromolecules. In: Ghuysen JM, Hakenbeck R (eds) Bacterial cell wall. New comprehensive biochemistry, vol 27. Elsevier, Amsterdam, pp.263-279

Nikaido H (1996) Outer membrane. In: Neidhart FC, Curtiss R III, Ingraham IL, Lin ECC, Low KB, Magasanik B, Reznikoff WS, Riley M, Schaechter M, Umbarger ME (eds) *Escherichia coli* and *Salmonella*: cellular and molecular biology, 2nd edn. ASM Press, Washington, pp 29-47

Singer SJ (1977) The fluid mosaic model of membrane structure. In: Abrahamson S, Pascher I (eds) Structure of biological membranes, Plec num Press, New York, pp 443-461

3 Periplasmic Space and Rigid Layer

3.1 The Periplasmic Space

The region between cytoplasmic and outer membrane of the Gram-negative bacteria is called the periplasmic space. It contains a concentrated gel-like matrix, the periplasm. Embedded in it is a rigid layer (peptidoglycan). Its width has been electron microscopically determined as 13-25 nm, a variation range that may be explained by differences in cultivation or preparation conditions. The total volume of the periplasmic space has been determined as 20-40% of the whole cell; this value, however, is not identical with that calculated from the width, which would allow a volume of only 8-16%. Because of its great importance for the vital processes of the bacteria, the existence of a space at least similar to the periplasmic is also thought to be present in Gram-positive bacteria (Dijkstra and Keck 1996).

The composition of the periplasm differs distinctly from that of the surrounding medium. It contains an aqueous solution of mono- and oligosaccharides, amino acids, peptides, soluble biosynthetic precursors of the peptidoglycan and other small molecules, but also degrading and detoxifying enzymes. Thus, at the membrane limiting the space to the outside, a Donnan potential of about 30 mV is created, the physiological importance of which is still unknown. Shielding of the cytoplasmic membrane, formation of a protonmotive force, regulation of the extent of opening the membrane pores and the influencing of surface organelles, such as, for instance, the flagella, are under discussion.

Besides the dissolved ones, there are also substances in the periplasm which are bound to one of the membranes or to the peptidoglycan.

It is difficult to determine the composition of the periplasm. A selective lysis, using, e.g. osmotic shock or treatment with chloroform, comprises only the suspended periplasmic components and even these not always completely and/or exclusively. The use of spheroplasts which are more easy to destroy may have variations in composition as a consequence. Some components can be specifically detected in situ, e.g. enzymes, by using diffusible substrates or proteins by labelling with diffusible cross-linking reagents. Histochemical and immunologic methods have also been described. It is recommended to always apply more than one method in parallel and to combine the results.

From the volume of the periplasm and the quantity and nature of the materials dissolved in it, it follows that the solution must be highly viscous and it occurs most likely in a gel-like state. This state as well as the sieving function of the peptidoglycan embedded in the periplasm drastically reduce the motility of the dissolved substances (to about 0.1% compared to an aqueous solution) and thus may support a formation of specific microenvironments with particular physiological assignments.

Since the cytoplasmic membrane is only marginally osmostable, cytoplasm and periplasm have to be nearly iso-osmotic. Therefore the bacteria possess fast-reacting regulatory mechanisms which keep the periplasmic osmotic pressure constant. Currently, the nature of the osmosensor is unknown. The osmotic stabilisation occurs in particular via the content of membrane-derived oligosaccharides (MDO), a mixture of oligosaccharides consisting of 6-12 glucose units in β-($1\rightarrow2$)- and β-($1\rightarrow6$)-linkage which are statistically substituted with sn-1-phosphoglycerol, 2-phosphoethanolamine and O-succinyl ester residues. It was found that osmotic differences between cytoplasm and periplasm can trigger the synthesis of MDO. The overall negative charge of the MDO represents one factor in the formation of the Donnan potential.

The proteins in the periplasm are basically assigned to three functional classes: enzymes (catabolic and detoxifying enzymes for protection against penetrated harmful substances as well as enzymes necessary for the biosynthesis of cell wall components and higher-molecular secretion products), chaperones and high-affinity binding proteins for vitally important substrates (amino acids, peptides, sugars, vitamins, coenzymes, nucleotides, inorganic ions and many others) which are important for the following transport across the cytoplasmic membrane. An important periplasmic protein, TonB, has already been described (see Sect. 2.3.2.3.3). All these proteins contribute to the function of the periplasm, i.e. represent an essential coordinating point for transports from the outside to the inside and vice versa. Some of these proteins are also present in Gram-positive bacteria in which they are anchored in the outer surface of the cytoplasmic membrane.

Outer and cytoplasmic membrane are not separated from each other at all sites of the cell wall. In electron micrographs, adhesion zones between both membranes (so-called Bayer´s patches, 100-200 per cell, zone width 20-100 nm) can be detected which are mainly found in exponentially growing cultures. Therefore it is supposed that these sites play a role in the biosynthesis of the cell wall or serve as periseptal anuli for the subcompartimentalisation of the periplasmic space as well as marking sites for a future cell division. However, there are opinions that the Bayer´s patches represent artefacts created during fixation of the preparations.

Bibliography

Dijkstra AJ, Keck W (1996) Peptidoglycan as a barrier to transenvelope transport. J Bacteriol 178:5555-5562

Duong F, Eichler J, Price A, Leonard MR, Wickner W (1997) Biogenesis of the Gram-negative bacterial envelope. Cell 11:567-573

Oliver DB (1996) Periplasm. In: Neidhart FC, Curtiss R III, Ingraham IL, Lin ECC, Low KB, Magasanik B, Reznikoff WS, Riley M, Schaechter M, Umbarger ME (eds) *Escherichia coli* and *Salmonella*: cellular and molecular biology, 2nd edn. ASM Press, Washington, pp 88-103

3.2 The Rigid Layer of Gram-Positive and Gram-Negative Bacteria

3.2.1 Significance, Isolation, Composition, Structure

Despite the general variability in composition and structure of the microbial cell wall, one construction element is astonishingly widespread: an exoskeleton structure of high mechanical stability which was developed 2 to 3 billion years ago. It is missing in only very few bacterial groups such as mycoplasma and in some archaea. In bacteria it is mainly named peptidoglycan, but often also called murein, more rarely mucopeptide, glycopeptide or basal structure. Peptidoglycans forms a sacculus consisting of polysaccharide chains cross-linked by peptide bridges. Thus, a net-shaped structure of high solidity is formed causing the mechanical stability of the bacteria and maintaining their shape. According to the kind of the bacteria, peptidoglycan can amount to 5-90% of the cell wall mass, a value that is higher in Gram-positive bacteria (40-60% in staphylococci) than in Gram-negative ones. Gram-positive peptidoglycan layers are distinctly thicker than those of Gram-negative bacteria. The intact cell wall of the Gram-negative bacteria resists pressures of up to 5 atm, that of the mechanically more stable Gram-positive ones (*Bacillus subtilis*) of up to 50 atm.

The lack of peptidoglycan caused, for example, by addition of penicillin to the cultivation medium, leads to spheroplasts under loss of the type-specific shape. These are stable (and are partially even able to multiply) only in solutions where the osmotic pressure corresponds to the internal pressure of the bacteria. Spheroplasts burst in more diluted solutions.

3.2.1.1 Isolation of the Peptidoglycan

The isolation of the peptidoglycan in its intact form causes difficulties. In both Gram-negative and Gram-positive bacteria further components of the cell envelope are covalently bound to it. It is difficult to selectively cleave the corresponding binding sites and to remove the additional components.

The preparation of the peptidoglycan starts either from intact bacteria or from the previously isolated cell walls. The isolation of the cell walls is performed by disintegration of the cells using ultrasonication, repeated freezing and thawing, strong shear forces, or grinding with quartz sand. The cell walls can be separated from other cell components by differential centrifugation.

In Table 3.1 some standard preparation procedures for the peptidoglycans of Gram-positive and Gram-negative bacteria are summarised. The methods used have to be relatively drastic in order to cleave the cell envelope components that are bound to the peptidoglycan. This is feasible because peptidoglycan is chemically quite stable. Nevertheless, application of most methods results in more or less significant modifications in the peptidoglycan, which may lead to misinterpretations during the investigations of both the chemical and biological properties.

Table 3.1. Isolation of peptidoglycans. (After Schleifer 1975)

Method	Disadvantage
Gram-positive bacteria	
Formamide, 150-180 °C, 30 min	Possibly formylation of OH and NH_2 groups
HNO_2, 0.8 M, 37 °C, 15 min	Splitting-off of unsubstituted primary amino groups
Trichloroacetic acid, 5-25%, 90 °C, 10 min	Partial acidic degradation
Trichloroacetic acid, 5-25%, 0 °C, several days	Partial acidic degradation
H_2SO_4, O.1 N, 60 °C, 24 h	Partial acidic degradation
NaOH, 0.1 N, 100 °C, 1 h	Hydrolysis of glycyl-peptide linkages, destruction of reducing terminal carbohydrate groups
NaOH, 0.1 N, 35 °C, 8 h	Hydrolysis of glycyl-peptide linkages, destruction of reducing terminal carbohydrate groups
Deoxycholate, 1%, 0 °C , 16 h	Extraction not sufficient
N,N-Dimethylhydrazine, 2%, 80 °C , 2 h	Extraction too extensive, destruction of amino sugars
Gram-negative bacteria	
Phenol/water (45:55), 68 °C , 5-35 min	Containing endotoxin
Sodium dodecyl sulfate, 4%, 100 °C	Containing endotoxin

In the case of *S. aureus*, the purified peptidoglycans contain glucosamine, muramic acid (= glucosamine which is etherified at O-3 with lactic acid), glycine, L- and D-alanine, L-lysine and D-glutamic acid, and sometimes also L-serine. The ratios of these components are not constant, but depend, for instance, on the composition of the cultivation medium. The peptidoglycans of *E. coli* strains are similarly composed, but contain no glycine, and *meso*-diaminopimelic acid (DAP) instead of L-lysine.

3.2.1.2 Chemical Structure of Peptidoglycan

In the following, the course of the structural elucidation of peptidoglycan in the 1960s is described, because this structure was very logically and clearly deduced from the experiments. With regard to the analytical possibilities of that time, the analysis of this cross-linked molecule consisting of a polysaccharide and a peptide moiety was a severe challenge. Its relatively uncomplicated fulfilment was due to the facts that the peptidoglycan is not irregularly constructed, but consists of identical and recurring structural units (repeating units), and that enzymes could be found that cleave the peptidoglycan either into its repeating units or the peptidoglycan or the repeating units at a single site; e.g. dividing polysaccharide and peptide moiety from each other.

3.2.1.2.1 Chemical Structure of the Polysaccharide Moiety. For a better understanding, a characteristic section of the structure of the peptidoglycan from Staphylococcus aureus may serve as an example to describe the individual reaction steps.

Action of endo-N-acetylmuramidase from *Streptococcus albus* or from *Chalaropsis B* on the peptidoglycan of *S. aureus* strain Copenhagen cleaved the polysaccharide chain between N-acetylmuramic acid and N-acetylglucosamine. In Fig. 3.1 the cleavage site is designated 1. Only glycopeptides were produced which had been exclusively linked via the peptide bridges. This peptide moiety could be split off in the intact arrangement by means of N-acetylmuramyl-L-alanine amidase from *Streptomyces* (cleavage site 2 in Fig. 3.1). In addition, two disaccharides were liberated: N-acetylmuramyl-β-(1→4)-N-acetylglucosamine and N,6-O-diacetyl-muramyl-β-(1→4)-N-acetylglucosamine. The elucidation of their chemical structure was carried out using classical chemical methods.

By using lysostaphin instead of endo-N-acetylmuramidase, the polysaccharide chain was cleaved between N-acetylglucosamine and N-acetylmuramic acid (3 in Fig. 3.1). Subsequent treatment with N-acetylmuramyl-L-alanine amidase liberated the two disaccharides N-acetylglucosaminyl-β-(1→4)-N-acetyl-muraminic acid and N-acetylglucosaminyl-β-(1→4)-N,6-O-diacetylmuraminic acid.

As may be concluded from these results, the polysaccharide chain consists of alternating N-acetylglucosamine and N-acetylmuramic acid both in β-(1→4)-linkage, the latter being O-acetylated to about 60% at C-6. Thus, the polysaccharide chain represents a kind of modified chitin which consists of long poly-β-(1→4)-acetylglucosamine chains. The difference between chitin and the peptidoglycan polysaccharide consists of a lactyl residue which in the latter is bound via an ether linkage to the 3-OH group of every second glucosamine residue (muramic acid). A second modification is the far shorter chain length of the glycan strands of the peptidoglycan polysaccharide, enabling a helical structure and rendering the peptidoglycan more flexible and elastic.

The peptide-free complete glycan strands could be obtained by enzymatic digestion of the peptidoglycan using an enzyme complex of myxobacteria, which hydrolyses the D-alanylglycine, glycylglycine and N-acetylmuramyl-L-alanine linkages (cleavage at 2 and 4 in Fig. 3.1).

In general, the structure of the polysaccharide moiety of all bacterial groups shows only little variability. The hydroxy groups at C-6 of the muramyl residue may be acetylated to different degrees; in addition, N-acetyl groups may be missing or modified. Mycobacteria contain N-glycolyl-muramic acid; thus, additional hydrogen bridges become possible, resulting in a further stabilisation of the structure.

Recently, the peptide bridges have been cleaved using an amidase and the obtained glycan chains have been separated by means of reversed-phase HPLC (Sect. 2.1.2). In the case of *E. coli*, the result was a very heterogeneous mixture of cleavage products with chain lengths of 1 to 30 disaccharide units, and, in addition, some unfractionated material with an average chain lengths of 80 disaccharide units was obtained. Thus, the chain-length distribution proves to be extreme. In the case of *E. coli*, an average chain length of only 9 disaccharide units for about 70% of the glycan chains is supposed, and of about 45 units for the remaining 30%. In

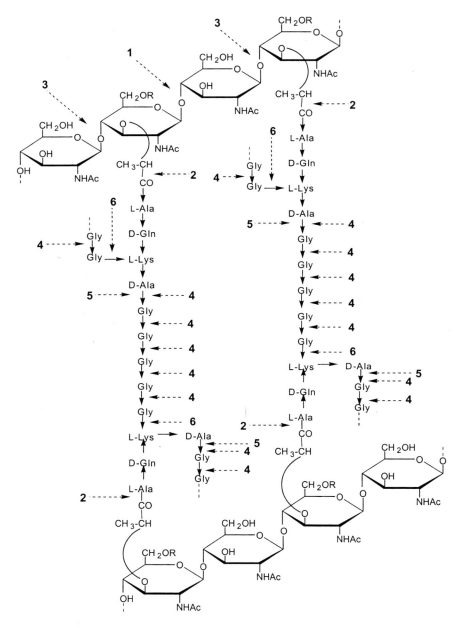

Fig. 3.1. Chemical structure and degradation scheme of the peptidoglycan from *Staphylococcus aureus*. The dotted arrows indicate the targets of the enzymes (see text)

addition, it has been detected that the reducing terminus of the glycan chain is not formed by the muramic acid itself, but by its 1,6-anhydro form.

3.2.1.2.2 Chemical Structure of the Peptide Moiety. The peptide is the more variable part of the peptidoglycan. It varies in dependence not only on the nature of the bacteria, but also on the cultivation conditions, e.g. in S. aureus. Up to now, altogether some 100 different peptidoglycan types have been identified, based to a great extent on differences in the peptide moiety.

On analysing the peptidoglycan of *S. aureus* strain Copenhagen, Schleifer (1975) found only about half the glycine content and slightly larger quantities of alanine than reported earlier by other authors, despite otherwise comparable analytical data. Schleifer supposed deviations in the glycine and alanine concentrations of the cultivation media as a reason for these differences. Peptidoglycan preparations from *S. aureus* analysed by other authors additionally contained serine.

In the case of *S. aureus*, the structural elucidation of the peptide moiety was also mainly based on enzymatic degradation steps. The peptide bridges between D-alanine and glycine in the intact peptidoglycan were cleaved by means of *Streptomyces* SA endopeptidases (5 in Fig. 3.1), and subsequently all glycine residues (five per lysine residue) were removed by the action of *Streptomyces* amidase (cleavage in 4 and 6).

Then the glycan chains were cut by *Streptomyces* F₁-endo-N-acetylmuramidase (cleavage in 1) and the obtained disaccharide tetrapeptide units were cleaved into the disaccharide and the tetrapeptide moiety by means of N-acetylmuramyl-L-alanine amidase (cleavage in 2).

The structure of the tetrapeptide (peptide subunit) was deduced from Edman-degradation as L-Ala-γ-D-Glu-L-Lys-D-Ala; the unusual linkage between glutamic acid via its γ-carboxyl group is noteworthy. The L-lysine possesses a free ε-amino group.

According to the degradation scheme, the tetrapeptide can be bound to the D-lactyl residue of the muramic acid only via the amino group of the L-alanine. The resulting D-L-D-L-D-sequence is interesting. With others, this prevents the formation of an α-helical peptide structure, and this gives the peptide much more flexibility.

As concluded from the *Streptomyces* SA-endopeptidase cleavage sites (cleavage in 5), the D-alanine has to be the starting point of a pentaglycine bridge. The cleavage of the glycine residues during treatment with the *Streptomyces* amidase (cleavage in 4 und 6) leads to the demasking of the ε-amino group of the lysine residue of the subsequent tetrapeptide; therefore the pentaglycine bridge has to terminate at this group. However, later it was discovered that this is not the only possibility. There are also cross-links starting from a diamino acid on both sides.

The peptidoglycan of *Staphyloccus epidermidis* strain Texas 26 shows a structure analogous to that of *S. aureus* strain Copenhagen; merely one of the glycine residues of the pentaglycine bridge is replaced by serine. However, not every glycine of the pentaglycine bridges can be replaced by serine. In 55% of the bridges the third glycine residue (counted from the N-terminus) was replaced, in 15% the first and the third one and in 10% the second, whereas 20% of the bridges contained exclusively glycine. The bridges of *S. epidermidis* strain 66 contain an L-alanine residue replacing one of the glycine residues.

Investigation of the peptidoglycan preparations which contain a lower percentage of glycine revealed that the pentaglycine chains in the bridges are not shortened to yield tetraglycine chains, but the number of pentaglycine bridges is reduced (from about 90% tetrapeptide units to about 60%). Due to the resulting fewer cross-links, such a peptidoglycan possesses a more dispersed structure.

As revealed by investigation of the *E. coli* peptidoglycan, its tetrapeptide units are directly linked to each other, i.e. without a bridge; either starting from D-alanine to the lysine-replacing *meso*-diaminopimelic acid or, more rarely, between two *meso*-diaminopimelic acid residues.

For a better understanding of the arrangement of polysaccharide and peptide in the peptidoglycan, a schematic presentation is given in Fig. 3.2. In the case of *E. coli*, about 50% of the peptides can for steric reasons not be involved in the cross-linking (helical structure of the polysaccharide, see below), and are present as monomers. The cross-linking peptides are predominatingly dimers (linkage of two glycan strands), besides which, about 5% trimers (linkage of three glycan strands) and 0.2% tetramers occur. In *S. aureus* the degree of cross-linking is distinctly higher due to a multiple-layer arrangement (Labischinski and Maidhof 1994).

Meanwhile, the structures have in principle been confirmed using modern analytical methods; however, it turned out that they are much less ideal and that microheterogeneities exist to a great extent. They relate to all parts of the peptidoglycan and concern especially the length of the glycan strands, the number of peptide subunits and the degree and the site of the cross-links. For example, in the case of *E. coli*, mostly 5 to 10 disaccharide units, rarely up to 80, with an average value of about 9 units are present in regions with only one layer (see below). In the case of *S. aureus* with its multilayered peptidoglycan, an average of about 15 disaccharide units was found. So far, there is only speculation as to their significance.

The glycan moiety of the peptidoglycan is identical in composition and structure in all bacteria (apart from insignificant modifications such as the degree of acetylation). The peptide moiety of Gram-negative bacteria is also quite uniform, whereas in Gram-positive bacteria it shows a marked variability in both the peptide subunit and the (not always present) bridging peptide. This is presumably caused by the fact that in Gram-positive bacteria the peptidoglycan is relatively unprotected, located at the cell surface and exposed to the defence mechanisms of the host. The modifications in composition and structure represent an attempt to resist these mechanisms. In Gram-negative bacteria, the outer membrane protects the peptidoglycan from host defence.

In the peptide subunit (see Table 3.2.), especially L-lysine but also D-glutamic acid and L-alanine can be replaced by other amino acids. The α-carboxyl group of the D-glutamic acid in position 2 may be present as an amide or be substituted by glycine, glycinamide or D-alaninamide.

Cross-linking of the peptide subunits may occur starting from either the D-alanine of one subunit to the diamino acid of another, as in the case of *S. aureus*, either via the bridging peptides, or directly. In the case of *E. coli*, the two peptide subunits are cross-linked via the D-alanine of one subunit linked to the *meso*-diaminopimelic acid of another, but also a direct cross-linking between *meso*-diaminopimelic acid residues of both peptide subunits is possible. Cleavage of the glycan chains with muramidase and separation of the cleavage products by means of reversed-phase HPLC (see above) resulted in an unexpectedly complex pattern of disaccharide oligopeptides. This confirms that in a given murein also the cross-linking is not uniform and that, instead, the above-mentioned different binding sites exist and the number of the glycan strands linked to each other via peptide bridges is also variable, however, it does not amount to more than four.

According to composition and binding, two groups (A and B) of bridge peptides can be recognised, which may be further divided into subgroups and variations.

In group A, the cross-linking within the peptidoglycan takes place between positions 3 and 4, in those of group B between positions 2 and 4 of the peptide subunits.

Figure 3.3 shows the most important subgroups of group A, Fig. 3.4. those of group B.

The approximately 100 peptidoglycan types found up to now can be classified into these two groups consisting of altogether six subgroups. The by far predominant part of these types is formed by peptidoglycans of the Gram-positive bacteria, whereas those of the Gram-negative bacteria belong mainly to subgroup A1 type, containing *meso*-diaminopimelic acid as diamino acid but forming no bridge peptide.

It was attempted to use the structure of the peptidoglycan as a taxonomic criterion for Gram-positive bacteria (Schleifer and Kandler 1972). Although successful attempts cannot be denied, difficulties appear due to the possible influence of the composition of the culture medium on the structure of the peptidoglycans.

3.2.1.3 The Spatial Structure of the Peptidoglycan

The initial idea of the spatial structure of peptidoglycan was that of a paracrystalline network (sacculus) enclosing the entire bacterium with presumably longitudinal polysaccharide strands arranged in ring-shape around the cell and its cross-linking by peptide bridges. This assumption seemed to be in accordance with the net-like structure distinctly perceivable in electron microscopic pictures.

One of the present ideas, based on the results of X-ray diffraction as well as of infrared and NMR spectroscopy, supposes an essentially more flexible irregular structure with gel-like properties (periplasmic gel). This model is, however, not generally accepted. The bulky lactyl-ether groups of the muramic acid molecules are assumed to force the glycan strands into a helical conformation with four to five disaccharide units per turn. This conformation reduces the number of cross-links by peptide bridges between the glycan strands which, to a great extent, are arranged in parallel and run vertically to the cell axis. The glycan strands themselves are only slightly flexible; thus, rotations of the individual hexosamine rings around the glycosidic bond and kinks in the chains are almost impossible. A small part of elasticity could come from switches of the boat configuration of the pyranose structure to the chair form under tension of the glycan chains.

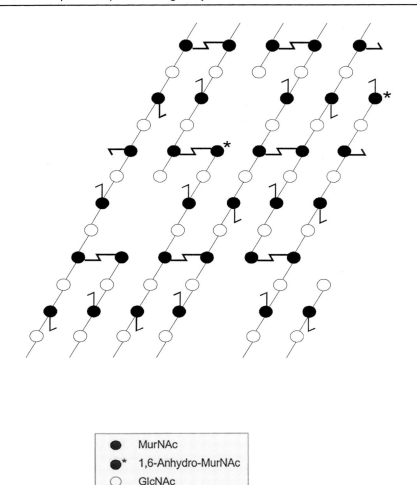

●	MurNAc
●*	1,6-Anhydro-MurNAc
○	GlcNAc
└	Peptide

Fig. 3.2. Schematic spatial structure of the peptidoglycan (Park JT (1996) The murein sacculus. In: Neidhart FC, Curtiss R III, Ingraham IL, Lin ECC, Low KB, Magasanik B, Reznikoff WS, Riley M, Schaechter M, Umbarger ME (eds) *Escherichia coli* and *Salmonella*: cellular and molecular biology, 2nd edn. ASM Press, Washington, pp 48-57, with permission)

The peptide is the more flexible region in the polymer. In a stress-free state, the peptide subunits wind nearly circularly round the polysaccharide helices; they are in contact with the glycane chains. In stress situations, an up to fourfold stretching of the peptide chains is possible. In this way, even Gram-positive bacteria are able to react adequately to osmotic variations. It was found that if the peptidoglycan strands are aligned in parallel, the interstrand spacing in *E. coli* sacculi increases by 12% with every 1 atm increase in (turgor) pressure.

Table 3.2. Possible amino acid variations in the peptide subunit (Schleifer and Kandler 1972; Labischinski and Maidhoff 1994)

Amino acid	Can be replaced by
L-Alanine	Glycine, L-serine
D-Glutamic acid	*threo*-3-Hydroxy-D-glutamic acid, glycine(-amide), D-alanine amide, cadaverine
L-Lysine	(Hydroxy-)*meso*-diaminopimelic acid, L,L-diaminopimelic acid, L-diaminobutyric acid, N_γ-acetyl-L-diaminobutyric acid, L-ornithine, L-hydroxylysine, L-lanthionine, L-homoserine, L-alanine, L-glutamic acid

This fact causes an elastic anisotropy of the *E. coli* cell amounting to a factor of 2 or 3 between the short and the long direction of the sacculus, in agreement with the much higher constancy of the cell diameter compared with its length. It presents additional evidence for the accuracy of the assumption that the glycan strands are aligned at right angles to the long cell axis.

In the case of a murein sacculus consisting of a single peptidoglycan layer (such as occurs in about 75-80% of *E. coli* strains, the remaining part consists of three layers; Barnickel et al. 1979); only every fourth lactyl residue is spatially arranged in such a way that the formation of a peptide bridge to the glycan strand adjacent at the same side is possible. The same number of residues makes a linking to the opposite adjacent strand possible, thus connecting in total 50% of the peptide residues with each other. In *E. coli*, the part of the murein sacculus formed by only one peptidoglycan layer has a thickness of 2.5 nm.

Koch (1998a) indicated the improbability of parallel glycan strands under the influence of internal cell pressure (stress-bearing wall). Under these conditions the smallest pores (tesserae, i.e. units having four corners) become hexagons (about 5 x 10^6 per cell, Fig. 3.5), demarcated by the saccharide chains and the peptide bridges, should be produced, which are fused together to form patches of a covalent structure. These patches are linked together by peptide bonds to cover the cell. An incomplete linking should conceivably lead to some larger pores. However, as shown by diffusion studies, there seem to be only few large imperfections in the peptidoglycan layer.

Much more favourable for cross-linking is a murein sacculus furnished from several peptidoglycan layers lying one above the other. In the case of *S. aureus* they amount to about 40. Thus, a degree of cross-linking of up to 90% is possible. The essentially longer peptide bridges compared to those in the peptidoglycan of *E. coli* are also responsible for this high degree of cross-linking, since they are also able to link lactyl residues to each other which are not immediately adjacent.

The assumption that the glycan strands are in parallel to each other and in general arranged cross to the cellular axis has been questioned. Already in 1979, Labischinski et al. proposed a model where the polysaccharide chains of adjacent peptidoglycan layers in Gram-positive cell walls are not arranged in parallel, but twisted towards each other.

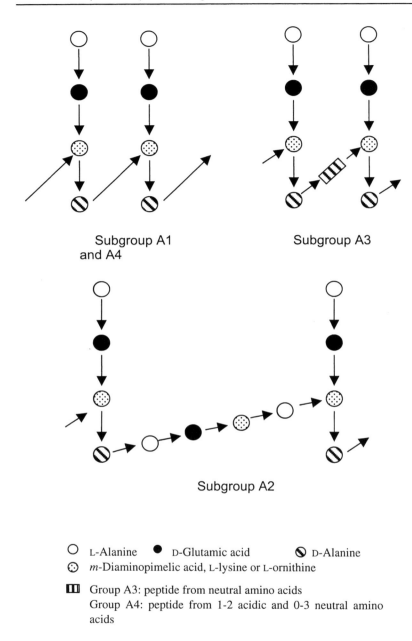

Subgroup A1
and A4

Subgroup A3

Subgroup A2

○ L-Alanine   ● D-Glutamic acid   ◐ D-Alanine
◉ *m*-Diaminopimelic acid, L-lysine or L-ornithine

▥ Group A3: peptide from neutral amino acids
Group A4: peptide from 1-2 acidic and 0-3 neutral amino acids

Fig. 3.3. Schematic structure of a peptide subunit from the group A peptidoglycans

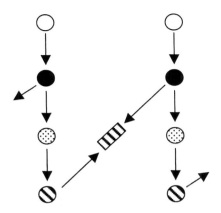

○   Glycine or L-serine

●   D-Glutamic acid or hydroxy-D-glutamic acid

☺   L-Lysine, L-homoserine, L-glutamic acid,
     L-alanine, L-ornithine or L-diaminobutyric acid

$$\begin{array}{ccc} & \alpha & \delta \end{array}$$

▥   at B1 → Gly$_{1\text{-}2}$→L-Lys←,

$$\begin{array}{ccc} \varepsilon & \alpha\ \delta & \alpha \end{array}$$
at B2 →D-Lys ←,→D-Orn← ,

$$\begin{array}{ccc} \delta & \alpha & \alpha \end{array}$$
→Gly→D-Orn←   or →D-DAB←

Fig. 3.4. Schematic structure of a peptide subunit from the group B peptidoglycans

Very recently, a basically new model of the peptidoglycan layer has been proposed (Dmitriev et al. 1999; see Chap. 7). It suggests that the glycan strands are arranged vertically to the cell wall plane and every strand is connected with four others by the peptide bridges placed within the cell wall plane. The thicker peptidoglycan layers of the Gram-positive bacteria are explained by the assumption of longer glycan strands.

The deviations of the peptidoglycan from the ideal structure due to microheterogeneities have diverse reasons. First, during bacterial growth, sites have to be present in the peptidoglycan where newly synthesised structural units may be attached. Therefore the bacterial cell walls contain autolytic enzymes that cleave the polysaccharide chains or the peptide bridges and thus create new receptor sites. These enzymes are present in an active form only at those sites where growth occurs; their activation is controlled by a regulatory mechanism. If peptidoglycan biosynthesis is inhibited, the unhindered continued cleavage of the

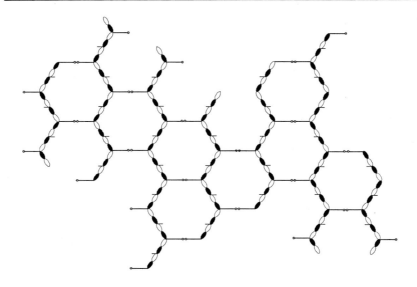

N-Acetylglucosamine, ○ ; *N*-Acetylmuramic acid, ● ; Peptide bridge, ───w───

Fig. 3.5. Hypothetical structure of a patch consisting of eight tesserea. The bonds to adjacent patches are indicated by arrows (Koch AL (1998b) How did bacteria come to be? Adv Microbial Physiol *40*, 353-399, with permission)

peptidoglycan leads first to the release of cytoplasmic material and finally to autolysis of the bacteria. A second reason for the deviations is the fact that, at least in Gram-positive bacteria, perfectly cross-linked peptidoglycans would represent a strong penetration barrier which prevents the uptake of nutrients as well as the secretion of metabolic products. Finally, the possibility has to be considered that part of the defective sites may be the result of inaccuracies during biosynthesis; thus, some free receptor sites are "overlooked" and not substituted.

Attempts have been undertaken to relate the bacterial shape to the structure of the peptidoglycan, however, the results are not unequivocal, since all three possibilities of mutual dependencies of composition and shape have been found, i.e. different shape in the case of identical composition, identical shape of two peptidoglycans in the case of different composition, different shape in the case of different composition.

If the biosynthesis of the peptidoglycans is prevented, e.g. by addition of penicillin to the medium, formation of spheroplasts results. They possess a globular shape even when they originate from rod-shaped bacteria. After removal of the penicillin, at first globular bacteria containing a peptidoglycan are formed and later rod-shaped ones. Thus, peptidoglycan biosynthesis is not connected with the cylindrical shape of the rods.

With regard to the interpretation of the analytical results, it has to be taken into account that all data represent average values obtained from the entire cell and that composition and structure of a peptidoglycan may be quite different at the pole of the bacteria than in its cylindrical part. There have been attempts to elucidate such differences. For this purpose, the peptidoglycans of *E. coli* spheroplasts (representing mainly the polar regions) produced by treatment with the antibiotic mecillinam or filaments (predominantly cylindrical region) produced by treatment with aztreonam, were degraded by means of muramidase, and the obtained muropeptides were separated by means of HPLC. No significant differences between the two peptidoglycans were found. On the contrary, after degradation using amidase, differences were detected, suggesting somewhat shorter glycan chains in the polar region. These results are contradictory to those obtained by the analysis of peptidoglycans of bacteria isolated from distinct stages of synchronous cultures in which no differences were found. It seems possible that the differences observed after treatment with antibiotics are due to that very influence.

Arthrobacter crystallopoietes, *Bacillus subtilis* and *E. coli* may appear in two different forms, a rod-shaped and a coccoid one. The investigation of composition and structure of the peptidoglycan, e.g. in the case of *A. crystallopoietes*, revealed no fundamental differences between the two forms. One of the differences found, however, is a higher average length of the polysaccharide chains (126 repeating units in the case of the rods, 34 in the coccoid form). This may be due to the activity of murolytic enzymes during chemical manipulations when preparing the bacteria for analysis, but, of course, it may exist in reality. In coccoid forms, a two- to fivefold amount of peptidoglycan has been found, suggesting a thicker peptidoglycan layer. In addition, differences in the quantity of other cell wall components have been detected. However, it is unclear what is the cause and what is the effect.

Differences in the peptidoglycans of spores and vegetative cells suggest an effect of the peptidoglycan structure on the shape of the bacteria. The peptidoglycans of spores are less stable than those of vegetative cells. They include about half the muramic acid residues as non-acetylated inner amide (Fig. 3.6), which thus is not disposable as binding site of the peptides. The inner amide additionally prevents the formation of hydrogen bridges stabilising the chain structure and thus renders the glycane chain more flexible. About half of the remaining 50% is substituted with only one L-alanine residue at its carboxyl group, and the remainder, i.e., only about 25% of all muramic acid, carries a complete, but very moderately cross-linked peptide. All this contributes to the reduced stability of the peptidoglycans of the spores.

3.2.1.4 Peptidoglycan as a Transport Barrier

On the one hand, the meshes of the intact peptidoglycan net have to be so close that they stabilise the cytoplasmic membrane against the internal pressure of the bacteria and prevent a forcing of membrane particles through the pores. On the other hand, they have to be wide enough not to prevent the transport of larger molecules, especially from inside to the outside.

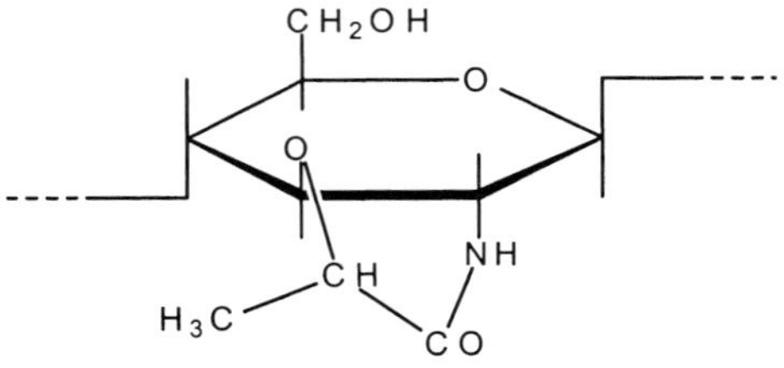

Fig. 3.6. Structure of the inner amide of a (1→4)-bound muramic acid molecule

The length of one disaccharide unit in the peptidoglycan, i.e. the distance between two peptide bridges, amounts to 1.03 nm. For the distance of the glycan chains, 1.25 nm are assumed in *E. coli*. Since even the multilayered and strongly cross-linked murein sacculus of *S. aureus* may swell and shrink considerably, the latter value has to be regarded as an average value with a rather great variation range. As the glycan chains are relatively short and in Gram-negative bacteria only every fourth disaccharide represents a starting site of one cross-linking peptide bridge in the same direction (see Fig. 3.2), the mesh size is distinctly higher than could be calculated from the data mentioned above. The diameter of the pores has been determined as approximately 2.2 nm and the permeability of the peptidoglycan for globular proteins as 55 kDa (Koch 1998b). In spite of the difference in wall thickness, the porosities of Gram-positive and Gram-negative bacteria are about the same.

Thus, the problem of penetration of larger molecules through the peptidoglycan layer arises, both for transport from inside to outside and vice versa. Impressive examples are the P ring of the flagellae (see the corresponding chapter), which is anchored in the peptidoglycan layer with a diameter of 26 nm, as well as the plasmid transfer where molecular masses of up to 210 kDa have to pass two cell walls. One explanation is given by the assumption of locally acting autolytic peptidoglycan hydrolases which may be determined by the gene cluster responsible for transport and cut holes of corresponding size into the network.

The fact that the peptidoglycan layer functions as a penetration barrier results in the existence of a space containing elevated concentrations of proteins and other compounds between cytoplasmic membrane and the peptidoglycan. This is even the case in Gram-positive bacteria and resembles the periplasmic space of the Gram-negative ones. However, most of these proteins are provided with a lipid moiety and thus could be fixed in the cytoplasmic membrane.

3.2.2 Biosynthesis of Peptidoglycan

The biosynthesis of a complicated macromolecule like peptidoglycan cannot take place in one single reaction pathway. Three clearly distinguishable stages can be recognised, which are located at different sites of the bacterial cell.

The first stage comprises the biosynthesis of *N*-acetylglucosamine 1-phosphate and from this UDP-*N*-acetylmuramyl pentapeptide. This process occurs in the cytoplasm. During the second stage, the complete disaccharide pentapeptide subunit and the bridging peptide bound to it are synthesised, which occurs at or in the cytoplasmic membrane. In the third stage, the new building block is integrated into the peptidoglycan lattice and cross-linked. This reaction takes place in the cell wall.

All genes participating in peptidoglycan biosynthesis have been identified, cloned and sequenced. They are mainly localised in two gene clusters termed *mra* (situated in the 2-min region of the *E. coli* chromosome) and *mrb* (in the 90-min region). The genes participating in the early steps of the biosynthesis, namely in the formation of UDP-*N*-acetylmuramic acid, are encoded by the *mrb* region. Those participating in later steps, namely the connection of all amino acids, the formation of lipid I and II as well as the cross-linking and the cell division processes, by the *mra* region. Comparable gene regions have also been found in other bacterial species, even in Gram-positive ones. This suggests evolutionary relationships in the peptidoglycan biosyntheses.

3.2.2.1 Biosynthesis of UDP-N-Acetylmuramyl Pentapeptide

The formation of UDP-*N*-acetylglucosamine via *N*-acetylglucosamine 1-phosphate has already been described for the case of LPS biosynthesis (Fig. 2.22). This compound participates in the biosynthetic pathways of both LPS and peptidoglycan.

The next intermediate, UDP-*N*-acetylglucosaminyl-pyruvat-enolether, is synthesised from UDP-*N*-acetylglucosamine and phosphoenolpyruvate (PEP). It is transferred into UDP-*N*-acetylmuramic acid (UDP-MurNAc) by cleavage of the double bond taking place in an NADPH-consuming reaction which is catalysed by a reductase, a flavoprotein (Fig. 3.7)

The investigation of the further course of the biosynthesis was possible because substances had been identified that interrupt the biosynthetic pathways at the individual stages which lead to an enrichment of the particular intermediate products. Isolation and analysis of these products, their application in further biosynthetic steps as well as isolation and investigation of the mechanisms of the enzymes necessary for the respective reaction step have to a great extent elucidated the biosynthetic pathways in *S. aureus*. In *Micrococcus luteus* (formerly *Micrococcus lysodeiktikus*) and *E. coli* similar pathways have been found.

The biosynthesis of the peptide moiety is different from that of most other cell wall protein components which use the template mechanism of the usual protein biosynthesis. The sequence of the single stages follows from order and strict substrate specificity in the enzymes catalysing this pathway.

In the first stage, the attachment of L-alanine to the lactoyl group of UDP-*N*-acetylmuramic acid takes place. Like the attachment of the following amino acids, this reaction requires ATP and Mg^{2+} or Mn^{2+}. ATP is needed for the carboxyl activation at all stages of this reaction chain. The acylphosphate intermediate thus formed reacts with the amino group of the amino acid to be integrated via a nucleophilic mechanism. The incorporation of L-alanine is inhibited by gentian violet, resulting in the enrichment of UDP-GlcNAc, UDP-GlcNAc-pyruvate-enolether, and UDP-MurNAc.

D-Glutamic acid necessary for the next steps is synthesised either by the action of glutamic acid racemase from the natural L-form (e.g. in *E. coli*) or by stereospecific transamination from α-ketoglutarate by D-alanine (e.g. *Bacillus* species).

If the diamino acid necessary for the third stage (L-lysine in the case of *S. aureus*, for those found in other bacteria see Table 3.2) is missing, UDP-MurNAc-L-Ala-D-Glu is enriched. The catalysing enzyme specifically binds the lysine residue to the γ-carboxyl group of the glutamic acid.

D-Alanine is not attached as a single amino acid, but as a dipeptide (D-Ala-X; X is in the case of *S. aureus* D-Ala; in other bacteria D-serine or D-lactate). Starting from L-alanine, D-alanine is synthesised under the influence of alanine racemase and then transformed into the dipeptide in an ATP-consuming reaction. The attachment of the dipeptide to the α-amino group of the terminal lysine also takes place under ATP consumption.

The synthesis of both the dipeptide and its attachment are inhibited by D-cycloserine or *O*-carbamoyl-D-serine; simultaneously, enrichment of UDP-MurNAc-L-Ala-D-Glu-L-Lys occurs. To explain the mechanism of the cycloserine effect, one starts from the steric similarities between D-alanine and D-cycloserine (Fig. 3.8). Whereas the alanine molecule may adopt several spatial structures due to the free rotability of its distinct atomic groups around the C-C and the C-N bond, the spatial structure of the cycloserine is stabilised by the ring. This is probably the reason for the fact that the Michaelis constant of D-alanine for both enzymes is about 100-fold higher than for D-cycloserine, i.e. the bond of D-cycloserine to the enzyme is 100-fold stronger. Thus, the antibiotic effect of cycloserine effects an interruption of the biosynthesis of the peptide moiety of the peptidoglycan and therefore of the whole peptidoglycan due to a competitive inhibition.

3.2.2.2 Biosynthesis of the Complete Subunit and of the Bridge Peptide

The synthesised UDP-*N*-acetylmuramylpentapeptide is metabolised under the influence of a translocase with the acceptor undecaprenyl phosphate (ACL, see Sect. 2.1.4.3.4). As an amphiphilic molecule, undecaprenyl phosphate is embedded in the cytoplasmic membrane with its hydrophobic region. The reaction occurs according to:

undecaprenyl-P + UDP-MurNAc-pentapeptide → undecaprenyl-PP-MurNAc-pentapeptide + UMP.

The reaction product is also called lipid I. UMP released during the reaction is retransformed into UTP under consumption of two molecules ATP and restarts the reaction cycle.

The equilibrium position of this reaction is located at the far left side. Therefore the products of this and the following reaction stages are present in very small quantities. This again leads to a turnover rate of lipid I of only a few seconds.

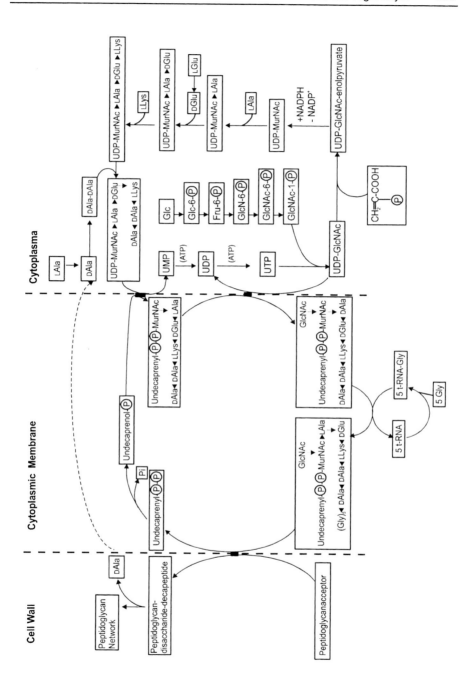

Fig. 3.7. Scheme of the peptidoglycan biosynthesis

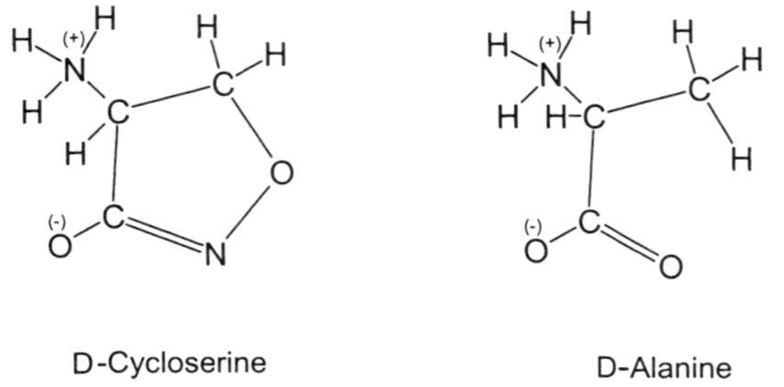

D-Cycloserine   D-Alanine

Fig. 3.8. Comparison of the spatial structures (projections) of D-cycloserine and D-alanine

To complete the subunit of the peptidoglycan, the incorporation of *N*-acetylglucosamine is still needed. This incorporation occurs in the next reaction stage starting with UDP-*N*-acetylglucosamine (Fig. 3.7) The reaction product is undecaprenyl-PP-(GlcNAc-)MurNAc-pentapeptide (lipid II). The reaction sequence of the last two stages has been confirmed by using ^{32}P-labelled substances; the liberated UMP derives from the muramyl peptide and the liberated UDP from UDP-*N*-acetylglucosamine. The transferase catalysing the reaction is bound at the inner surface of the cytoplasmic membrane.

In *S. aureus*, the bridging peptide is synthesised in a next biosynthesis complex, which clearly differs from biosynthesis of the peptide subunit in its sensitivity towards ribonuclease. This indicates a participation of glycyl-t-RNA in the reaction. Three different types of glycyl-t-RNA have been found which all take part in peptidoglycan biosynthesis; one of them is inactive in protein biosynthesis and is not bound to ribosomes in the presence of one of the glycine codons. The role of the t-RNA in this synthesis remains unclear. It is certain, however, that the corresponding anticodon has nothing to do with the biosynthesis.

Glycine is activated under ATP consumption, bound to t-RNA and then transferred to lipid II. From the simultaneously decreasing number of ε-amino groups in the diamino acids in lipid II the binding site of the glycine residues can be deduced. In the case of *S. aureus*, the incorporation of the five glycine residues occurs step by step and in opposite direction to the case of protein biosynthesis. For the integration of a seryl residue into the peptide bridge, the presence of a seryl-t-RNA is necessary. The incorporation depends on the preceding one of glycine, thus, the localisation of the seryl residue is not arbitrary.

Defined t-RNA species are also necessary for the incorporation of L-threonine into the bridge peptide of *Micrococcus roseus* and of L-alanine into that of *Arthrobacter crystallopoietes*.

Finally, the α-carboxyl group of the glutamic acid is amidated. It is not certain whether this takes place before or after the incorporation of *N*-acetylglucosamine and the bridge peptide, respectively. The amidation is ATP-dependent and uses free NH$_4$OH.

In the biosynthetic pathway described, the pre- and intermediate stages up to that of the UDP-*N*-acetyl-muramyl pentapeptide exist in such high concentrations in the bacteria that under normal conditions, feedback inhibitions are not expected.

On the contrary, lipid I is present in extremely small quantities (less than 700 molecules per cell). This is caused by the fact that the equilibrium of the furnishing reaction from undecaprenyl phosphate and UDP-MurNAc-pentapeptide (see above) is situated far on the left side, i.e. it depends greatly on the concentration of the first reaction component. This may indicate that in *E. coli* the ratio between the three cell wall components peptidoglycan, LPS and ECA is regulated in this way, as undecaprenyl phosphate is also necessary for the biosynthesis. However, it is equally possible that this regulation prevents a free diffusion of lipid I and II in the membrane; thus, both are exclusively present in an enzyme-bound state. One consequence of the low concentrations of lipid I is its high turnover rate to lipid II. This step depends on the concentration of UDP-GlcNAc which, in turn, depends on protein synthesis. This means that also in this way a regulation of the whole biosynthesis pathway can take place.

3.2.2.3 Transfer to the Growing Terminus of the Polysaccharide Chain

The peptide moiety of lipid II is then bound to an acceptor, namely the growing glycan chain in the peptidoglycan (Fig. 3.7). It is still unclear whether the binding takes place at the reducing or the non-reducing terminus of the chain. This may differ in individual bacterial species. In parallel to the liberation of the lipid moiety, an 1,6-anhydro formation of the muramic acid may occur which is located at the reducing terminus of the glycan chain. In a consecutive reaction the liberated undecaprenyl pyrophosphate is cleaved into phosphate and undecaprenyl phosphate, the latter of which again participates in the biosynthesis cycle.

Since the growing peptidoglycan is located outside the cytoplasmic membrane, and the biosynthesis of the subunit occurs at least partially in the interior, membrane transport must take place. Its mechanism is still unknown to a great extent.

For the integration of the completed building blocks which are capable of cleaving intact peptidoglycan enzymes also play a role, e.g. between MurNAc and GlcNAc, between D-Ala and MurNAc, as well as between L-Lys and Gly. It is possible that these enzymes create receptor sites for subunits that are newly incorporated (see Sect. 3.2.2.5).

3.2.2.4 Cross-Linking

After the integration of a new building block into the growing glycan chain, it must also be cross-linked to the peptide moiety. Since the reaction occurs outside the cytoplasmic membrane, ATP as an energy source is not available. However, the reaction presumably takes place close to this membrane, because otherwise it would be out of reach of membrane-bound transpeptidases. The required energy is obtained from the cleavage of the terminal D-alanine of the peptide subunit, the total reaction thus represents a transpeptidation (Fig. 3.9). A prerequisite for the process of cross-linking is a spatially favourable position of both reactants towards each other.

In numerous cell walls, a D,D-carboxypeptidase (see Table 3.3) has been detected, the function of which is cleavage of the D-alanyl-D-alanine linkage. The enzyme presumably regulates the extent of cross-linking in the peptidoglycan by removal of the energy necessary for cross-linking via cleavage of this linkage. In *E. coli*, mainly tetrapeptide units (-L-Ala-D-Glu-L-Lys-D-Ala) serve as acceptor molecules, but tri- and pentapeptides may also do so. The tripeptides are in most cases formed from the pentapeptides by stepwise cleavage of both D-alanine residues, at first by means of a D,D-carboxypeptidase (pentapeptide → tetrapeptide) and subsequently by an L,D-carboxypeptidase (tetrapeptide → tripeptide). These enzymes are present in Gram-negative bacteria in higher concentrations than in Gram-positive ones; a correlation to the essentially stronger thickness of the murein layer in the latter case is obvious. In the presence of higher concentrations of D-amino acids or glycine, this reaction does not result in a tripeptide, but in an exchange of lysine-bound D-alanine with the added amino acid. In this way, no further tripeptide units are created and the biosynthesis of the peptidoglycan ceases. Thus, the well-known growth inhibition of bacteria by glycine or D-amino acids may be explained.

As a result of cross-linking, bridges consisting mainly of two tetrapeptide units (tetra-tetra bridges) are formed. In the course of bacterial maturation, cleavage of terminal D-alanine residues may occur by means of a L,D-carboxypeptidase resulting in the formation of tetra-tri bridges. They are important for length growth, whereas the tetra-tetra bridge is needed for bacterial division (see also Sect. 8.4).

In the case of *S. aureus*, the reaction mechanism is slightly different because of the presence of the pentaglycine bridge (Fig. 3.7). This bridge is formed by stepwise addition of glycine residues from t-RNA-Gly to the disaccharide pentapeptide. The consecutive cross-linking occurs between the amino group of the terminal glycine residue and the carboxyl group of the D-alanine.

It has been doubted whether the transglycosylation, i.e. the incorporation of a new building block into the growing glycan chain, in fact takes place before the transpeptidation. After all, some of the enzymes were shown to be bifunctional (see

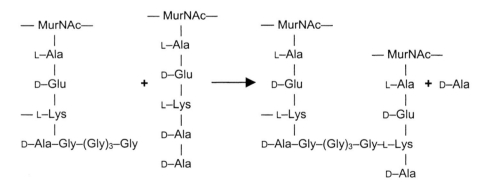

Fig. 3.9. Scheme of transpeptidation in *E. coli*

Table 3.3). The results allow the conclusion that both reactions may occur in the first place. Transpeptidation, however, might even be preferred.

3.2.2.5 Regulation of Peptidoglycan Biosynthesis

In general, the growth of rod-shaped bacteria represents a sequence of length growth and transversal division. The omission of one of the two steps results in the formation of abnormal cell shapes (minicells and filaments, respectively). Both growth stages are characterised by different mechanisms of peptidoglycan synthesis which may be differently inhibited by antibiotics. During length growth, new building blocks are inserted into an already existing macromolecule. On the contrary, during transversal division a completely new synthesis takes place, presumably by incorporation of smaller and smaller concentric peptidoglycan rings into a template. Both processes have to be chronologically coordinated. In *E. coli*, for example, the velocity of length growth is, in fact, retarded during septum formation, and in periods of length growth the velocity of peptidoglycan synthesis decreases in the septal area.

The biosynthesis of peptidoglycan is a complex and complicated process. Not only bacterial shape and diameter have to be preserved, but after having obtained the correct length, the septation necessary for the bacterial division has to be initiated exactly in the central region of the cell wall, and the velocity of the biosynthesis of the peptidoglycan has to be adapted both to the growth phase (logarithmic, stationary) and to the biosynthesis of other cell envelope components. In some cases, cell wall thickenings have to be regulated. The formation of spores and presumably also the transition of one bacterial form to another one (see Sect. 3.2.1.3) requires a modification of the incorporation mechanisms. All this has to be controlled by a complex regulatory system.

The ratio between cell elongation and septum formation which is important for growth and multiplication of the bacteria, is differently regulated in individual bacterial species.

Different hypotheses have been developed to explain the mechanism of cell wall growth of rod-shaped bacteria, one of which is the "belt model". It postulates that during length growth of the *E. coli* cell, the building block is presumably completely synthesised and subsequently inserted as a whole into an acceptor site created by the action of hydrolytic enzymes and finally cross-linked. The enzyme complexes necessary for the incorporation process are supposed to migrate into one direction around the cell and thus allow a regular length growth. Especially in Gram-negative bacteria, the creation of such mechanically weak regions in the peptidoglycan being exposed to the full internal bacterial pressure is not unproblematic, because they may lead to membrane disruption and, thus, to the release of cell material. Since this is obviously not the case, theories have been developed that try to explain the incorporation under maintenance of stability. The "make-before-break" theory supposes that the single-stranded building block is moved under the peptidoglycan layer and linked to two adjacent strands. Only after this are the bonds between the old strands enzymatically cleaved. The "in-between" theory modifies the preceding one. It assumes the building block to be triple-stranded and postulates a linking to two parallel strands of the old peptidoglycan which are separated from each other by a third. After this the third strand is removed enzymatically (Fig. 3.10A). The region finally is stretched by the internal pressure of the bacteria. By this mechanism, one old strand is replaced by three new. A third assumption says that the units to be incorporated are so small that the temporarily limited defects in the peptidoglycan network have no destabilising effect. Two enzymes regulating the incorporation process are the penicillin-binding proteins 2 and 4 (PBP2 and PBP4, see Sect. 3.2.2.6).

A further hypothesis that explains the mechanism of cell wall growth of rod-shaped bacteria is the "surface stress" theory. This hypothesis postulates that the cell wall is enlarged randomly. When an undecaprenol-bound disaccharide pentapeptide, a peptidoglycan acceptor region, and the transglycolase activity of the statistically distributed PBP1B are proximal (see Fig. 3.7), the growing chain is increased by one disaccharide at its N-acetyl glucosamine end. At the same time, chain-linking PBP2 diffuses close enough to a nascent chain and links the chains in a way consistent with cylindrical growth. As the diameter must be kept constant, the dimensions are controlled by physical forces operating on the growing sacculus. Models were developed showing that elongation under these conditions can indeed take place spontaneously as long as the poles are metabolically inert and rigid and, in addition, of fixed and appropriate radius. The above mentioned "in-between" theory is also regarded as appropriate for this model.

Even more complicated than their elongation is the septation of the cells resulting in the formation of two new pole caps. This demands both an exact spatial and temporal control. More than 15 different genes participate in this process. An enzyme essentially regulating this process is represented by PBP3, a periplasmic protein which is N-terminally anchored in the cytoplasmic membrane. It acts both as a transpeptidase and a transglycosidase. Formation of filaments occurs if it is missing. Due to its action, two closely connected cross-wall layers are formed from outwards by the incorporation of smaller and smaller concentric peptidoglycan rings. A possible mechanism resembles that of the "in-between" theory. The only difference consists in the fact that the third strand (see above) is not immediately liberated after incorporation of the triple-stranded building block (Fig. 3.10B). This occurs only after the integration of the following triple-stranded building block at the first one. In continuing this process, the peptidoglycan grows inwards in the form of a double layer. After completion, both layers are completely separated under the control of the EnvA protein and then represent the new pole caps. During this separating process, quite a number of low molecular peptidoglycan fragments are released. In Gram-negative bacteria, the outer membrane grows from outwards to the same extent as the separation of both peptidoglycan layers proceeds inwards; it becomes anchored in the peptidoglycan via Braun´s lipoprotein and OmpA.

In Gram-positive bacteria, peptidoglycan biosynthesis is simpler insofar as the murein is multilayered and no precautions have to be taken to keep the cell stable towards its own pressure. In this case, the corresponding innermost layer is synthesised in a compressed form and linked to the already existing murein layer (inside-to-outside growth). During growth, the outer peptidoglycan layers are enzymatically removed step by step from the outside until the "new" layers become more and more stressed and expand.

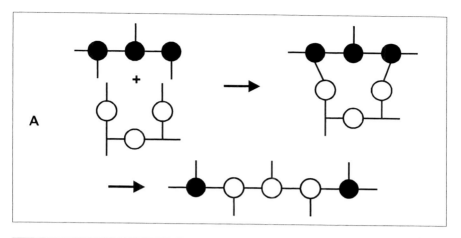

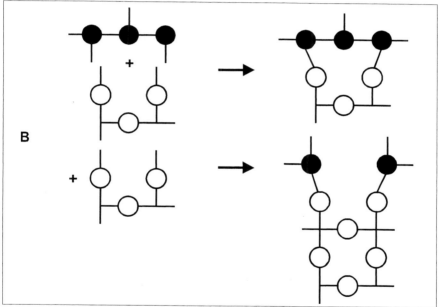

old polysaccharide in chain direction
new polysaccharide in chain direktion
peptide

Fig. 3.10 A and B. Schematic representation of cell elongation (**A**) and division (**B**) at the peptidoglycan level according to the in-between-theory. For details see text

The process of growth of coccoid bacteria is also complicated, similarly to that of the rod-shaped ones. A model has been developed for *Streptococcus faecalis* in which cross-wall formation, cell elongation and thickening are assumed to occur starting from a single zone located in the centre of the bacterium (Fig. 3.11). The elongation of the peptidoglycan occurs at two regions arranged in parallel adjacent to each other and protruding into the cell. This process is overlapped by stretching of the cell. It depends on the ratio of the velocities of the peptidoglycan elongation (v_1) and the cell stretching (v_2), whether cell elongation ($v_2 > v_1$) or cell division ($v_1 > v_2$) occurs. After the division into daughter cells is almost completed, their division starts vertically to the first direction, i.e. the zone in the cell envelope where the biosynthesis takes place remains spatially unchanged (Fig. 3.12).

Besides its formation, a degradation of the murein sacculus occurs, which depends very much on the nature of the bacteria. In the case of *Bacillus subtilis* and *E. coli* it amounts to up to 50% per generation, in *Bacillus megaterium* 30% and in *Streptococcus faecalis* it has not been proven at all. A great part of the cleavage products is taken up again by the cell. Glycosidases, peptidases and amidases take part in this process. For this, three reasons are thought probable, namely the creation of acceptor sites for the incorporation of new peptidoglycan subunits as well as for cell wall components covalently bound to the peptidoglycan, temporal and spatial regulation of the cellular shape and the "recycling" of the building blocks.

Up to now, the biosynthetic pathways have been mainly investigated in the Gram-negative *E. coli* and the Gram-positive *S. aureus*. They are mostly identical with the exception of the bridging peptide synthesis in the case of *S. aureus*, and it is supposed that these processes occur in many other bacteria according to an essentially equal scheme.

3.2.2.6 Possibilities to Influence the Biosynthesis of Peptidoglycan

Peptidoglycan biosynthesis may be inhibited by quite a number of substances; some examples have already been mentioned in the explanation of biosynthesis mechanisms. Of special importance for the fight against infectious diseases are antibiotics, some of them effecting the biosynthesis of peptidoglycan. In the following, the mechanism of action of representative examples is described.

The transfer of enolpyruvate from phosphoenolpyruvate to UDP-*N*-acetylglucosamine is inhibited by fosfomycin. This antibiotic substance, a low molecular compound of the structure depicted in Fig. 3.13, reacts with the cysteine residue of the transferase and thus initiates the inhibition. Its transport through the cytoplasmic membrane occurs via the L-α-glycerolphosphate transport system.

The action mechanism of cycloserine has already been discussed. Amphomycin utilises a similar mechanism. In its presence, UDP-MurNAc-pentapeptide, UDP-MurNAc-L-Ala and UDP-MurNAc become enriched.

The transfer of GlcNAc-1-P or MurNAc-1-P residues from the corresponding uridine nucleotides to ACL is inhibited by tunicamycin.

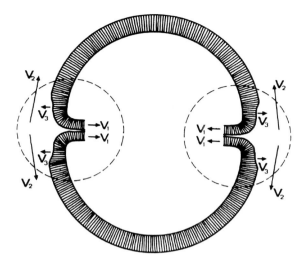

Fig. 3.11. Scheme of peptidoglycan growth (at *Streptococcus faecalis*). V_1 Direction and velocity of the linear peptidoglycan growth; V_2 direction and velocity of the bacterial growth, V_3 direction and velocity of the peptidoglycan thickening

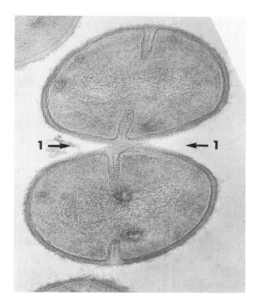

Fig. 3.12. Alternating of division plane in *S. aureus* (Labischinski H, Maidhof H (1994) Bacterial peptidoglycan: overview and evolving concepts. In: Ghuysen JM, Hakenbeck R (eds) Bacterial cell wall. New comprehensive biochemistry, vol 27. Elsevier, Amsterdam, pp 23-38, with permission)

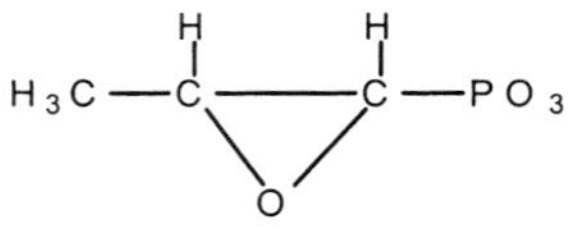

$$H_3C-\overset{\overset{\displaystyle H}{|}}{C}\underset{\underset{\displaystyle O}{\diagdown\diagup}}{}\overset{\overset{\displaystyle H}{|}}{C}-PO_3$$

Fig. 3.13. Structure of fosfomycin

The antibiotics of the vancomycin group (vancomycin, ristocetin A and B, ristomycin and actinoidin) inhibit the incorporation of GlcNAc-MurNAc-pentapeptide bound to undecaprenyl-PP into the acceptor, i.e. the growing peptidoglycan, presumably by attachment to the D-alanyl-D-alanine terminus.

Bacitracin inhibits the enzymatic dephosporylation into undecaprenyl phosphate of the undecaprenyl pyrophosphate released during the transfer of GlcNAc-MurNAc-pentapeptide into the growing peptidoglycan and thus prevents the transition of newly synthetised MurNAc-pentapeptide from UDP to the lipid and therefore the start of a new biosynthetic cycle.

For the effect of the β-lactam antibiotics, the so-called penicillin-binding proteins (PBP) in the cell envelope play an essential role. The PBP possess enzymatic functions and are involved in the biosynthesis steps occurring after transport of the growing peptidoglycan precursors across the cytoplasmic membrane. The binding capability of PBP to β-lactam antibiotics connected with a serine residue induces inhibition of the biosynthesis. The PBP found in E. coli have been investigated in detail, and are listed in Table 3.3. All the higher molecular PBP (1 - 3) are characterised by an enzymatic double function (transglycosidase and transpeptidase); both activities may be inhibited differently. They all accomplish important tasks during the formation of the murein sacculus. In all cases, the penicillin-binding serine of the higher molecular PBP is located near the transpeptidase domain. PBP can be found in both Gram-negative and Gram-positive bacteria. The resistance to a number of β-lactam antibiotics very common in Gram-negative bacteria, however, is frequently related to the missing penetrability of the outer membrane.

The PBP1 complex (see Table 3.3) is involved in cell elongation; it may be inhibited by penicillin and cephaloridin. For the PBP1B partial complex a function during incorporation of the GlcNAc-MurNAc-pentapeptide into the growing peptidoglycan is supposed. Together with the RodA protein, PBP2 is essential for the peptidoglycan biosynthesis during cell elongation and is responsible for the maintenance of the cellular shape during biosynthesis. In E. coli, its loss leads to the formation of spherical cells. The mechanism of PBP2 action is still unclear; it is possible that, besides the main function during cell elongation, it also plays a role in cell division. It is inhibited by mecillinam. As already mentioned, PBP3 is responsible for septum formation. Its transpeptidase property is located at its C-terminus, the transglycosidase property at its N-terminus. The presence of the FtsW protein promotes its influence. It is blocked by cephalexin, piperacillin and furazlocillin.

A resistance mechanism towards β-lactam antibiotics, which is particularly well developed in Gram-positive bacteria (e.g. in the methicillin-resistant S. aureus, MRSA), consists in the biosynthesis of a penicillin-binding protein with very low affinities for these antibiotics (PBP2A), but with the remaining ability to complete

Table 3.3. Penicillin-binding proteins in *E. coli*

Number	Molecular mass	Physiological function	Enzyme activity
1A	93 500	Cell elongation	Transglycosylase and transpeptidase
1B-α	94 100	Cell elongation (and division)	Transglycosylase and transpeptidase
1B-β		Cell elongation (and division)	Transglycosylase and transpeptidase
1B-γ	88 800	Cell elongation (and division)	Transglycosylase and transpeptidase
1B-δ		Cell elongation (and division)	Transglycosylase and transpeptidase
2	70 867	Determination of the cell shape	Transglycosylase and transpeptidase
3	63 850	Cell division	Transglycosylase and transpeptidase
4	49 568	Cell wall lysis and maturation	D-Carboxypeptidase and D,D-endopeptidase
5	41 340	Cell wall maturation	D-Carboxypeptidase
6		Cell wall maturation	D-Carboxypeptidase

the peptidoglycan biosynthesis, though with reduced capacity for cross-linking. The *mecA* gene produced by fusion of a PBP-encoding gene with a penicillinase gene and which can also be localised on a plasmid, encodes a protein C-terminally corresponding to PBP, but N-terminally contains a smaller domain with penicillinase properties.

Bibliography

Barnickel G, Labischinski H, Bradaczek H, Giesbrecht P (1979) Conformational energy calculations on the peptide part of murein. Eur J Biochem 95:157-165

Dijkstra AJ, Keck W (1996) Peptidoglycan as a barrier to transenvelope transport. J Bacteriol 178:5555-5562

Dmitriev BA, Ehlers S, Rietschel ET (1999) Layered murein revisited: a fundamentally new concept of bacterial cell wall structure, biogenesis and function. Med Microbiol Immunol 187:173-181

Höltje JV (1998) Growth of the stress-bearing and shape-maintaining murein sacculus of *Escherichia coli*. Microbiol Mol Biol Rev 62:181-203

Koch AL (1998a) Orientation of the peptidoglycan chains in the sacculus of *Escherichia coli*. Res Microbiol 149:689-701

Koch AL (1998b) How did bacteria come to be? Adv Microbial Physiol 40:353-399

Labischinski H, Maidhof H (1994) Bacterial peptidoglycan: overview and evolving concepts. In: Ghuysen JM, Hakenbeck R (eds) Bacterial cell wall. New comprehensive biochemistry, vol 27. Elsevier, Amsterdam, pp 23-38

Labischinski H, Barnickel G, Bradaczek H, Giesbrecht P (1979) On the secondary and tertiary structure of murein. Eur J Biochem 95:147-155

Park JT (1996) The murein sacculus. In: Neidhart FC, Curtiss R III, Ingraham IL, Lin ECC, Low KB, Magasanik B, Reznikoff WS, Riley M, Schaechter M, Umbarger ME (eds) *Escherichia coli* and *Salmonella*: cellular and molecular biology, 2nd edn. ASM Press, Washington, pp 48-57

Schleifer KH (1975) Chemical structure of the peptidoglycan, its modifiability and relation to biological activity. Z Immunforsch 149:104-117

Schleifer KH, Kandler O (1972) Peptidoglycan types of bacterial cell walls and their taxonomic implications. Bacteriol Rev 36:407-477

Yao X, M. Jericho, Pink D, Beveridge T (1999) Thickness and elasticity of Gram-negative murein sacculi measured by atomic force microscopy. J Bacteriol 181:6865-6875

4 Further Cell Wall Components of Gram-Positive Bacteria

In addition to peptidoglycan, the cell walls of Gram-positive bacteria contain mainly proteins and polysaccharides. With the exception of mycolata and similar bacteria they are generally poor in lipids and lack the outer membrane. The lipid moiety of the lipoteichoic acids is not a constituent of the cell wall, but embedded in the cytoplasmic membrane.

The cell walls of the mycolata show a quite different assembly; in particular, they are rich in (glyco-)lipids. Since these bacteria clearly differ in general from other Gram-positive bacteria, their cell walls are discussed in a separate section.

4.1 Polysaccharides (Except Capsular Polysaccharides)

Gram-positive bacteria contain in their cell wall substantial amounts (10-60%) of polysaccharides of a mostly anionic character which are anchored either at the peptidoglycan or in the cytoplasmic membrane. Among the former are teichoic acids, teichuronic acids and anionic polysaccharides, among the latter are lipoteichoic acids (LTA) and lipoglycans. They are all important for bacterial vitality. In some genera they are of taxonomic significance.

4.1.1 The Teichoic Acid Family

Teichoic acids were first described in 1958. They proved to be bound to the peptidoglycan (cell wall teichoic acids) and could be divided into two groups, namely the glycerol teichoic acids and the ribitol teichoic acids (polyglycerol or ribitol phosphates). Three years later, a group of substances similar to teichoic acids was detected which were, however, free from phosphates and obtained their acidic character through uronic acids. Therefore they were called teichuronic acids. Again 3 years later, another group of teichoic acids was found which was fixed in the cytoplasmic membrane. They were first termed membrane teichoic acids, but later lipoteichoic acids, due to their lipid moiety.

The classical procedure to isolate teichoic acids is the extraction of the bacteria with ice-cold 5-10% trichloroacetic acid and subsequent precipitation with ethanol. Another method is the extraction with 0.1 M NaOH at 20-22 $^{\circ}$C. In both cases, a cleavage of labile bonds occurs. A milder procedure consists in the extraction with phenylhydrazine or *N,N*-dimethylhydrazine. Purification occurs by means of fractionated precipitation and subsequent ion exchange chromatography.

Teichoic acids are built up of repeating units like other cell wall polysaccharides. The elucidation of their structure can be carried out by means of the procedures described for LPS and the capsular polysaccharides. Frequently, phosphomono- and diesterases are used in order to cleave the polymers and to isolate fragment structures.

4.1.1.1 Teichoic Acids

Teichoic acids are cell wall polymers which represent polyalditol phosphates containing diester linkages and which may additionally have sugar (phosphates) in the main chain as well as D-alanyl or glycosyl residues as side chains. They form two main groups: the poly-(polyolphosphates) and the poly-(glycosylpolyolphosphates). Their attachment to peptidoglycan occurs via a special linkage unit attached to the *O*-6 of the *N*-acetylmuramic acid.

4.1.1.1.1 Poly-(Polyolphosphate)-Teichoic Acids.

In most cases, teichoic acids of this type contain glycerol or ribitol, but also mannitol, erythritol or arabinitol may appear as polyol component.

In Gram-positive bacteria, teichoic acids of the poly-(glycerolphosphate) type are most common. There are two isomers, namely the poly-(1,3-glyceroldiphosphate) type and the poly-(2,3-glyceroldiphosphate) type. In Fig. 4.1 typical chain regions showing the individual linkages are depicted.

As substituent R, mainly glucose, galactose or an *N*-acetyl-amino sugar occur, but also *O*-linked D-alanine and L-lysine as well as *O*-acetyl residues are found, often in non-stoichiometric amounts. In *Streptomyces* sp. VKM Ac-1830, the two unsubstituted isomers (R = H) were identified in a proportion of 2:1. One example for a substituted poly-1,3-(glycerol phosphate) teichoic acid is that of *Streptococcus lactis* 7944 containing an *N*-acetyl-glucosamine residue in glycosidic linkage at C-2. *Glycomyces harbinensis* contains a poly-2,3-(glycerol phosphate)-teichoic acid, the OH groups of which are completely substituted at C-1 with α-D-glucopyranosyl residues.

In the case of the teichoic acids of the poly-(ribitol phosphate) type, two basic isomers have been found as well, namely the poly-(1,5-ribitol diphosphate) type present in many Gram-positive bacteria and the more rarely occurring poly-(3,5-ribitol diphosphate) type (Fig. 4.2).

The free OH groups may be substituted, mainly with glucose or with an *N*-acetylamino sugar, but also with D-alanine. In *Staphylococcus aureus,* a D-alanine is ester-linked to the C-2 of the 1,5-ribitolteichoic acid (about 50% substitution) and an *N*-acetylglucosamine to C-4. Here, the chain comprises about 40 repeating units.

The following two examples give an impression of the diversity of substituents in case of poly-(ribitol phosphate)-teichoic acids: in *Nocardiopsis albus*, pyruvic acid is situated in a (2→4)-ketalic linkage at ribitol and thus generates additional

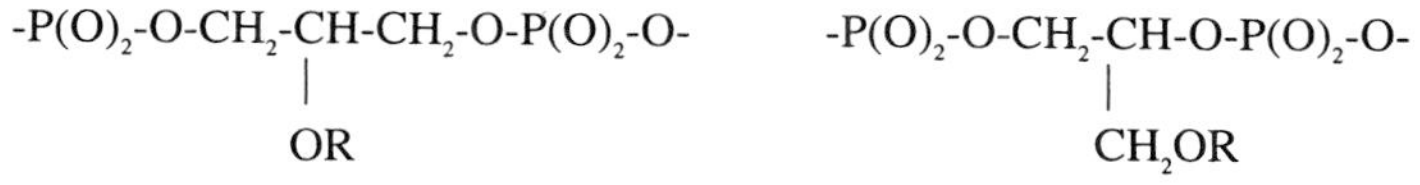

-P(O)₂-O-CH₂-CH-CH₂-O-P(O)₂-O-   -P(O)₂-O-CH₂-CH-O-P(O)₂-O-
              | |
             OR   CH₂OR

  1,3-Glycerol bisphosphate   2,3-Glycerol bisphosphate

Fig. 4.1. Characteristic sections of the poly(glycerol phosphate) teichoic acid chains.
For significance of *R* see text

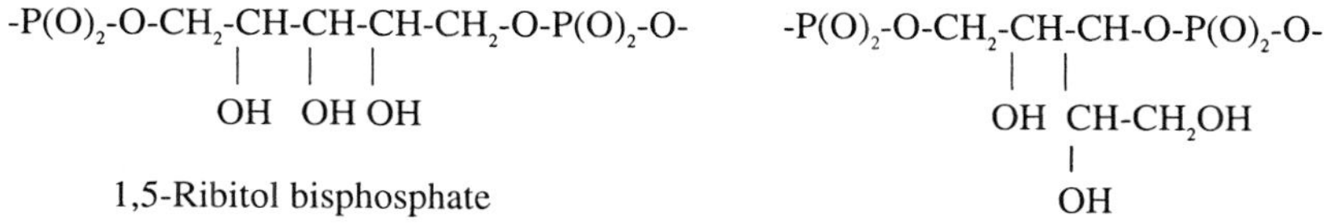

-P(O)₂-O-CH₂-CH-CH-CH-CH₂-O-P(O)₂-O-   -P(O)₂-O-CH₂-CH-CH-O-P(O)₂-O-
              | | | | |
             OH OH OH   OH CH-CH₂OH
                                                          |
  1,5-Ribitol bisphosphate   OH

                                               3,5-Ribitol bisphosphate

Fig. 4.2. Characteristic sections of the poly(ribitol phosphate) teichoic acid chains

negative charges, and in *Agromyces fucosus* a relatively hydrophobic oligomer α-L-Rha*p*-(1→3)-α-L-Rha*p*-(1→3)-β-D-Glc*p*NAc-(1→2)-α-L-Rha*p*-(1→ has been detected, which is bound to O-2 or O-4 of the ribitol.

The basic chain of poly-(mannitol phosphates) contains the (1→6)-isomer which, in, e.g. *Brevibacterium iodinum*, is substituted in a 4,5-ketalic linkage by pyruvic acid and to about 50% at C-2 with α-D-glucopyranose.

The (1→4)-poly-(erythritol phosphate) of *Glycomyces tenuis* is a good example for poly-(erythritol phosphates). It consists of approximately 23 units, about five of which are substituted with β-D-*N*-acetyl-glucopyranosyl residues.

4.1.1.1.2 Poly-(Glycosylpolyolphosphate)-Teichoic Acids.

Polymers of this kind may have been developed by incorporation of one glycosyl residue between phosphate and polyol in one of the above described polyglycerol or polyribitol teichoic acids. Poly-(glycosylglycerol phosphate) teichoic acids are more widespread than poly-(glycosylribitol phosphate) teichoic acids.

The monosaccharide of the poly-(glycosylglycerol phosphate) teichoic acids is bound via its OH group at C-6 to the phosphodiester group and via its glycosidic oxygen to O-1 of glycerol, in some cases also to O-2. One example is given in Fig. 4.3 depicting the structure of the teichoic acid of *Actinoplanes philippinensis* VKM Ac-647 that contains galactose as glycosidically linked component and a mannosyl residue bound in a (1→4) linkage as a side chain.

Poly-(glycosylribitol phosphate) teichoic acids may contain several monosaccharides in their repeating units such as -6-β-D-Gal*f*-(1→3)-β-D-Gal*p*-

$$-P\text{-}6\text{-}\beta\text{-}D\text{-}Galp\text{-}(1{\rightarrow}1)\text{-}sn\text{-}Gro\text{-}3\text{-}$$
$$4$$
$$\uparrow$$
$$1$$
$$\alpha\text{-}D\text{-}Manp$$

Fig. 4.3. Structure of the teichoic acid from *Actinoplanes philippinensis* VKM Ac-647. P $PO_2^{(-)}$

(1→6)-β-D-Gal*f*-(1→6)-β-D-GalNAc*p*-(1→3)-α-D-Gal*p*-(1→1)-Ribitol-5-P-   of *Streptococcus oralis* C 104.

4.1.1.1.3 Heterogeneity of Teichoic Acids.

Like LPS and capsular polysaccharides, the different teichoic acid types are also not absolutely regularly assembled, but show structural (micro-)heterogeneities. In the case of the ribitol phosphate polymer of *Bacillus subtilis* possessing a glucose side chain, only part of the molecules could be precipitated using an antiserum specific for terminal glucose, and even in the precipitate the glucose: ribitol ratio was <1. It has to be supposed that in this preparation ribitol phosphate chains substituted either completely, partially, or not at all with glucose occur simultaneously. The same is true for the teichoic acids of *Nocardiopsis dassonvillei* where glycerol phosphate and *N*-acetyl-β-D-galactopyranosyl-glycerol phosphate residues occur simultaneously in comparable quantitities. In the case of a glycerol phosphate-polymer of *Lactobacillus plantarum* N.I.R.D. C106 with a glucosidic side chain glucose, glucosyl-(1→2)-glucose and glucosyl-(1→3)-glucose could be detected besides glycerol phosphate. All three different polysaccharides could be separated by means of concanavalin A.

4.1.1.1.4 Linkage Units to Peptidoglycan.

Originally, it was supposed that teichoic acid chains are bound to peptidoglycan via a simple phosphodiester bridge. However, this could not be confirmed in most cases. On the contrary, it was discovered that binding units are inserted between peptidoglycan and teichoic acid, in which glycerol phosphate represents a common component, e.g. in the 1,5-ribitol teichoic acid of *Staphylococcus aureus* which is bound to the 6-hydroxyl group of the muramic acid of the peptidoglycan via -(OCH$_2$-CHOH-CH$_2$O-P)$_3$-3-GlcpNAc-(1→4)-GlcpNAc-1-P-.

4.1.1.2 Teichuronic Acids

In Gram-positive bacteria cultivated under usual conditions, teichuronic acids are mainly found only in low quantities besides the main component, teichoic acid. If the bacteria are cultivated under phosphate-deficient conditions, the production of teichoic acids is stopped and the formation of teichuronic acid is essentially reinforced. Thus, the uronic acid residues in the teichuronic acids take over the

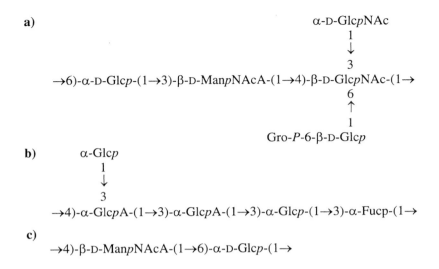

Fig. 4.4 a-c. Structure of the repeating units of the teichuronic acids from *Bacillus megaterium* AHU 1375 (**a** and **b**) and from *Micrococcus luteus* (**c**). P   PO$_2^{(-)}$; *ManpNAcA*   2-acetamido-2-deoxy-mannopyranuronic acid

functions of the teichoic acid phosphate residues, especially the binding of metal cations.

In *Bacillus megaterium* AHU 1375, the structure of the teichuronic acids is quite complicated. It contains two chains similar to each other with the repeating units shown in Fig. 4.4, however, chain a in Fig. 4.4 contains 24-28 units and chain b 85-95 units. The binding of chain a to the peptidoglycan occurs via a disaccharide β-ManpNAcA-(1→4)-GlcpNAc.

The teichuronic acids of *Micrococcus luteus* are more simply composed (Fig. 4.4.c). Its repeating unit represents a disaccharide containing *N*-acetyl-mannosaminuronic acid.

In the case of the latter, the polysaccharide chain is bound to the *O*-6 of the muramic acid of the peptidoglycan via a trisaccharide consisting of a reducing *N*-acetylglucosamine residue and two *N*-acetyl-mannosaminuronic acid residues as well as a phosphodiester bridge.

4.1.1.3 Lipoteichoic Acids

LTA represent partially substituted poly-(polyolphosphates), the repeating units of which do not essentially differ from those of the corresponding teichoic acid. However, instead of units linked to the peptidoglycan they possess a lipid component by means of which they are fixed in the cytoplasmic membrane. Their biosynthesis is not stopped in the case of phosphate deficiency, contrary to that of the cell wall teichoic acids.

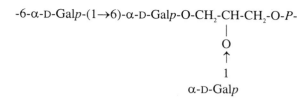

-6-α-D-Gal*p*-(1→6)-α-D-Gal*p*-O-CH₂-CH-CH₂-O-*P*-

$$-6\text{-}\alpha\text{-}D\text{-}Galp\text{-}(1{\rightarrow}6)\text{-}\alpha\text{-}D\text{-}Galp\text{-}O\text{-}CH_2\text{-}CH\text{-}CH_2\text{-}O\text{-}P\text{-}$$

α-D-Gal*p*

Fig. 4.5. Structure of the repeating unit of the lipoteichoic acid from *Lactococcus garvieae*. *P* PO₂⁽⁻⁾

→6)-β-D-Glc-(1→3)-α-GalNAc4N-(1→4)-α-D-GalNAc-1→3-β-D-GalNAc-(1→1)-Ribitol-5-ⓟ-

Fig. 4.6. Structure of the repeating units identically for teichoic acid and lipoteichoic acid of *Streptococcus pneumoniae*; each of the two GalNAc-residues contains an -O-PO₂⁽⁻⁾-O-CH₂-CH₂-N⁽⁺⁾(CH₃)₃ group at the C-atom 6. Gal*N*Ac4N 2-acetamido-2,4-dideoxy-galactose

The poly-(glycerol phosphate) type is most common. As in the case of teichoic acids, its hydrophilic moiety consists of 16-40 poly-(glycerol phosphate) residues linearly 1,3-linked (Fig. 4.1). The substituents R depicted there are comparable to those described for teichoic acids. In *Enterococci*, mono- to tetra-(1→2)-α-D-glucopyranosyl residues have been found as substituent R.

A second LTA type is represented by poly-(glycosylpolyol phosphate). Their hydrophilic part is similar to the corresponding teichoic acids. In the case of *Lactococcus garvieae*, the chain is composed of nine to ten repeating units (Fig. 4.5).

The repeating units of the lipoteichoic acid of *Clostridium innocuum* are structured more simply, containing only one galactopyranose residue in the main chain instead of the disaccharide depicted in Fig. 4.5. The secondary OH group of the glycerol is either unsubstituted (25%) or substituted with glucosamine cation (50%) or *N*-acetylglucosamine (25%).

Pneumococci represent an exception since their teichoic acid and lipoteichoic acid possess the same complex structure in the polyol moiety. In the special case of *Streptococcus pneumoniae* polymers, five to eight tetrasaccharidic ribitol phosphate units are present. These tetrasaccharides consist of two *N*-acetyl-galactosamine residues, each of which carries one phosphocholine residue at C-6, one residue of the rare 2-acetamido-4-amino-2,4,6-trideoxy-D-galactopyranose (Gal*p*NAc4N) and one glucose residue (Fig. 4.6).

Like teichoic acids, lipoteichoic acids show a more or less distinct heterogeneity which manifests itself not only in diverse chain lengths, but also in different substitutions. The lipoteichoic acid of *Clostridium innocuum* described above represents one example.

The hydrophobic part of the lipoteichoic acid is in most cases a glycolipid bound via a phosphodiester linkage. It derives from free lipids of the cytoplasmic membrane. In Fig. 4.7 some structures of this binding anchor are listed. In *Staphylococcus aureus* every fifth to eighth lipid molecule of the outer monolayer of the cytoplasmic membrane represents such a lipid anchor.

a)   α-Glc/Gal-(1→2)-α-Glc-(1→3)-O-CH$_2$-CH-CH$_2$

                                 | |

                             OR OR

b)   β-Glc-(1→6)-β-Glc-(1→3)-O-CH$_2$-CH-CH$_2$

                                 | |

                             OR OR

c)   β-Glc-(1→6)-α-Gal-(1→2)-α-Glc-(1→3)-O-CH$_2$-CH-CH$_2$

                                              | |

                                           OR OR

Fig. 4.7 a-c. Structures of binding anchors of lipoteichoic acids. *R* Different fatty acids, **a** with a terminal Glc, occurring in *Streptococcus, Lactococcus, Enterococcus, Leuconostoc*; **a** with a terminal Gal, occurring in *Streptococcus, Lactobacillus, Listeria*; **b** occurring in *Staphylococcus, Bacillus*; **c** occurring in *Lactobacillus*

4.1.1.4 Functions of the Teichoic Acids

All three members of the teichoic acid family are of great importance for the vital functions of the bacteria. Even under limiting cultivation conditions (e.g. restriction of Mg^{2+}, K^+, nitrogen or SO_4^{2-}) no perceivable reduction of their biosynthesis takes place. The importance of teichoic acids for bacteria is reflected by the fact that in the case of phosphate deficiency in the culture medium, teichuronic acids, in which uronic acids take over the acidic function, are synthesised instead of teichoic acids. Thus, the polysaccharide chains with their high negative charge density are essential for the principal function of teichoic acids, which is most likely the binding of metal cations (mainly Mg^{2+}, but also Ca^{2+} and K^+). It is recalled that the core region of the LPS of Gram-negative bacteria has a similar function. Also, such polysaccharide chains may possess protective functions.

Bacteria obviously do not require the same charge densities at all growth phases and in all surface regions. Regulation of charge densities is most likely possible via neutralisation of the charges by means of amino acids (D-alanine, D-lysine) or non-acetylated amino sugars. In some cases, the bacteria contain several acidic polysaccharides of diverse charge densities. It is conceivable that they are located at different sites on the surface.

Teichoic acids play an important role in the regulation of the activity of autolytic enzymes necessary for growth and division of the cell (e.g. *N*-acetylmuramyl-L-alanin-amidase) via direct interactions. Due to their distinct hydrophilicity, they provide the bacteria with good suspendibility. However, teichoic acids may also have unfavourable properties, a number of which may lead to destruction of the bacteria.

Teichoic acids represent effective phage receptors. If an extracted and purified receptor is added to the system phage/bacteria, a decrease in the magnitude of the lysis created by the phages follows due to the competition between free and bound receptor. This is frequently the case if isolated teichoic acids are added. The teichoic acid-peptidoglycan complex present in the cell wall, however, is often more effective since, due to the peptidoglycan matrix, it creates more favourable steric conditions for the phage adsorption. If the bacteria are cultivated under phosphate deficiency, they become resistant to these phages, and the teichuronic acids synthesised in this case have no receptor function.

The immunogenic efficiency of teichoic acids is also important. Like other polysaccharides, they are frequently only weakly effective in the isolated form. Immunisation with bacteria, on the contrary, induces a strong antibody production. The best antigenic determinants of teichoic acids are the branching sugar molecules. On the other hand, D-alanine bound as side chain is less efficient, presumably since it is rapidly cleaved in the host.

The poly-glycerol and –ribitol chains are also immunogenic, even if they are substituted with sugar or D-alanine molecules. Presumably, non-substituted regions in these basic chains or those having become unmasked by cleavage of D-alanine are effective in this process. A whole number of serologic cross-reactions between Gram-positive bacteria are caused by antibodies directed against these regions.

Due to their amphiphilic nature, lipoteichoic acids show characteristics which are missing in teichoic and teichuronic acids. In the cell wall-bound form, they influence the physical properties of the cytoplasmic membrane; also regulating functions are discussed. In the released form, they resemble LPS. For instance, they build up micelles and are able to bind to host cells and to trigger secondary reactions such as the release of interleukins and tumor necrosis factor α. During complement activation (see Sect. 8.3.4), both their negative charges and their micellar structure play a supporting role. Yet they are in general distinctly less active than LPS. For instance, an ability to induce pyrogenicity or lethal toxicity is expressed only poorly, if it is present at all.

The release of lipoteichoic acids from the bacteria occurs continuously during bacterial growth and may be reinforced by antibiotics. It takes place via an intermediate stage where the hydrophilic part is still located in the cell wall, whereas the hydrophobic part extends outwards. Thus, the bacterial cells receive a more hydrophobic surface, enabling them to adhere to host cells.

4.1.2 Acidic Polysaccharides

Acidic polysaccharides include sugar phosphate polymers, anionic polysaccharides and lipoglycanes.

Sugar phosphate polymers may consist of repeating units of 1-8 monosaccharides linked to each other via phosphodiester bridges, e.g. →6)-α-D-Glc$p$-(1→3)-α-D-Gal$p$NAc-1-P- in *Micrococcus* sp. 1. They differ from other similarly structured capsular polysaccharides in their stable binding to the cell wall.

Anionic polysaccharides contain as side chains acidic groups such as glycerol phosphate, organic acids (pyruvic and succinic acid) or sulfate. The latter provides the polysaccharides with an extremely strong, negative charge. The polysaccharide of *Bacillus cereus* AHU 1356, for example, possesses a repeating unit with the structure →4)-α-D-Glc$p$NAc-(1→3)-α-L-Rha$p$-(1→3)-α-D-Gal$p$-(1→4)-[Gro-1-P-2-]-α-L-Rha$p$-(1→.

Lipoglycans are distinguished from the polymers above especially by the fact that they are membrane-bound via a hydrophobic region, and that they are not always composed of repeating units. They consist of linear or branched homo- or heteropolysaccharides and may carry a monomeric glycerol phosphate as side chain. One example is represented by the lipoglycan of *Bifidobacterium bifidum*. It consists of a linear β-(1→5)-D-galactofuranan which is substituted at C-6 by glycerol phosphate. This region is attached to a linear β-(1→6)-D-glucopyranan which, in turn, is glycosidically linked to a hydrophobic region consisting of a partially acylated galactose in glycosidic linkage to a diacylglycerol:

Gro-1-*P*-6-D-Gal*f*-[(1→5)-{Gro-1-*P*-6-}-β-D-Gal*f*]$_n$-[(1→6)-β-D-Glc*p*]$_m$-(1→6)-β-D-Gal*p*-(1→1)-GroRR',

in which n = 7-14 and m = 7-10, R and R´ represent long-chain fatty acids, and the galactose linked to glycerol is esterified up to about 30% with long-chain fatty acids at its C-3.

Lipoglycans also comprise the lipoarabinomannan described further below.

4.1.3 Biosynthesis

The biosynthesis of the discussed polysaccharides occurs in cyclic processes as in the case of LPS and capsular polysaccharides. However, some differences to these, as well as among each other, have been found. For instance, teichoic acids are synthesised as a whole, whereas the biosynthesis of the LPS regions lipid A-core on the one hand and O-specific polysaccharide on the other occurs separately (see Sect. 2.2.5).

4.1.3.1 Biosynthesis of Teichoic Acids

During the first synthesis stages, the activated monosaccharide derivatives are produced in the same manner as described in Section 2.2.5.1. The syntheses of activated glycerol (CDP-Gro) and other glycerol derivatives are described in Section 2.1.4.3, to which the synthesis of the activated ribitol occurs in analogy.

The biosynthesis of substituted chains is demonstrated by, e.g. that of *Bacillus subtilis* 168. At first the binding unit is synthesised (see Fig. 4.8) starting with the transfer reaction UDP-GlcNAc + L → L-P-P-GlcNAc + UMP. L represents a carrier molecule which is at least similar to ACL (Sect. 2.1.4.3.3).

During the following steps, ManNAc and subsequently glycerolphosphate are attached according to an analogous mechanism, thus producing the binding unit. For the formation of the unsubstituted chain, the activated polyols are the only substrates. Under the influence of a polymerase, the chain is lengthened by the attachment of new building blocks to the terminal glycerol of the acceptor molecule via a phosphate residue. Subsequently, a part (about 50%) of the free OH groups in the glycerol is substituted with glucose (from UDP-Glc). The growing or mature chain is transported from the inner to the outer monolayer of the cytoplasmic membrane by means of a two-stage ABC transporter (Sect. 8.1.2.2), then a further portion (about 25%) of the free OH groups in the glycerol is esterified with D-alanine taken from the lipoteichoic acids. Finally, the mature molecule is linked to O-6 of the muramic acid of peptidoglycan via a phosphodiester brigde.

The genes necessary for biosynthesis are located as a cluster in the region of approximately 308°, some of them are depicted in Fig. 4.8. This arrangement is not universal.

4.1.3.2 Biosynthesis of Teichuronic Acids

Corresponding to the biosynthesis of teichoic acids, that of teichuronic acids starts with the synthesis of the binding unit attached via a diphosphate to a carrier molecule L similar to ACL. In the case of *Micrococcus luteus*, this molecule is ManNAcA-ManNAcA-GlcNAc-P-L. It is developed by the gradual attachment of three monosaccharide units, starting from the corresponding UDP derivatives. After this follows the gradual and alternating attachment of glucose and *N*-acetyl-mannosaminuronic acid, also starting from the corresponding UDP derivatives. After having achieved the final chain length, the transfer to the primary alcohol group of the muramic acid of peptidoglycan occurs.

4.1.3.3 Biosynthesis of Lipoteichoic Acids

The biosynthesis of poly-(glycerolphosphate) lipoteichoic acids starts with that of the lipid anchor. In the case of *Staphylococcus aureus*, 1,2-diacylglycerol is transformed with two molecules of UDP-Glc into β-Glc-(1→6)-β-Glc-(1→3)-diacylglycerol. The latter reacts with one molecule phosphatidylglycerol to form Gro-P-β-Glc-(1→6)-β-Glc-(1→3)-(Acyl)$_2$Gro. This substance is detectable in the cytoplasmic membrane in changing quantities and serves as a lipid anchor for the following reaction steps. Starting from phosphatidylglycerol (PtdGro), glycerolphosphate residues are gradually attached to it up to the desired chain length according to

Gro-P-β-Glc$p$-(1→6)-β-Glc$p$-(1→3)-(Acyl)$_2$Gro + n PtdGro →

(Gro-P-)$_n$-β-Glc$p$-(1→6)-β-Glc$p$-(1→3)-(Acyl)$_2$Gro + n (Acyl)$_2$Gro.

The released diacylglycerol is transformed into phosphatidic acid, consuming ATP, and subsequently into phosphatidylglycerol, and then returns to the cyclic biosynthesis process.

Possible substituents may be incorporated during or after chain elongation. Hexoses are activated as undecaprenol phosphates via the nucleoside diphosphates, D-alanine is attached to the D-alanine carrier-protein via the AMP derivative and is thus transmitted.

Genetic encoding and exact loci of biosynthesis are still unknown.

4.1.3.4 Control of the Biosyntheses

The relative amounts of synthesised polysaccharides of the Gram-positive cell wall are subject to control mechanisms as in the case of other cell wall components.

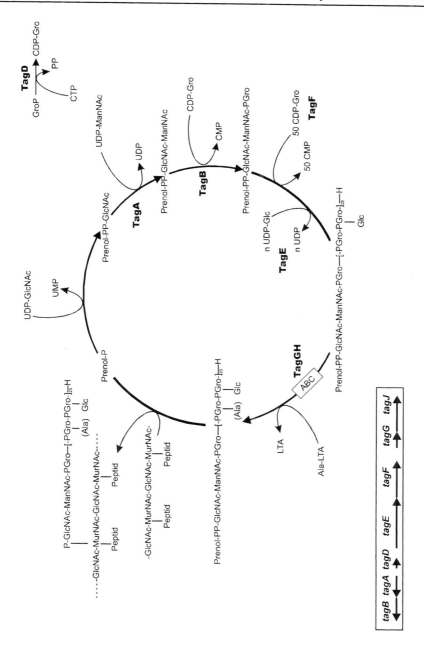

Fig. 4.8. Biosynthesis of the teichoic acid of *Bacillus subtilis* 168 (Fischer W (1997) Lipoteichoic acid and teichoic acid biosynthesis. Targets of new antibiotics? Biospektrum Sonderband 47-50, with permission). *P* phosphate

Especially worth mentioning are the controls of the ratios peptidoglycan/teichoic acid on one hand and teichoic acid/teichuronic acid on the other. The former ratio is regulated via the lipid carrier. The synthesis of peptidoglycan may be inhibited by the addition of the biosynthetic precursors of teichoic acids (and the other way round). The ratio teichoic acid/teichuronic acid is regulated by the concentration of phosphate. CDP-Glycerol-pyrophosphorylase obviously plays an important role in this process.

Bibliography

Delcour J, Ferain T, Deghorain M, Palumbo E, Hols P (1999) The biosynthesis and functionality of the cell wall of lactic bacteria. Antonie van Leeuwenhoek Int J Gen Mol Microbiol 76:159-184

Fischer W (1994) Lipoteichoic acids and lipoglycans. In: Ghuysen JM, Hakenbeck R (eds) Bacterial cell wall. New comprehensive biochemistry, vol 27. Elsevier, Amsterdam, pp 199-215

Fischer W (1997) Pneumococcal lipoteichoic and teichoic acid. Microb Drug Resist 3:309-325

Fischer W (1997) Lipoteichoic acid and teichoic acid biosynthesis. Targets of new antibiotics? Biospektrum Sonderband 47-50

Naumova IB, Shashkov AS (1997) Anionic polymers in cell walls of Gram-positive bacteria. Biochemistry (Moscow) 62:809-840

Ward JB (1981) Teichoic and teichuronic acids: biosynthesis, assembly, and location. Microbiol Rev 45:211-243

4.2 Proteins

The cell walls of Gram-positive bacteria are richer in proteins than those of the Gram-negative ones. More than 25 different proteins may be present only on the cell surface. According to their functions, two types of proteins can be distinguished, namely those located in a space similar to the periplasmic space of the Gram-negative bacteria, which have biosynthetic functions, and proteins appearing further outside which interact with the bacterial environment. Since the former have been mentioned in the discussion of the biosyntheses of the other cell wall components, this section will describe only some cell surface proteins which are important for infection processes caused by pathogenic bacteria.

4.2.1 Isolation and General Structure

The isolation and structural elucidation of cell wall proteins can be carried out using the same methods as in the case of Gram-negative bacteria. Differences exist in the higher stability of the Gram-positive cell wall and, therefore, mechanical decomposition demands more drastic methods. Enzymatic decompositions, on the contrary, are more easily carried out because the outer membrane is lacking. There is no universal methodology for the release of cell wall proteins of Gram-positive bacteria, since their fixation in the cell wall may be quite different. Staphylococcal protein A, for example, is covalently bound to the peptidoglycan, whereas streptococcal M-protein is mainly associated with lipoteichoic acids. Nevertheless, extraction of the latter is difficult and requires acids or detergents. In addition, destruction may occur during the procedure. In both cases, it is advantageous to degrade the cell wall enzymatically beforehand.

Three categories of cell surface proteins can be differentiated: (1) those that anchor at the C-terminal end of the molecule, (2) those that bind by their charge or by hydrophobic interactions and (3) those that are fixed via their N-terminal region (lipoproteins).

The proteins of the first group have a number of common structural characteristics. They usually consist of three regions, a C-terminal anchor region, a wall-associated region, and a surface-exposed region.

In contrast to the proteins remaining after biosynthesis in the Gram-negative cell wall, the corresponding ones of the Gram-positive bacteria often contain a sorting signal. This signal consists of a short sequence of up to seven mainly charged amino acids at the C-terminus, followed inwards by 15-22 mainly hydrophobic amino acids sufficient to span the cytoplasmic membrane. The hydrophobic region is followed by the highly conserved sequence Leu-Pro-X-Thr-Gly- (preserved in more than 100 surface proteins of Gram-positive cells; X is any amino acid). This motif is situated immediately at the outer surface of the cytoplasmic membrane. It is thus supposed that the charged tail and the hydrophobic region slow down the release of the C-terminus of the mature proteins from the membrane and induce cell wall binding of the proteins. During sorting, the Leu-Pro-X-Thr-Gly sequence is cleaved between threonine and glycine and the liberated carboxyl group of threonine may be linked to a cellular substrate like the amino group of the pentaglycine cross-bridge of the peptidoglycan. Therefore, the hydrophobic domain and the charged tail are, for instance, missing in M-proteins isolated from the streptococcal cell wall.

The cell wall-associated region is characterised by a high percentage of proline/serine residues distributed more or less regularly. The length and primary sequence of the latter region depend on the nature of the bacteria. This region should be positioned within the peptidoglycan layer.

The surface-exposed region is responsible for the biological effects. Areas consisting of almost identical building blocks coupled one after the other (tandem repeats) occur frequently. Their size varies and may amount to 2 to more than 100 amino acid residues.

The second group of cell surface proteins, i.e. those anchoring by their charge or by hydrophobic interactions, mainly consists of enzymes (isomerases, dehydrogenases, kinases, mutases, enolases). They can be removed from the cell wall by chaotropic agents.

Finally, N-terminal-anchored (lipo-)proteins have only recently been identified in Gram-positive bacteria. They may play a role as adhesins in the infection process.

4.2.2 Streptococcal Cell Wall Proteins

Among the longest known bacterial cell wall proteins are the M-proteins. In intact Streptococci they are localised on long filaments together with acylated and deacylated lipoteichoic acid and represent both efficient virulence factors and high-protective antigens. Members of the M family of proteins are molecules that can bind a variety of host components like albumin, immunoglobulins, kininogens, plasminogen and fibrinogen. Group A Streptococci containing M-protein resist the phagocytosis of normal sera, and presence of anti-M-protein antibodies renders them sensitive. This activity is type-specific, thus, in this way, nearly 100 specific M-protein types could be distinguished within group A streptococci.

The name M-protein is mainly applied to anti-phagocytically active members of the M family, whereas the others are called M-like. Only the former precipitate fibrinogen. Recently, it has been proposed to apply the expression M-protein to all types and to distinguish them according to functionally neutral terms (Emm, Mrp, Enn).

Under in vitro conditions, the formation of M-protein can be reversibly switched off.

The complete or at least partial sequences of a number of emm or emm-like *Streptococcus pyogenes* genes coding for M-protein biosynthesis have been elucidated. The genes are always located at the same locus of the chromosome. The sequence analysis shows that the proteins contain four repeat blocks (A, situated at the N-terminal end of the molecule, B, C, D) besides the above-mentioned C-terminal anchor region. The C-terminal region of the molecule including the sorting signal is almost constant while the N-terminal regions of the mature M-proteins are variable, the repeat block A hypervariable. The sequence diversity of the variable regions may be created by point mutations, small insertions and deletions, and intragenic recombinations between tandem repeat domains. A possible horizontal gene transfer is discussed.

The regions containing repeat sequences and being situated outwards from the cell in all cases develop elongated dimeric α-helical coiled-coil structures similar to each other with mainly hydrophobic amino acids on the surface of the helices. The central rod-like region formed thus contains the repeat blocks A, B and C.

The majority of the cell surface proteins are multifunctional. In M-proteins, the biological activity of the individual tandem repeats is different. The A-repeats bind albumin, the B-repeats fibrinogen and the C-repeats factor H of complement and keratinocytes. The regions are mainly located not in the constant, but, astonishingly, in the variable part of the molecule.

The serological diversity of the M-protein is the result of the variability of corresponding regions in the emm-like genes. The exchange of those genes within the group A-strains is responsible for the considerable antigenic variations. Above that, a gene transfer to Streptococci from other groups is possible.

The role of M-proteins as virulence factors is not unequivocal. As already mentioned, their antiphagocytic effect is connected with their ability to bind to molecules found in body secretions, including immunoglobulins (IgA and IgG), albumin, fibronectin and fibrinogen. This binding initiates a conformational signal on the bacterial cell surface to perform a specific function. The binding of M-protein to complement factor H could impair factor C3b. Today it is still believed that further bacterial components are important for the creation of this phagocytosis resistance. In addition, it has been supposed that the ability of M^+ strains to bind to host surfaces depended directly on the protein. However, this does not seem to be the general case. It is possible that the intermediate stage of lipoteichoic acids release (see Sect. 4.1.1.4), where the hydrophilic part is still attached in the cell wall but the hydrophobic part points outwards, is mediated by M-protein. This hydrophobic part possibly mediates the binding of the bacteria to the host cells. Further functions are the creation of an inflammatory response and the mediation of the invasion into deeper tissue regions.

A number of group A-streptococci are able to evoke opalescence in sera. The serum-opacity factor (OF) is an apoproteinase and up to now not well defined; however, it allows the division of the streptococci into two subgroups (OF^+/OF^-). The formation of OF is initiated by the emm-like gene. In OF^- types, the chromosomal region contains only one single emm-like gene, in contrast to a triplet of these genes at the same site in OF^+-types.

The T-proteins of streptococci can be extracted from approximately 95% of all serogroup A strains, occasionally also from other groups. They mostly appear only after treatment of the bacteria with trypsin and serve as an additional typing characteristic. T-Proteins were identified by immunoelectron microscopy as components of the outer streptococcal cell wall. In this method, particular cell wall components are marked with (for example ferritin-) labelled specific antibodies and the ultrathin sections are examined under the electron microscope. The proteins were found to be located on short filaments a little below the M-proteins close to the electron-dense cell wall structure. They have no influence on the virulence of Streptococci. It is possible that the molecular structure of individual T-proteins is relatively diverse, and that, thus, T-proteins resemble each other only in their resistance to trypsin.

There are also enzymes anchored in the cell wall such as neuraminidase, which play a role in virulence or as β-N-acetylhexosaminidase in *Streptococcus pneumoniae* and β-D-fructosidase and dextranase in oral streptococci.

4.2.3 Staphylococcal Protein A and Similar Proteins

Protein A was first described in 1940 and can be found in almost all *Staphylococcus aureus* strains. With a portion of 6-7%, the cell wall of strain Cowan I is especially rich in protein A. Even though a small part of the protein may be extracted with phosphate buffer at about 90 °C, the bulk cannot be isolated by means either of detergents or trichloroacetic acid. On the contrary, cell lysis by lysostaphin is successful. Purification of protein A can be simply managed by its extraordinary property to form precipitates in a pseudo-immunoreaction with the F_c fragment of γ-globulin. This can be shown in the double diffusion test according to Ouchterlony. Using antiprotein A serum, a double line can be observed. One line is the result of the normal antigen-antibody reaction, the second is caused by the pseudo-immunoreaction. This property enables the isolation of protein A from crude cell extracts by its binding on a IgG-Sepharose column at neutral pH, washing the column with neutral buffer, and eluting the protein at pH 3. Also, IgG or, after fragmentation with papain, its F_c fragment can be isolated with immobilised protein.

Some staphylococcal strains are not able to incorporate their protein A into the cell wall. They release it into the culture medium from where it can be obtained, e.g. by means of IgG-Sepharose. Using erythrocytes coated with anti-erythrocyte antibodies in a mixed culture of protein A-carrying and -non-carrying staphylococci, the former can be isolated by binding to the F_c part and the cosedimentation thus caused. The non-carrying bacteria remain in the supernatant.

Staphylococci may as well be coated with specific antibodies via the F_c fragment without affecting their serologic specificity. If such coated Staphylococci come into contact with microorganisms against which the antibodies are directed, a coagglutination takes place. There are test systems based on this principle.

Protein A has a molecular mass of about 42 kDa. Using partial trypsin cleavage and isolation of the F_c-reactive cleavage products it was shown (Sjödahl 1977) that the protein A molecule contains four F_c-reactive areas at the N-terminus, each of which consists of about 60 amino acid residues with a similar structure. According to recent investigations, however, the gene carries codations for five similar tandem repeats, and the region contains about 175 amino acid residues. Presumably, the structure developed from the ancestoral gene by several gene duplications. The C-terminus is constructed in analogy to the general structural principles described above.

The possibility of cleavage of protein A as of other proteins from the Gram-positive cell wall by means of lysostaphin indicates their binding in this wall. It occurs at the free NH_2-group of a non-cross-linked pentaglycine bridge (see Sect. 3.2.1.2.2; Fig. 4.9) via threonine-COOH which is released after cleavage of the linkage between Thr and Gly of the Leu-Pro-X-Thr-Gly sequence.

The IgG-F_c-binding capacity is not limited to staphylococcal protein A, e.g. streptococci of the groups A, C and G are also able to produce similarly reacting proteins. In total, at least five types have been found. Type I is represented by the staphylococcal protein A, type III by the streptococcal protein G. Group II is very heterogeneous and divided into subgroups. Comparable to the proteins of this group are also some emm-like proteins of *Streptococcus pyogenes*.

Additionally, proteins have been described that dock at the F_c fragment of IgA or which bind the F_{ab} region of IgA, IgE, IgG and IgM in a non-immune mechanism.

According to crystallographic investigations, the binding regions of the different F_c fragment-binding proteins are not identical. The binding seems to occur via

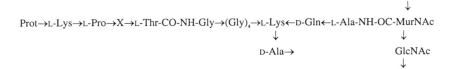

Fig. 4.9. Binding of protein A to the peptidoglycan. *Prot* protein; *X* amino acids

hydrophobic contacts between residues of two particular α-helices of the protein A-binding unit and residues in the intersection of the C_H2 and C_H3 domains in the F_c region of IgG.

The influence of Ig-binding proteins on the virulence of the strains has not yet been definitively elucidated. Isolated protein A has been described as immunogenic and toxic; also hypersensitivity reactions have been observed. It is possible, however, that these effects are at least partially due to impurities in the preparation. The reaction of protein A with the F_c fragment of IgG is of importance since it leads to the blockade of the antibody region responsible for the activation of the complement system and thus to the impairment of the immune defense. Even inhibitions of the complement system by protein A and protein A-antibody complexes have been observed.

4.2.4 Clumping Factor of Staphylococci

Staphylococcus aureus possesses a surface protein that binds fibrinogen and in this way leads to clumpings (clottings) in the blood plasma. Due to its functional similarities to coagulase which belong to the secretory proteins it was supposed that the clumping factor represented a cell bound coagulase. This could not be confirmed. The primary structure of the clumping factor is distinctly different from that of coagulase and corresponds basically to other cell wall proteins (see above). However, a sequence of 150 times repeated Asp-Ser-dipeptides is incorporated in the proline-rich region at the C-terminus. This acidic region could react with a basic region at the C-terminus of the fibrinogen γ-chain consisting of 15 amino acid residues.

Coagulase as well as clumping factor may induce the formation of a fibrin layer round the bacteria. This is recognised by the host as non-foreign and thus allows the bacteria to grow unhindered. In addition, the clumping factor provides the bacteria with adhesive properties. Both effects represent virulence factors, the role of which in the infection process has not yet been unequivocally elucidated.

4.2.5 Further Proteins of the Gram-Positive Cell Wall

As is described below (Sect. 8.3), the attachment of the bacteria to outer or inner surfaces is a precondition to colonising a host. Many proteins of the Gram-positive cell wall are able to bind specifically to components of host surfaces, such as, e.g. fibronectin-binding proteins (FnBP) of staphylococci and streptococci or the collagen-binding protein of staphylococci. Other cell wall proteins bind to soluble proteins of the host organism, such as (immuno-) globulins, complement components, fibrinogen, kininogen, prothrombin and salivary glycoproteins. Finally, some cell wall proteins represent enzymes that cleave, for example, large nutrients into transportable sizes (dextranase in the case of streptococci, casein peptidase in lactococci).

Staphylococcus aureus strains may contain one or two similar FnBPs. They possess a structure which is typical for cell wall proteins and contain a region consisting of tandem repeats in each of which a highly acidic sequence Glu-Glu-Asp-Thr is essential for the binding of fibronectin. The binding sites in FnBP of *Streptococcus pyogenes* resemble those of *Staphylococcus aureus*, possibly since their genes developed from a common precursor. Fibronectin is a very large multifunctional glycoprotein situated ubiquitously in both dissolved and insoluble matrix form in the human organism and also on its surfaces. It binds a wide variety of compounds such as collagens, fibrin, heparin and actin. It is thought that the bacteria could use this molecule as a target for binding.

Regarding its basic structure, the collagen-binding protein also does not essentially differ from those described so far. It has a large unique N-terminal region followed by a set of 187 amino acid repeats, a proline-rich region, and a classic cell wall sorting signal. Collagens are very frequently occurring long-fibrous, high-molecular scleroproteins present in connective tissue or in the protein-containing bone matrix and, together with fibronectin among others, important for anchoring processes. Both binding proteins mentioned enable the bacteria to colonise almost any region of the host organism and thus to induce the infection process.

4.2.6 Biosynthesis

The biosynthesis of cell wall and secretion proteins of Gram-positive bacteria, their penetration of the cytoplasmic membrane, and the cleavage of the N-terminal signal sequence occur similarly to those of the corresponding proteins of the Gram-negative bacteria (see Sect. 2.3.2.6). In both cases, cell wall and secretion proteins are synthesised according to similar mechanisms. The signal sequence in Gram-positive bacteria also consists mainly of a core of 15-20 hydrophobic amino acids which is flanked by positively charged residues at the N-terminal end.

The differences between Gram-negative and Gram-positive bacteria begin with the sorting of the proteins. In Gram-positive bacteria, an additional signal is mostly necessary for the incorporation of proteins in the cell wall. The fixation may take place by covalent or stable non-covalent binding to other cell wall components. To initiate the binding, the presence of the hydrophobic/charged end (see above) is necessary, which is later enzymatically cleaved.

In contrast to the biosynthesis of teichoic acids, which is coupled to that of the peptidoglycan, the synthesis of *Staphylococcus aureus* protein A occurs independently. This indicates that its incorporation is not carried out into one of the precursors of the peptidoglycan biosynthesis, but into the mature peptidoglycan. However, this is hard to imagine, since the peptidoglycan is highly cross-linked and contains only a few free pentaglycine amino groups. Therefore, binding to a precursor seems to be more probable here as well, with a subsequent incorporation into the cell wall by means of a transglycosylation reaction.

The mechanisms of biosynthesis in proteins which are anchored in the cytoplasmic membrane, especially the incorporation of the lipid anchor, resemble those of lipoprotein biosynthesis in Gram-negative bacteria.

The biosynthesis of Gram-positive cell wall proteins can be inhibited by antibiotics attacking the ribosome, such as chloramphenicol, streptomycin or puromycin. In this case, the biosynthesis of the peptidoglycan proceeds still one generation further. Vancomycin, on the contrary, inhibits the biosynthesis of the peptidoglycan, but not that of proteins. Peptidoglycan and proteins are thus synthesised independently of each other. Even after the inhibition of the peptidoglycan biosynthesis by vancomycin, an incorporation of protein A is still possible, i.e. it takes place in already produced peptidoglycan. This is a contrast to the incorporation of newly synthesised teichoic acids.

Bibliography

Kehoe MA (1994) Cell-wall-associated proteins in Gram-positive bacteria. In: Ghuysen JM Hakenbeck R (eds) Bacterial cell wall. New comprehensive biochemistry, vol 27. Elsevier, Amsterdam, pp 217-261
Navarre WW, Schneewind O (1999) Surface proteins of Gram-positive bacteria and mechanisms of their targeting to the cell wall envelope. Microbiol Mol Biol Rev 63:174-229
Sjödahl J (1977) Structural studies on the four repetitive Fc-binding regions in protein A from *Staphylococcus aureus*. Eur J Biochem 78:471-490
Ton-That H, Faull KF, Schneewind O (1997) Anchor structure of staphylococcal surface proteins. A branched peptide that links the carboxyl terminus of proteins to the cell wall. J Biol Chem 272:22285-22292

4.3 Cell Wall Components of Mycolata

Mycolata belong to the Actinomycetes and contain mycolic acids as a common taxonomic marker (Barry et al. 1998). Among them are species of the genera listed in Table 4.1 as well as the not mentioned *Dietzieae*. They mostly represent harmless bacteria such as saprophytes; some of them, however, are very dangerous, e.g. *Mycobacterium tuberculosis*.

The cell walls of Mycolata differ from those of other Gram-positive bacteria in their elevated content of glycolipids with a high lipid portion. Their hydrophilic moiety consists in part of rather unusual linear or branched homo- or hetero-polysaccharides. In mycobacteria, two special layers of the cell wall control the

$$| \longleftarrow \text{Meromycolate moiety} \longrightarrow | \longleftarrow \alpha\text{-Branch} \longrightarrow |$$

$$\textbf{R-X-(CH}_2)_{13,15}\textbf{-Y-(CH}_2)_{16,18}\textbf{-CHOH-CH-(CH}_2)_{21,23}\textbf{-CH}_3$$
$$|$$
$$\textbf{COOH}$$

Fig. 4.10. Basic structure of mycolic acids from Mycobacteria. For the significance of *R*, *X*, and *Y* see Fig. 4.12

permeability to substances, i.e. an outer lipid barrier and a so-called capsule consisting of proteins and polysaccharides. In the first case, the mycolic acids together with an assortment of free lipids containing fatty acids of medium and short chain lengths form a pseudo-outer membrane.

4.3.1 The Mycolyl Arabinogalactan Complex

This complex represents the largest part of the Mycolata cell walls. It is covalently bound to peptidoglycan and thus represents a component of a high-molecular cell wall skeleton which is responsible for the unusual stability of these bacteria.

4.3.1.1 The Mycolic Acid Moiety

4.3.1.1.1 Properties, Structure. Mycolic acids are high-molecular fatty acids alkylated in α-position and hydroxylated in β-position, and occur in the cell walls partly in free, loosely associated form, partly bound to polysaccharide (especially to arabinogalactan). They range to 80 carbons in total chain length and represent the main components of the cell walls. The free mycolic acids may be extracted from the cells with chloroform/methanol, whereas alkaline conditions are necessary for the isolation of the bound ones. They can be separated from each other by two-dimensional thin-layer chromatography (TLC) on silica gel or by high-performance liquid chromatography (HPLC). The mycolic acids of the mycobacteria are essentially bigger than those of other species (Table 4.1). The main chain of the mycolic acids, the meromycolate moiety (Fig. 4.10) is the variable region of the molecule. Apart from the chain length, the chain structure of the α-branch is conservative.

The elucidation of the chemical structure has been facilitated by the fact that mycolic acids may be cleaved by pyrolysis into the meroaldehyde and the α-branch as shown in Fig. 4.11. For this reason, the carboxyl group and the α-C-atom of mycolic acids are regarded as part of the α-branch. The analysis of both cleavage products was carried out by using the methods mentioned in Section 2.1.2. It revealed that the α-branch is composed of a straight-chained, unbranched, unsubstituted and saturated hydrocarbon chain, in which merely the number of C-atoms may vary. The meromycolate moiety is, in fact, also straight-chained and unbranched, unsubstituted, and saturated in three larger, "constant" regions (Fig. 4.12). However, it may contain *cis*- or *trans*-double bonds, cyclopropane and epoxy rings in the variable regions termed X and Y, as well as methyl, methoxyl and oxo groups at different sites. The extreme chain length of the meromycolate moiety with a linearly structured inner part (between CHOH group and Y region) as well as the linear structure of the likewise rather long (>20 C-atoms) α-branch of the mycobacterial mycolic acids facilitate a dense parallel packing of the two hydrocarbon chains. On the other hand, the *cis*-double bonds and the substituents induce kinks in the packings (Fig. 2.1). Both are actually located at a considerable distance from the carboxyl group, especially in the case of oxygen substituents that are at least 35 C-atoms apart.

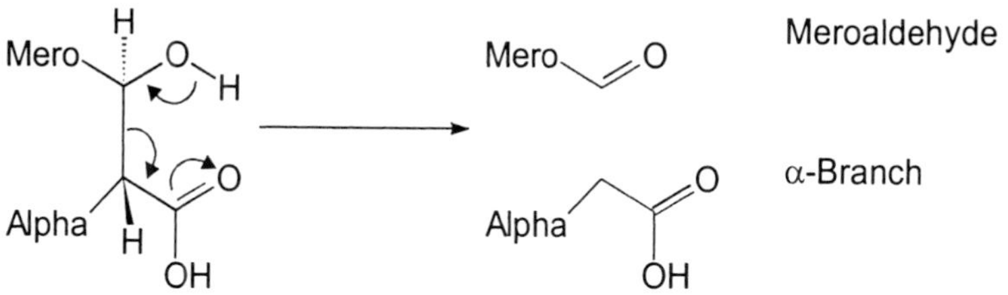

Mero — H, O, H

Mero — O   Meroaldehyde

Alpha — H, O, OH

Alpha — O, OH   α-Branch

Fig. 4.11. Pyrolytical fragmentation of mycolic acids (Barry CE III, Lee RE, Mdluli K, Sampson AE, Schroeder BG, Slayden RA, Yuan Y (1998) Mycolic acids: structure, biosynthesis and physiological functions. Prog Lipid Res *37*, 143-179, with permission)

Mycolate type	R	X		Y
α'	-	CH_3CH_2-		Y_1
	$CH_3-(CH_2)_{15-19}-$	H—C=C—H, CH₂—	(cis)	Y_1, Y_2, Y_3
α	$CH_3-(CH_2)_{15-19}-$	H—C—C—H (CH₂ ring), CH₂—	(„cis")	Y_1, Y_2, Y_3
Epoxy	$CH_3-(CH_2)_{15,17}-$	CH_3 —CH— C—C (O epoxide) —H, H	(„trans")	Y_1, Y_2, Y_3, Y_4
Keto	$CH_3-(CH_2)_{15,17}-$	CH_3 O —CH—C—CH_2—		Y_1, Y_2, Y_3, Y_4
Methoxy	$CH_3-(CH_2)_{15,17}-$	CH_3 O—CH_3 —CH—CH—CH_2—		Y_1, Y_2, Y_3, Y_4
Wax ester	$CH_3-(CH_2)_{15,17}-$	CH_3 O —CH—O—C—CH_2—		Y_1, Y_2, Y_3, Y_4

Y_1:	H—C=C—H, CH₂—	(cis)	Y_3:	H—CH(CH₃)—C=C—H	(trans)	
Y_2:	H—C—C—H (CH₂ ring), CH₂—	(„cis")	Y_4:	H—CH(CH₃)—C—C (CH₂ ring)—H	(„trans")	

Fig. 4.12. Structures of mycolic acids from Mycobacteria (Brennan and Nikaido (1995), with permission, from the *Annual Review of Biochemistry*, Volume 64 © 1995 by Annual Reviews www.AnnualReviews.org). For the localisation of *R*, *X*, and *Y* in the chains see Fig. 4.10. The designations α for the mycolate-type on the one hand and as α-branch on the other have been chosen independently of each other and have no connections to each other

The individual genera of Mycolata differ characteristically from each other in the total number of C-atoms, the number of double bonds and of substituents in the meromycolate moiety as well as in the number of C-atoms in the α-branch (Table 4.1.). Even within the genera, the mycolic acids may vary sufficiently to serve as taxonomic markers which, for instance, can be used for comparing their TLC or HPLC patterns.

A considerable microheterogeneity within a single type is caused by alternations between two C-atoms (Fig. 4.10) of chain lengths of the "constant" regions in the meromycolate moiety as well as of the α-branch in the mycolic acid. In the case of a "mycolic acid" isolated from *Mycobacterium smegmatis*, more than 100 such closely related molecular species could be identified. If the possible *cis-/trans-*isomerism depicted in Fig. 4.12 is considered in addition, a single "mycolic acid" amounts to more than 500 different molecular species.

4.3.1.1.2 Biosynthesis.

4.3.1.1.2 Biosynthesis. The probable biosynthetic pathway of mycolic acids consists of four stages:

1. Synthesis of the hydrocarbon chain of the α-branch including C-atom 2 and the carboxyl group of the whole mycolic acid.
2. Synthesis of the basic hydrocarbon chain of the meromycolate moiety.
3. Modification of the meromycolate chain.
4. Condensation of both molecular parts.

The biosynthesis of the two hydrocarbon chains (steps 1 and 2) takes place according to the same mechanisms described in Section 2.1.4 up to a chain length of C_{24}. Then the biosynthetic pathways fork. ACP, CoA or a membrane-bound protein serve as activators (R_2 in Fig. 4.13) for the synthesis of the meromycolate chain. For biosynthesis of the α-branch and the completion of the molecule, the so-called mycolyl phospholipid is responsible (Myc-PL, a β-D-mannopyranosyl-octahydroheptaprenol phosphate to which the substrate is bound at O-6; R_1 in Fig. 4.13). After the completion of the α-branch, its C-atom is activated by carboxylation in α-position to the carboxyl group.

The modification of the meromycolate chain starts with it elongation and the insertion of two *cis*-double bonds, a distal and a proximal one (Fig. 4.13). This reaction takes place before condensation with the α-branch and is probably connected with the chain elongation. It is not yet clear whether a β-hydroxyacyl dehydrase or a desaturase affects this process. The introduction of a *C*- or *O*-methyl group as well as a methylene bridge to form a cyclopropane ring occurs by starting from (*S*-adenosyl-)methionine. The first step of the reaction consists in a transfer of a methyl cation to the double bond. The resulting cationic intermediate may be deprotonated to the cyclopropane ring or to a *trans*-double bond with an adjacent methyl group or it may react with water to an α-hydroxymethyl compound which may further react to an α-oxo- or α-methoxy compound (Fig. 4.14).

The linking of both parts of the molecule presumably occurs via Claisen condensation. For this purpose, the carbonyl-carbon of the mero acid has to react as an electrophile with the activated nucleophilic carbon atom of the α-branch.

Table 4.1. Variations in the mycolic acid-size in dependence of the genus. (Barry et al. 1998)

Genus	Total number of C-atoms	Number of double bonds	C-Atoms in the α-branch
Corynebacteria	22-36	0-2	14-18
Gordona	48-66	1-4	16-18
Mycobacteria	60-90	2*	20-26
Nocardia	44-60	0-3	12-18
Rodococcus	34-48	0-4	12-18
Tsukamurella	64-78	1-6	20-22

* According to Lee et al. (1996) mycolates from *M. tuberculosis* do not contains double bonds.

Subsequently, the reaction product still has to be decarboxylated and its oxo-group reduced to the secondary alcohol group. The resulting mycolic acid ester is transported into the cell wall, where it is bound either to arabinogalactan (AG) to form AG-mycolate or to trehalose to form the cord factor.

4.3.1.2 The Arabinogalactan Moiety

This polysaccharide is rather unusual as both arabinose and galactose are present in furanosidic form. The basic chain consists of a linear poly-β-D-galactofuranose chain (about 32 residues) in alternating (1→5)- and (1→6)-linkages, to which three poly-(1→5)-α-D-arabinofuranose chains with each about 27 sugar residues are bound to O-5 of Gal*f* residues that are substituted at O-6. The polyarabinose chains are partially substituted by further α-D-arabinofuranose units branching from O-3. The non-reducing ends are capped by a branched hexa-arabinose motif that contains all Ara*f* residues being substituted at O-2 and which carries mycolic acids (Fig. 4.15). The disaccharide structure α-L-Rha*p*-(1→3)-D-Glc*p*NAc-1-*P* is located at the reducing end and binds the complete molecule to O-6 of some of the muramic acids of the peptidoglycan via the phosphate residue.

The biosynthesis starts with a transfer of GlcNAc-1-*P* from UDP-GlcNAc and subsequently of Rha, originating from dTDP-Rha, to a polyisoprenoid carrier. After this, the incorporation of Gal*f* (from UDP-Gal*f*) and then Ara*f* (from β-D-Ara*f*-1-monophosphoryl decaprenol) follows.

4.3.1.3 The Complete Molecule

The arabinogalactan part is covalently bound to peptidoglycan and substituted by mycolic acids and thus represents the connecting piece between both. The

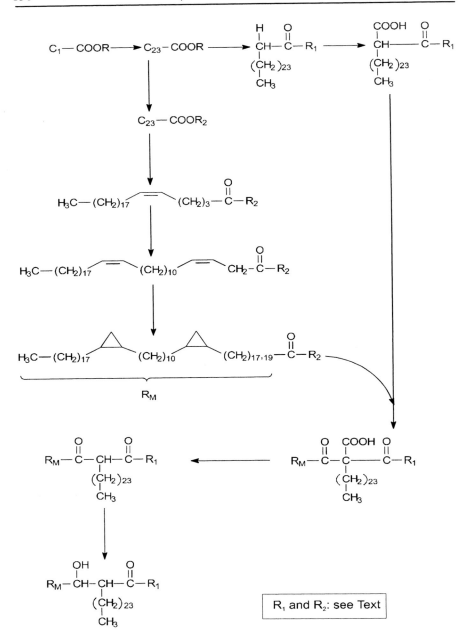

Fig. 4.13. Presumable biosynthetic pathway of the mycolic acid of *Mycobacterium tuberculosis* (after Brennan and Nikaido 1995)

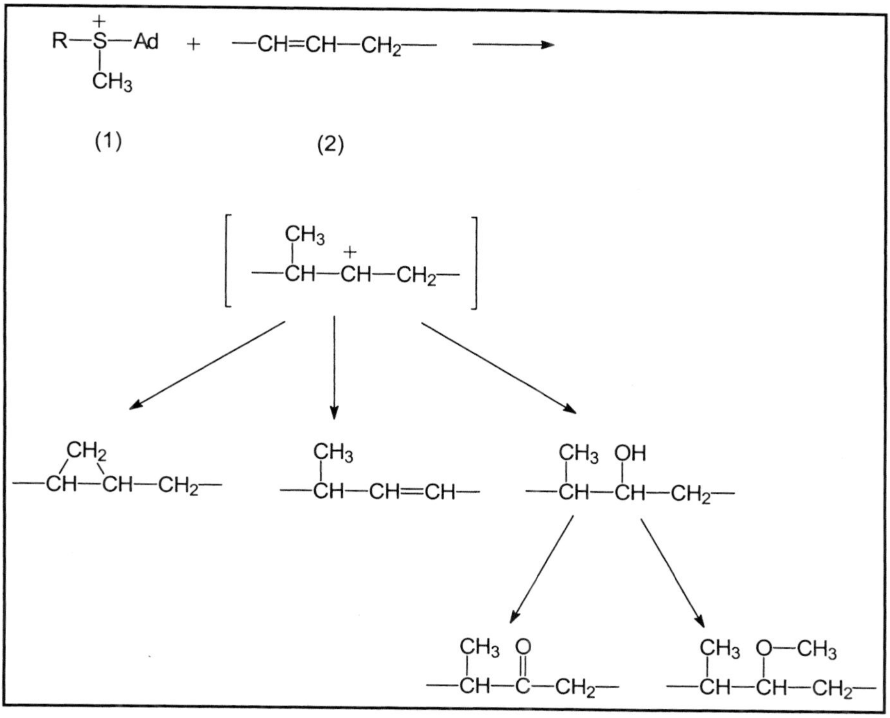

(1) S-Adenosylmethionine.
(2) Detail of a meromycolate chain.
See Text for further explanations.

Fig. 4.14. Modification reactions of the meromycolate chain

binding of arabinogalactans to the peptidoglycan occurs via the described bridge Rha-(1→3)-GlcpNAc-1-P, and the mycolic acids are ester-linked to about two thirds of the primary alcohol groups of the hexaarabinose motifs. Thus, a molecular arrangement results which renders the cell wall very stable and very difficult to be penetrated (see Sect. 7.2). The diffusion barrier caused by these molecules is 100- to 1000-fold less permeable to hydrophilic substances than that of E. coli.

4.3.2 Lipoarabinomannan

Besides the mycolyl arabinogalactan macromolecules, mycobacteria contain in the cell wall a lipoarabinomannan of a likewise very complex and not yet completely elucidated structure (Fig. 4.16). Basically, it consists of three regions:

1. A phospholipid part which is substituted by region 2 via inositol and in most cases esterified with one molecule palmitic acid and one molecule tuberculostearic acid at the glycerol.
2. A poly-(1→6)-linked mannan branching at O-2 of each residue with mannose (the mannan core).
3. A heterogeneous poly-arabinofuranan consisting of an α-(1→5)-linked main chain with branches at O-3 which is bound to O-2 of the second distal mannose residue; the non-reducing ends may be capped by mono-, di- and trimers of mannose, the latter two comprising α-(1→2)-linkages.

In the polysaccharide regions of lipoarabinomannan of some mycobacteria, succinate and other substituents are ester-linked providing the molecule with an acidic character.

4.3.3 Extractable Lipids

4.3.3.1 Lipooligosaccharides (LOS)

The lipooligosaccharide of *Mycobacterium kansasii* is described as example. Its sugar backbone is formed by an α,α′-trehalose esterified with long-chain fatty acids, such as 2,4-dimethyl tetradecanoic acid, at O-3, O-4 and O-6, and substituted by an oligosaccharide at O-2′. The oligosaccharide consists of a tetra-

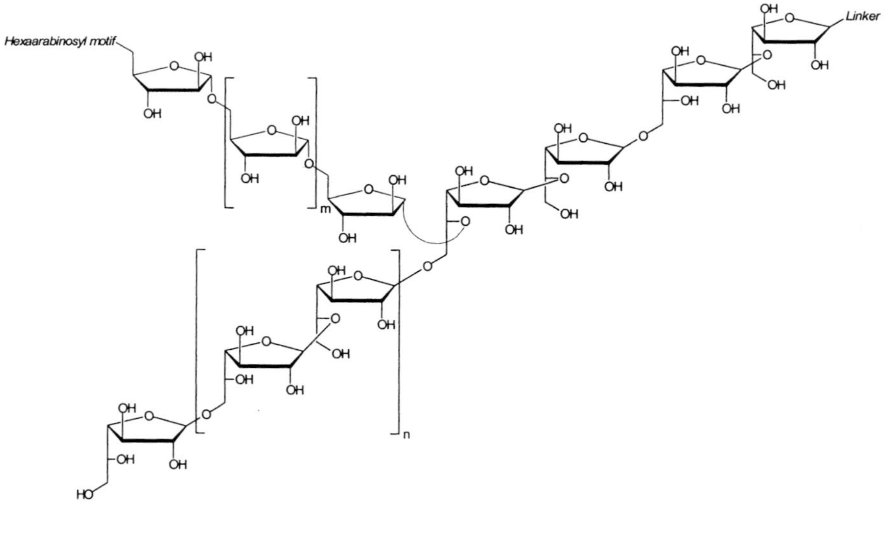

Fig. 4.15. Structure of the arabinogalactan moiety

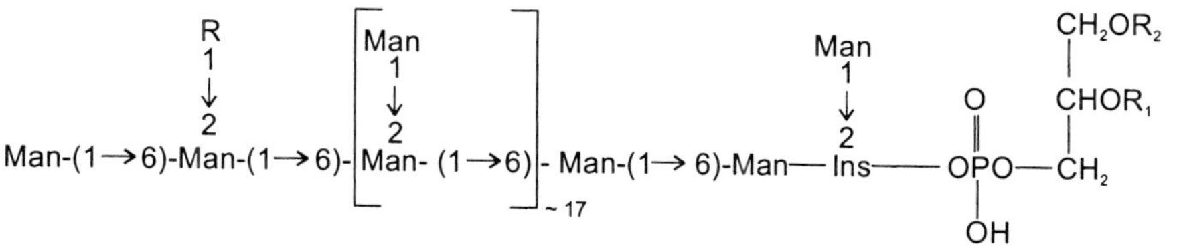

$$
\begin{array}{c}
R \\ 1 \\ \downarrow \\ 2
\end{array}
\quad
\left[\begin{array}{c}
\text{Man} \\ 1 \\ \downarrow \\ 2
\end{array}\right]
\quad
\begin{array}{c}
\text{Man} \\ 1 \\ \downarrow \\ 2
\end{array}
$$

Man-(1→6)-Man-(1→6)-|Man- (1→6)|- Man-(1→6)-Man—Ins——OPO—CH₂ ... with CH₂OR₂ / CHOR₁ branch, ~17

Fig. 4.16. Approximated structure of the lipoarabinomannan from *Mycobacterium tuberculosis*. R Ara-(1→2)-[Ara-(1→5)]$_{-22}$-(1→, with some (1→3)-branches by in part further branched oligo-(1→5)-Ara; at the non-reducing end in part in -(1→5)-linkage an oligo-(1→2)-mannan consisting of 1-3 residues. *Man* α-D-Man*p*; *Ins* myo-Inositol; *Ara* mainly α-D-Ara*f*; R_1, R_2 long-chain fatty acids, e.g. palmitic acid or tuberculostearic acid

glycose unit containing bound xylose, 3-*O*-methyl rhamnose, fucose and *N*-acylkansosamine (Fig. 4.17).

Also the cord factor, 6,6´-dimycolyl-α,α´-trehalose (Fig. 4.18), can be regarded as member of this class of compounds. It has been detected in *Mycobacterium tuberculosis*, but is also present in other mycobacteria. The name cord factor was given since strains containing it join together into cord-like formations in the culture medium.

4.3.3.2 Phenolic Glycolipids (PGL)

Phenolic glycolipids and the glycopeptidolipids described below are also termed mycosides. Mycoside A (*Mycobacterium kansasii*), B (*Mycobacterium bovis*) and G (*Mycobacterium marinum*) belong to the phenolic glycolipids. Their basic body is formed by a phenol residue that is substituted in *p*-position with a long (about 30 C-atoms) *n*-alkane chain which is hydroxylated twice and simply *C*- and *O*-methylated each (Fig. 4.19). Both hydroxyl groups are esterified with a typical C_{34} fatty acid, i.e. mycocerosic acid. The oligosaccharide bound to the phenolic hydroxyl group consists of one to four relatively hydrophobic monosaccharides. 2-

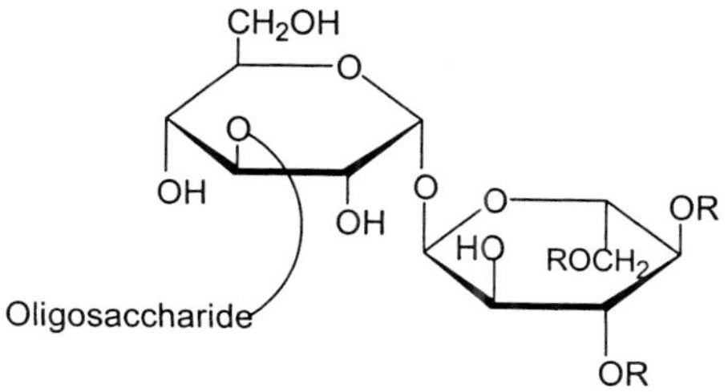

Fig. 4.17. Structure of the lipo-oligosaccharide from *M. kansasii*. *R* Long-chain fatty acid residue

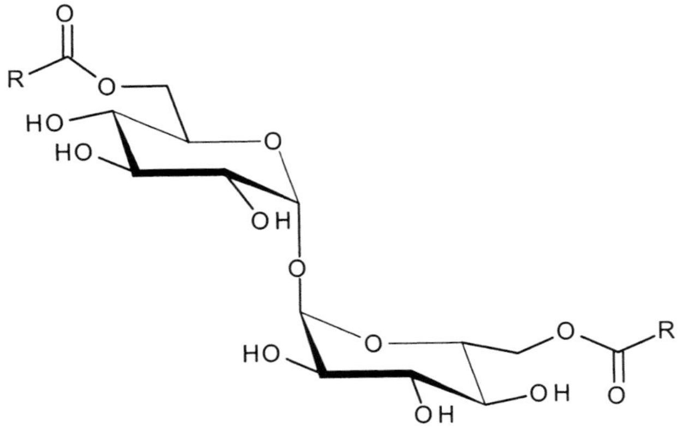

Fig. 4.18. Structure of the cord factor of *M. tuberculosis. RCOO* mycolyl residue

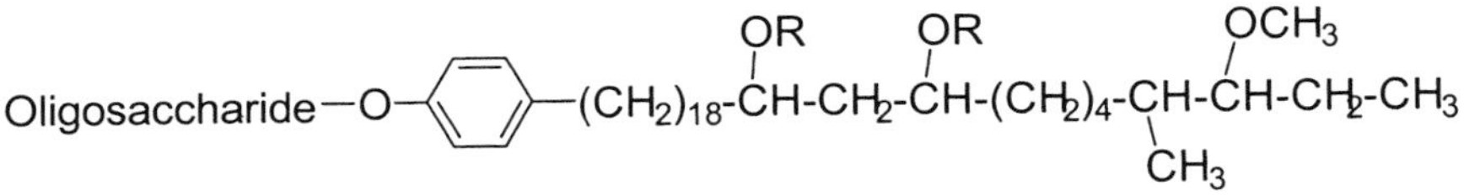

Fig. 4.19. Structure of a phenolic glycolipid. R long-chain fatty acid residue

O-Methylfucose as well as 2-*O*-methyl-, and 3,4-di-*O*-methyl-rhamnose have been found in mycosid A, and 2-*O*-methyl-rhamnose in mycoside B.

4.3.3.3 Glycopeptidolipids (GPL)

The glycopeptidolipids are also called mycoside C. The basic body of the molecule depicted in Fig. 4.20 as an example is formed by the peptide D-Phe-D- *allo*-Thr-D-Ala-L-alaninol. The hydroxyl group of alaninol is substituted with 3,4-di-*O*-methyl-L-rhamnose, the hydroxyl group of threonine carries a residue α-L-rhamnopyranosyl-(1→2)-6-deoxy-L-talopyranose. The phenylalanine is amidated, for example with palmitic acid. Especially the non-reducing terminal monosaccharides are highly variable and can be formed, e.g. by amido sugars, sugar acids, branched-chain and pyruvylated monosaccharides. These residues are often immuno-dominant.

Glycopeptidolipids are rather widespread in mycobacteria, and often represent the most important surface antigen. They can be divided into more than 30 serotypes due to their different chemical structures.

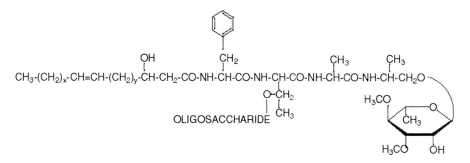

Fig. 4.20. Structure of a glycopeptidolipid

4.3.4 Waxes and Sulfolipids

The waxes of mycobacteria are esters of long-chain diols (phthiocerol A, 3-*O*-methyl-4-methyl-9,11-dihydroxy-n-$C_{32,34}$-alkan, and the phthiocerol B shorter by one C-atom) with mycocerosic acid (an *n*-C_{28}-fatty acid methyl-branched at C-atoms 2,4,6 and 8).

Sulfolipids represent a closely related family of trehalose-2′ sulfates esterified with long-chain fatty acids in positions 2,3,6,6′; they occur only in *Mycobacterium tuberculosis*. The following fatty acids could be detected: palmitate/stearate, phthioceranate (= 2,4,6,8,10,12,14-heptamethyl-*n*-C_{30}-1 carbonic acid) and 17-hydroxy-phthioceranate. It is supposed that these molecules support the intracellular survival of the bacteria.

4.3.5 Proteins

The cell walls of mycobacteria contain in general rather small amounts of proteins. About their functions, see Sect. 8.3.6.3.

Bibliography

Barry CE III, Lee RE, Mdluli K, Sampson AE, Schroeder BG, Slayden RA, Yuan Y (1998) Mycolic acids: structure, biosynthesis and physiological functions. Prog Lipid Res 37:143-179

Brennan PJ, Nikaido H (1995) The envelope of Mycobacteria. Annu Rev Biochem 64:29-63

Lee RE, Brennan PJ, Besra GS (1996) *Mycobacterium tuberculosis* cell envelope. Cur Top Microbiol Immunol 215:1-27

5 Cell Wall Components of Archaea

Archaea represent a group of microorganisms which often live under extreme conditions and obviously originate from a very early evolutionary era of the earth (extremophiles). Although they actually belong to the prokaryotes and morphologically even resemble the "classical" bacteria, they differ quite extraordinarily from them in many respects. Therefore they are at present taxonomically classified as a third domain (archaea) besides (eu-)bacteria (bacteria) and eukaryotes (eucarya). In older nomenclatures, archaea are called archaebacteria.

The cell walls of archaea show very interesting structural variants especially for the adaptation to their partially extreme living conditions. Like those of Gram-positive bacteria, they have a monoderm cell structure, i.e. they are surrounded by a single membrane. Thus, their cell walls are more closely related to each other than to those of the Gram-negative bacteria, which have a diderm structure.

The classification of the archaea is not yet unanimous; various authors differentiate quite distinctly. According to their 16S rRNA phylogeny, archaea can be divided into two kingdoms (see Noll 1992): euryarchaeota and crenarchaeota. A further division occurs into physiological groups.

Euryarchaeota are subdivided into:
- The extreme halophiles, some of which may grow in alkaline milieus (3.5 M NaCl at pH 9.5 are optimal for *Natronococcus occultus*).
- The methanogenes, strict anaerobes reducing CO_2 to methane.
- The extreme thermophiles, some of which reduce sulfate, other elementary sulfur to sulfide.

Crenarchaeota are subdivided into:
- The acido-thermophiles living at temperatures above 70 $^{\circ}$C and at pH below 4 (*Sulfolobus acidocalderaus* at pH 1.5) and representing obligate or facultative aerobes; they are able to oxidise elementar sulfur to H_2SO_4.
- The strict anaerobes living at temperatures even above 100 $^{\circ}$C and at pH 4.5-6.5 and reducing elementary sulfur to H_2S.

Opposite to Gram-positive or Gram-negative bacteria, the cell walls of archaea are neither formed according to similar construction schemes nor do they contain any common cell wall polymer. As shown in Table 5.1, a number of different cell wall components can be found which, however, show quite diverse compositions and structures. Table 5.1 does not include the glutaminylglycan of Natronococcus.

Like bacteria, archaea can also be differentiated into Gram-positive and Gram-negative prokaryotes, however, the taxonomic importance of this differentiation and the molecular background of the reaction are less significant, even though perceptible (see Table 5.1).

Table 5.1.: Composition of cell walls of Archaea (König 1988)

Genus	GRAM	HP	MC	PM	PS	SL
Halococcus	+	+		+		
Methanobacterium	+	+		+		
Methanothermus	+			+		+
Methanosarcina	+/-		+			+
Halobacterium	-					+
Methanococcus	-					+
Methanospirillum	-				+	+
Methanothrix	-				+	+
Thermoplasma	-					
Pyrodictium	-					+
Sulfolobus	-					+
Thermoproteus	-					+

HP heteropolysaccharide; *MC* methanochondroitin; *PM* pseudomurein; *PS* protein sheath; *SL* S layer. In the upper part of the table Euryarchaeota are listed, in the lower part Crenarchaeota.

The cell-stabilising component in the cell walls of the Gram-positive archaea is mostly formed by a polymer similar to the bacterial murein (peptidoglycan) and termed pseudomurein; in the case of the Gram-negative ones, it is the S-layer that takes over this function. *Thermoplasma* completely lack the cell wall; the stabilization of the cells occurs by means of a combination of a glycoprotein and a lipoglycan embedded in the cytoplasmic membrane. Both molecules are rich in mannose, and the glycoprotein is strongly branched at the non-reducing end, which consists of mannose, whereas the lipoglycan is linear.

In the following, some cell wall components of the archae are discussed in detail, independently of the taxonomic classification of the germs.

5.1 Pseudomurein

Pseudomurein represents the cell-stabilising component, mainly in Gram-positive archaea. It is a polymer which is quite similar to murein (peptidoglycan; Sect. 3.2.1.2), but differs from it as follows:
- Instead of the *N*-acetylmuraminic acid of murein, *N*-acetyltalosaminuronic acid is bound.
- The binding of the peptide bridge to the disaccharide does not occur via a lactic acid residue, but to the carboxylic group at position 6 of the *N*-acetyltalosaminuronic acid.
- A highly variable part of D-glucosamine is replaced by D-galactosamine.
- The glycosidic linkages in the polysaccharide strands are β-(1→3)- instead of β-(1→4)- in the murein.
- Contrary to murein, the peptide bridge does not contain D-amino acids, but relatively unusual peptide linkages (via γ-COOH or ϵ-NH$_2$).

The similarities between murein and pseudomurein concern not only their structure and properties, but also the methods for their isolation and structural elucidation.

The structure of the pseudomurein is represented in Fig. 5.1. The amino acids depicted in brackets are able to replace the predominantly occurring ones. Such a replacement may already occur after the addition of increased concentrations of the respective amino acid into the culture medium.

The spatial structure of the pseudomurein examined by means of X-ray structural analysis is comparable to that of murein. On the other hand, the modifications in composition and structure of pseudomurein compared to those of the murein are important enough to make the archaea resistant to both antibiotics such as penicillin, cycloserine or vancomycin and to enzymes such as lysozyme and proteases that attack α-peptide linkages.

The biosynthesis of pseudomurein is very similar to that of the murein, however, not identical, with respect to both the mechanism and the localization of the individual steps in different regions of the cell envelope (cf. Sect. 3.2.2.1).

The formation of UDP-N-acetyltalosaminuronic acids occurs in the cytoplasm, probably by epimerization at C-2 and by oxidation of the primary alcohol group of UDP-N-acetylgalactosamine during the synthesis of the disaccharide UDP-D-Glc$p$NAc-(3←1)-β-L-Tal$p$NAcA, which forms the polysaccharide strands from the UDP-activated monosaccharides.

Parallel to this process and also in the cytoplasm, the peptide bridge shown in Fig. 5.2 is formed. For this purpose, alanine, lysine (with its ϵ-amino group), glutamic acid (with its γ-carboxylic group to the α-amino group of lysine) and partially again alanine are stepwise attached to the UDP-glutamic acid (UDP bound to the α-amino group of the glutamic acid).

By cleavage of the UDP-Glu-linkage, the peptide is attached to the carboxyl group at position 6 of the N-acetyltalosaminuronic acid of the disaccharide UDP-D-Glc$p$NAc-(3←1)-β-L-Tal$p$NAcA, and the UMP bound at the glucosamine is replaced by undecaprenyl monophosphate (undec-P; see ACL in Sect. 3.2.2.2), in order to enable the molecule to penetrate the cytoplasmic membrane (Fig. 5.3).

Outside this membrane, the disaccharide pentapeptide is integrated into the growing glycan chain as in the case of murein biosynthesis. The following cross-linking to complete the pseudomurein could likewise occur by transpeptidation. The terminal L-alanine could play the same role in this process as the terminal D-alanine in the biosynthesis of murein, i.e. it could be split off to gain the linkage energy for the connection of the lysine of building block I with the glutamic acid of building block II. However, the mechanism of this process is not clear.

5.2 Methanochondroitin

The cell walls of *Methanosarcina* frequently consist of a 20-200-nm-thick layer of acidic polysaccharides which may be solubilised by treatment with sodium borohydride at 60 °C for 18 h and then possess a molecular mass of about 10 kDa. Both insoluble and soluble material predominantly consist of N-acetylgalactosamine and glucuronic acid, and of smaller amounts of glucose. Some strains also contain galacturonic acid.

In the soluble cell wall polysaccharide of *Methanosarcina barkeri* a structure [-β-D-GlcpA-(1→3)-β-D-GalpNAc-(1→4)-β-D-GalpNAc-]$_n$ has been identified. It is similar to the saccharide moiety of chondroitin sulfate which is the reason why the

←1)-β-D-GlcNAc-(3←1)-β-L-TalNAcA-(3←
6
↓
Glu (Asp)
↓γ
Ala (Thr, Ser)
↓ε
Lys←γ-Glu→(δ-Orn)
↑
Lys←γ-Glu
↑ε
Ala (Thr, Ser)
↑γ
Glu (Asp)
↑
6
→3)-β-L-TalNAcA-(1→3)-β-D-GlcNAc-(1→

Fig. 5.1. Structure of the pseudomurein from Methanobacteria

UDP-Glu-γ→Ala→ε-Lys→Ala
↑γ
Glu

Fig. 5.2. Structure of the activated bridge peptide

Undec-PP-D-GlcNAc-(3←1)-β-L-TalNAcA-(3←
6
↓
Glu
↓γ
Ala
↓ε
Lys←γ-Glu
↓
(Ala)

Fig. 5.3. Structure of the disaccharide-pentapeptide building blocks for the pseudomurein biosynthesis (with terminal Ala: building block I, without terminal Ala: building block II). *P* phosphate

whole class of molecules is termed methanochondrotin.

The biosynthesis of the methanochondrotin occurs in a way essentially different to that of the structurally comparable acidic polysaccharides of the bacteria. Lipid carries play a role only at a relatively late stage; before that, nucleoside diphosphate-activated intermediates are used.

During the first stage, UDP-Gal$p$NAc-(3←1)-Glc$p$A (= UDP-chondrosine) is produced from UDP-Gal$p$NAc and UDP-Glc$p$A and reacts with a further molecule, UDP-Gal$p$NAc to form UDP-Gal$p$NAc-(4←1)-Gal$p$NAc-(3←1)-Glc$p$A. Only then does the exchange of UDP to undecaprenyl monophosphate (undec-P), the penetration of the cytoplasmic membrane and finally the polymerisation take place.

5.3 Heteropolysaccharides

Halococcus forms 50-60-nm-thick cell walls consisting of complex acidic heteropolysaccharides. These are extraordinarily stable and demand drastic methods (0.5 N NaOH, 60 °C, 18 h) to become at least partially solubilised. The structure is remarkably complex (Fig. 5.4); according to present knowledge, three domains can be distinguished. Domain I and III are relatively high-molecular and contain chain branches. They are connected with each other by the straight-chained domain II. The linkage of domains I and II is unusual since the only amino acid in the molecule, the glycine, is bound to a glucosamine. Equally unusual is the very large quantity of sulfate residues in all three domains which render the molecule extremely acidic and have an influence on the stability of the cell wall even under extreme environmental conditions.

5.4 Glutaminylglycan

Natronococcus (see above) contains a rigid cell wall polymer. It is formed by linear chains of about 60 poly-γ-L-glutamine residues which may be divided into three regions (A, B, C, Fig. 5.5). The glutaminic acid residues are unsubstituted in region A, whereas in region B they are partially substituted with oligosaccharide X via an amide bond to their α-carboxyl group, and in region C partially with oligosaccharide Y.

Surface polymers consisting of poly-γ-glutaminic acid chains are not a domain of archaea, but are also present in bacteria. It is recalled that, for example, a poly-γ-glutaminic acid capsule (however, in its D-configuration) represents a virulence factor in *Bacillus anthracis*.

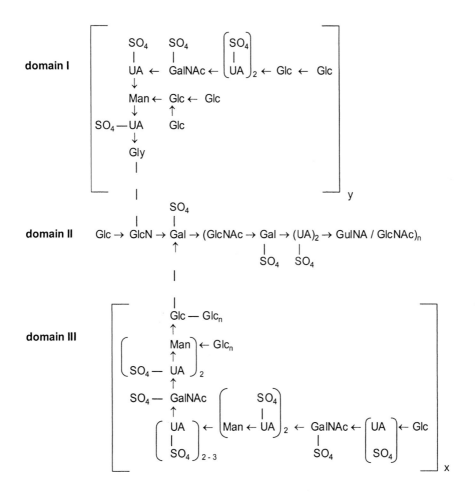

Fig. 5.4. Structural elements of the heteropolysaccharide from *Halococcus morrhuae*. *UA* unknown uronic acid (after Kandler and König 1993)

5.5 Lipids

The cell wall lipids of archaea differ characteristically from those of bacteria in both structure and properties. Firstly, their hydrocarbon chains are basically constructed of isopentanyl units (Fig. 5.6). Secondly, these chains are not ester- but ether-linked to a hydroxyl group of glycerol, which means that the fatty acids are replaced by long-chain alcohols. Such ether lipids are much more resistant to hydrolysis and are thus better adapted to the extreme environmental conditions of archaea. Thirdly, the lipids do not contain any C-C-double bonds. Fourthly, the hydrocarbon chains are not bound to the *sn*-1- and -2-position, but to the *sn*-2- and -3-position of the glycerol molecule.

The construction of isopentanyl units with their only short methyl side chains, and the lack of mainly *cis*-double bonds provide the archaea ether lipids with a rather high packing density. Thus, the membranes formed are still present in the liquid-crystalline state at quite high surrounding temperatures. In some lipids, cyclopentane rings are integrated into the chain to keep the fluidity constant to a large extent. To give examples, Fig. 5.7 shows the structure of archeol (the core lipid of the archaea) and nonitolcaldarchaeol. The latter may have been synthesised by head-to-head condensation of two identical partial molecules. Because of its chain length and the glycerol residues at both ends, it is thought to represent a transmembrane lipid.

The ether instead of ester linkage of the hydrocarbon chains with glycerol renders the molecule resistant to hydrolysis even under extreme environmental conditions. The *sn*-1-OH group of glycerol is not always present in a free state but is involved in the synthesis of sulfolipids, lipoglycans and phosphoglycolipids. In the case of sulfolipids, the sulfate residue is bound to O-1 of glycerol in the form - $PO-O-CH_2-CHOH-CH_2OSO_3H^-$.

Also in the case of lipoglycans and the phosphoglycolipids, the binding of the oligosaccharide to this site occurs directly or via a phosphate residue.

The lack of C-C-double bonds in their lipids is essential for archaea, since it provides them with stability towards oxidative degradation.

At first, the biosynthesis of the lipids containing the archaeol structure occurs similarly to that described in Fig 2.11 via the mevalonate pathway and geranyl pyrophosphate, and subsequently to a C_{20} product (presumably geranyl-geranyl-pyrophosphate). By condensation with dihydroxyacetone, this produces a compound called pre-diether which is structurally similar to the archeol but contains no double bonds. The further reaction steps to form the lipids mentioned above are listed in Fig. 5.8.

Region A   Region B   Region C

[γ-Gln→γ-Gln]→[γ-Gln→γ-Gln]→[γ-Gln→γ-Gln]
                                ↑   ↑
                                X   Y

X = α-Glc-(1→4)-α-Glc-(1→3)-β-GalNAc-(1→3)-β-GalNAc-(1→

Y = [4)-β-GalNAc-(1→]₄4)-β-GalNAc-(1→3)-[3)-α-GlcNAc-(1-α-]₄3)-α-GlcNAc-(1→

Fig. 5.5. Possible structure of the oligosaccharide substituents of the glucaminylglycan from Natronococcus

5.6 S-Layers and Sheaths

These two cell wall components are wide-spread in archaea. Since they also can be found in bacteria, however, they are described in Section 6.2.

Bibliography

Kandler O, König H (1993) Cell envelopes of archae: structure and chemistry. In: Kates M, Kushner DJ, Matheson AT (eds) The biochemistry of archae (archaebacteria). Elsevier, Amsterdam, pp 223-259

König H (1988) Archaebacterial cell envelopes. Can J Microbiol 34:395-406

Noll KM (1992) Archaebacteria (archae). In: Lederberg J (ed) Encyclopedia of microbiology, vol 1. Academic Press, New York, pp 149-160

Ratledge C (1996) Lipids, microbial. In: Meyers RA (ed) Encyclopedia of molecular biology and molecular medicine, vol 3. Verlag Chemie, Weinheim, pp 426-442

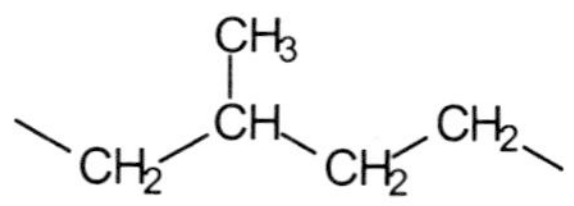

Fig. 5.6. Structure of the isopentanyl unit

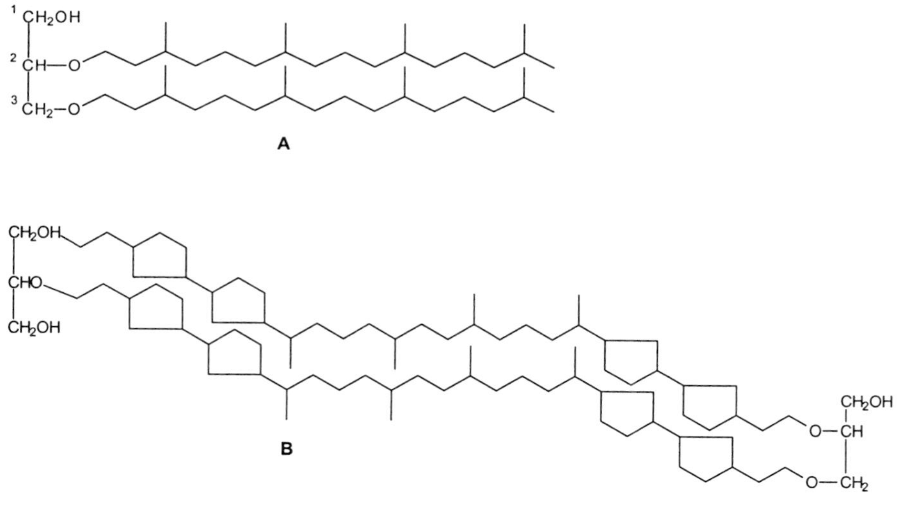

Fig. 5.7 A, B. Structure of archeol (**A**) and nonitolcald archaeol (**B**)

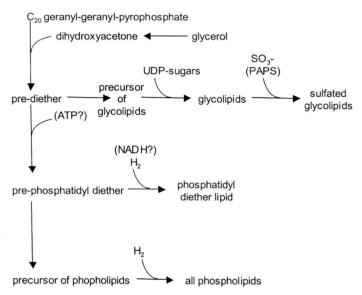

Fig. 5.8. Proposed pathway for the biosynthesis of the archeol-containing lipids of Archaea. *PAPS* phosphoadenosine phosphosulfate (Ratledge C (1996) Lipids, microbial. In: Meyers RA (ed) Encyclopedia of molecular biology and molecular medicine, vol 3. Verlag Chemie, Weinheim, pp 426-442, with permission).

6 Components Outside the Cell Wall

6.1 Capsules and Slime Layers

6.1.1 General Characterisation

Many bacteria are surrounded by an extracellular layer, the bacterial capsule. This compartment is a gel-like highly hydrated mesh mostly formed by polysaccharides (capsular polysaccharide = CPS). It covers the underlying cell wall components and is at least physically connected with it. Capsules are found in both Gram-positive and Gram-negative bacteria and create an additional protection, which in many cases results in an elevated virulence. Encapsulated bacteria may cause severe diseases: e.g. meningitis, septicemia, pyelonephritis, osteomyelitis, pneumonia or septic arthritis. Capsules may vary in their thickness from one strain to the other and even in different isolates of the same strain; on the one hand it can be extremely thin, on the other, it may surpass that of the cell wall by more than 1 order of magnitude.

Most extensive encapsulation is often observed with fresh isolates and may decrease on sub-cultivation under in vitro conditions. This is in accordance with the fact that the capsule actually protects bacteria against the defence mechanisms of the host but is not necessary in all phases of growth and multiplying. The capsule-forming capacity can also be kept stable in vitro over longer periods by means of exterior signals, e.g. an increased concentration of carbohydrates (glucose) or addition of serum to the culture media. However, it is not certain in all cases that the capsule material thus formed is still identical to the natural one.

Slimes are similar to capsules with respect to their physical structure and their chemical composition. Therefore, capsules are not easy to differentiate from slimes. A distinctive feature of capsules is their fixation to the cell wall, mostly through ionic or hydrophobic interactions, while slimes are excreted by the bacteria. They may form so-called slime walls, but also become released into the surrounding medium.

Despite this relatively simple and clear definition, the differentiation of the capsule from other bacterial cell surface components frequently causes difficulties. In the past, the proof of acidic polysaccharides was regarded as a relatively reliable criterion for the presence of a capsule. For several reasons this is no longer the case. Firstly, there are additional acidic cell wall polysaccharides such as the teichoic acids. Secondly, "normal" neutral cell wall polysaccharides, e.g. the O-specific polysaccharide of LPS, possess an acidic character in quite a number of bacteria. Thirdly, there are capsules formed by neutral polysaccharides, such as that produced by LPS of *E. coli* O111. Fourthly, the excretion products of bacteria, such as the slime substances released into the medium, often consist of acidic polysaccharides. Fifthly, capsules have been described which represent proteins and are neutral or basic (e.g. the poly-γ-glutamat capsule of *Bacillus anthracis* or the pretentious K88 and K99 "capsules" of *E. coli* mimicked by a very dense fimbriation).

Already in the 1920 it was found that capsules hamper the "normal" serologic reactions of the bacteria such as the O-agglutination of Enterobacteriaceae or the reaction with specific phages. However, the capsular components were found to be antigens or bacteriophage receptors in their own right. The first of these antigen found in *E. coli* was termed A-antigen. Kauffmann ascertained the correlation between this antigen as well as further ones, called B- and L-antigens, and the presence of capsules. Therefore he summarised them as K-antigens (from the German: Kapsel). Later, this term was applied to the analogous antigens of other bacterial groups, both Gram-positive and Gram-negative. The sera gained by immunisation with encapsulated bacteria are important for the K-typing of the bacteria. It is emphasised, however, that all K-antigens certainly indicate capsular structures, but not all capsules give rise to K-antigens.

In many cases, injection of encapsulated bacteria into appropriate host organisms leads to the production of specific (protective) antibodies. The serological typing of *E. coli* in particular, but also that of other Gram-negative and Gram-positive (*Staphylococcus aureus*) bacteria, revealed that (O-), K- and the H-antigens, the latter determined by fimbriae, are not arbitrarily combined in Nature.

6.1.2 Detection of Capsules

Like the differentiation of the capsule from other cell wall components, the unequivocal proof of its existence is often difficult. This is especially the case if microcapsules, pseudocapsules or slime walls exist. Several procedures are regarded as rather reliable. The first is the light microscopic evidence in a preparation containing Indian ink, mostly after staining of the bacteria for example with carbolfuchsin. The capsule then appears as a white halo between the pink bacteria and a black background (see Fig. 1.4). When a specific anticapsular antiserum is used, the immune precipitation around the cell can be easily seen and the capsule appears thicker (Kapselquellungs reaction). For a more detailed analysis of capsules, electron microscopy is well suited. Due to its instability during the sample preparation, the capsules have to be fixed, with either a negatively charged electron-dense protein, such as ferritin, or with a metal oxide. In immune electron microscopy, anticapsule antibodies are used for the demonstration of a capsule. In all these cases it is the contrasting protein that is visualised. In the immuno gold version of the immunoelectron microscopy, the antibodies are attached to spheres of a gold colloid with defined particle size. Instead of antibodies, lectins may be used. The methods, however, are limited, and there are examples where both the positive and the negative reaction had been caused by other cell components and was thus false.

Therefore it is recommended to include further procedures additionally. The following incomplete listing gives examples:

- Agglutination of the bacteria with anticapsule sera.
- Non-appearance of agglutination with sera directed against components of the cell wall (e.g. anti-O-sera).
- Prevention of the lysis by cell wall-specific phages.
- Lysis by capsule-specific phages.
- Mucous colony growth (mainly when slimes are produced).
- Increased virulence.
- Application of gene probes which, however, indicate only the principle ability to form a capsule but not the formation itself.

It is emphasised again that none of these reactions in itself represents an unequivocal proof for the existence of a capsule.

6.1.3 Isolation and Chemical Analysis of the Capsule Material

Due to their exposed position, the capsules may easily be removed from the cells. Even an extraction with isotonic saline, possibly by means of ultrasonication, a homogenizor or elevated bath temperatures may be sufficient. However, the procedures are rather unselective and require extensive purification operations. Extraction of the bacteria with 45% aqueous phenol at 65-68 $^{\circ}$C, which has already been described for the isolation of LPS (Sect. 2.2.3.1), is also frequently employed. The CPS is found in the ultracentrifugation supernatant of the aqueous phase together with LPS and nucleic acids. Further purification of acidic capsules may be achieved by fractional precipitation with cationic detergents (e.g. cetavlon = cetyl trimethyl ammonium bromide), followed by repeated precipitation with ethanol from solution in saline. Gel permeation chromatography, ion exchange chromatography, and/or preparative electrophoresis are also used, often as final steps in the purification. The solutions of CPS are generally quite viscous.

For the characterization of isolated CPS (K-antigens), a number of serological techniques are available: the agglutination of antigen-coated erythrocytes (passive hemagglutination), the complement reaction with antigen-coated erythrocytes, or a number of precipitation reactions in agarose gel with or without prior electrophoretic separation. In the latter technique, immuno electrophoresis, the capsular antigens migrate towards the anode and are then precipitated as K antigens with specific K antiserum. A more unspecific precipitation with cetavlon is also possible.

Since CPS often contain acidic components, which may be characteristic, they can be detected by chemical analysis of a suitable bacterial extract. This is, however, not a reliable method, since the presence of other cell components which contain these acidic components cannot easily be excluded.

The negative charge allows CPS to be separated in gel electrophoresis. Subsequent specific staining procedures show that the CPS preparations of most encapsulated bacteria are populations of polysaccharide chains with different length.

Like most bacterial polysaccharides (e.g. the O-specific polysaccharides of LPS), CPS are composed of repeating units. This relatively simple structural feature is of advantage in the application of NMR spectroscopy for the elucidation of the chemical structure. Today, this method, together with mass spectrometry, has practically replaced the classical methods of carbohydrate analysis.

6.1.4 The Capsules of Gram-Negative Bacteria

6.1.4.1 Escherichia coli

The first subdivision of E. coli capsular antigens was into types: A, B and L. The agglutination with homologous O-, K- and OK-sera, and the different heat sensitivity of this agglutination were regarded as the main differences for this classification. Later, the individual K-antigens were defined by an affixed Arabic number (K1, etc). Most E. coli capsules are predominantly formed by acidic polysaccharides; today about 80 capsules of the polysaccharide type are known. The acidic character of the CPS is generated by (N-acetylamino) uronic acids, but also by onic acids (e.g. Kdo) or by acidic substituents like pyruvic acid (in ketal linkage) or phosphoric acid.

Earlier, for the grouping of K-antigens, predominantly serological and biochemical criteria were applied (Jann and Jann 1997). Very recently, however, a new scheme was proposed using mainly genetic parameters for this purpose (Whitfield and Roberts 1999). This scheme is in use today.

One group of strains produces capsules even at low temperatures, whereas another group does not (see below). The capsules of these two groups clearly differ from each other in more than one feature (see Table 6.1).

The capsules of group I strains are to a great extent the same as those classified earlier in type A. They are detectable in only a few O-serotypes and may be divided into two subgroups, namely IA (containing no amino sugar in the CPS) and IB (containing amino sugars). The gene cluster encoding their biosynthesis is termed cps and localised adjacent to the wba-linked his-operon of the bacterial chromosome (see Fig. 2.27) responsible for the biosynthesis of the O-specific polysaccharide of the LPS. In some cases a connection with the wzy gene has been found.

The capsules of group II may appear together with many O-antigens. They differ markedly from CPS of group I in their monosaccharide composition and contain a further spectrum of acidic components. Their biosynthesis is encoded by the kps gene cluster located near the serA gene. They may likewise be divided into two subgroups due to their thermoregulation (yes/no).

Some CPS (K3, K10, K11, K54, K96, K98) chemically resemble those of group II, but on the other hand show some differences. Even though they are relatively heterogeneous among each other, they might be summarised as group III.

Table 6.1. Capsular groups I and II of *E. coli* (Jann and Jann 1997)

Characteristic	Group	
	I	II
Acidic component	Glucuronic acid Galacturonic acid Pyruvate	Glucuronic acid Neuraminic acid Kdo *N*-Acetylmannosaminuronic acid Phosphate
Expressed below 20 °C	+	-
Coexpression with O-antigens	O8, O9, O20	Mny O-antigens
Lipid at the reducing end	Core lipid A	Phosphatidic acid
Chromosomal determination (near to)	*wba* (*his*), *wzy* (*trp*)	*kpsA* (*ser*)
Examples	K27, K28, K29, K42, K55	K1, K5, K7, K12, K92, K100

For a possible group III see text.

Table 6.2. Classification of *E. coli* capsules. (Whitfield and Roberts 1999)

Characteristic	Group			
	1	2	3	4
Former K-group	IA	II	I/II or III	IB (O-antigen capsules)
Coexpressed with O-antigens	Limited range (O8, O9, O20, O101)	Many	Many	Often O8, O9, but sometimes none
Terminal lipid moiety	Lipid A core in K_{LPS}; unknown for capsular K antigen	Phosphatidic acid	Phosphatidic acid?	Lipid A core in K_{LPS}; unknown for capsular K antigen
Genetic locus	*cps* near *his* and *wba*	*kps* near *serA*	*kps* near *serA*	*wba* near *his*
Thermoregulated (i.e. not expressed below 20 °C)	No	Yes	No	No
Examples	K30	*K1, K5*	*K10, K54*	K40, O111
Similar to	*Klebsiella, Erwinia*	*Neisseria, Haemophilus*	*Neisseria, Haemophilus*	Many genera

A classification scheme proposed very recently (Whitfield and Roberts 1999) is based solely on genetic and biosynthetic criteria (Table 6.2). It distinguishes four different groups designated with Arabic numbers. Groups 1 and 4 are limited to a few O-types, groups 2 and 3 are more widespread. The biosynthesis of the group 1 CPS is determined by the *cps*-locus (see above), the one of group 4 CPS by the *wba*-locus. In this regard, there are no differences between groups 2 and 3 CPS; the structure of both is determined by the *kps*-locus (close to *serA*).

CPS expression is subject to regulatory systems which respond to external signals. There are, however, differences between the groups, e.g. in their thermoregulation. Whereas CPS of groups 1 and 4 are not thermoregulated and thus may also be produced below 20 °C, those of group 2 are subject to thermoregulation. It was shown that the lower growth temperatures in the latter group interferes with the transcription process and with the processing of the CPS-specific messenger RNA.

In contrast to previous assumptions, it could be shown that CPS are linked to a hydrophobic moiety via their reducing end. This moiety anchors the polysaccharides in the outer membrane, especially by interaction with the lipid A of LPS.

Many CPS from groups 1 and 4 are linked to the core lipid A region, the latter of which represents the hydrophobic moiety. Such CPS may thus be regarded as LPS-like. In the strains of group 4, this similarity is underlined by the fact that the chemical structure is also encoded by the *wba*-locus.

Strains of group 1 may contain an LPS form termed K_{LPS} besides their O-antigenic LPS. K_{LPS} contain only one to a few K-repeating units besides the core lipid A region. In addition, a high-molecular CPS consisting of the same repeating units can be found; however, the nature of its binding to the cell surface is unknown.

The group 4 capsules are known as O-antigen capsules. They were first described in serotype O111, where about half of the neutral antigenic polysaccharides is present in the smooth LPS fraction and the rest in an unlinked capsular form. Additional examples are represented by the serotypes O26, O55, O100 and O113. In the absence of a second LPS-linked O-antigen, the K-antigen of a strain may be regarded as an O-antigen, e.g. in the case of the O32 and the K87 as well as the O104 and the K9 antigens.

In strains of group 2 and presumably also of group 3, phosphatidic acid has been detected as a lipid anchor; however, it appears in only scarcely half of the CPS molecules, the rest are unsubstituted. In some strains, Kdo provides the linking molecule between the lipid and the polysaccharide. It seems to be possible that the unsubstituted part is ionically bound to the cell surface, or that both LPS and CPS may interchelate and interact with each other on the bacterial cell surface.

For the structural presentation of CPS, a display of repeating units and the joining linkage is sufficient. Table 6.3 shows the repeating structures of some representative CPS from several Gram-negative and Gram-positive bacterial strains.

The CPS of K1, K92 and K5 are especially interesting. The two former consist of *N*-acetyl-neuraminic acid (NeuNAc, sialic acid) residues linked to each other by (2→8)- (K1; identical with the CPS of *Neisseria meningitidis* type B) or alternately (2→8)- and (2→9)-bonds (K92). Since the embryonal form of the neural cell adhesion molecule (NCAM) consists in its carbohydrate moiety of α-(2→8)-linked

Table 6.3. Structure of KPS from several Gram-negative and Gram-positive bacteria

K-Antigen	Repeating unit
E. coli K1 *N. meningitidis* B	→8)-α-NeuNAc-(2→
E. coli K5	→4)-β-GlcA-(1→4)-α-GlcNAc-(1→
E. coli K12	→3)-α-Rha-(1→2)-α-Rha-(1→5)-β-Kdo-(2→ 　　　　　　　　　　　　　　　　|7/8 　　　　　　　　　　　　　　　　O-Ac
E. coli K27	→6)-Glc-(1→3)-GlcA-(1→3)-Fuc-(1→ 　↑1,3 　Gal
E. coli K40	CO-NH-threonine | →4)-β-GlcA-(1→4)-α-GlcNAc-(1→
E. coli K92	→8)-α-NeuNAc-(2→9)-α-NeuNAc-(2→
E. coli 100	→3)-β-Rib-(1→2)-ribitol-5-O-PO(OH)-O-
Salmonella Vi	→4)-α-GalNAcA-(1→ | O-Ac
H. influenzae B	→3)-β-Rib-(1→1)-ribitol-6-O-PO(OH)-O-
Klebsiella K5	O-Ac 　　　　　　|2 →4)-β-GlcA-(1→4)-β-Glc-(1→3)-Man-(1→ 　　　　　　　　　　　　||4,6 　　　　　　　　　　　　CH₃-C-COOH
P. aeruginosa	→4)-β-ManA-(1→ 　|2 　O-Ac
S. aureus Typ 2	→4)-β-GlcNAcA-(1→4)-β-GlcN(*N*-acelylalanyl)AcA-(1→
S. aureus Typ 8	→3)-β-ManNAcA-(1→3)-α-FucNAc-(1→3)-β-FucNAc-(1→ 　|4 　O-Ac
Streptococcus B TypIII	→4)-β-Glc-(1→6)-β-GlcNAc-(1→3)-β-Gal-(1→ 　　　　　　↑1,4 α-NeuNAc-(2→3)-β-Gal
S. pneumoniae 19A	→4)-β-ManNAc-(1→4)-α-Glc-(1→3)-α-Rha-1-O-PO(OH)-O-
S. pneumoniae 19F	→4)-β-ManNAc-(1→4)-α-Glc-(1→2)-α-Rha-1-O-PO(OH)-O-
M-Antigen (slime)	O-Ac 　　　　　　| →3)-β-FucNAc-(1→4)-α-FucNAc-(1→3)-β-Glc-(1→ 　↑1,4 β-Gal-(3←1)-β-GlcA-(4←1)-β-Gal 　　　　　　　　　　||4,6 　　　　　　　　　　R *R* CH₃-C-COOH, CH-CH₃ or CH₂

Where the subscript notation in the final row reads: R CH$_3$-C-COOH, CH-CH$_3$ or CH$_2$

neuraminic acid residues, these capsules are not recognised as foreign by the host and no antibodies are produced. K5 has the same structure as *N*-acetyl-heparosan, a precursor of heparin. Here as well, a molecular mimicry exists which, however, is not as perfect as in the case of K1. *E. coli* strains which possess one of the mentioned K-antigens often show an elevated virulence. Thus, >75% of the *E. coli* strains causing neonatal sepsis or meningitis contained the K1-antigen (Robbins et al. 1974), and infections with *E. coli* strains carrying K5 are also quite frequent.

Some CPS contain Kdo as acidic component. This substance was originally regarded as a characteristic component of LPS (see Sect. 2.2.4.2). In the meantime, the substance has been found to occur in CPS rather often and also in larger quantities (up to 40%), in some cases even as an immuno-dominant component. However, in LPS, Kdo is always present in α-anomeric linkage, whereas in most CPS it is β-linked.

Several capsular polysaccharides contain glycerol phosphate or ribitol phosphate. Thus, they do not represent polysacharides in the original sense and resemble teichoic acids. Other capsule antigens containing phosphate as the acidic component have a basic polysaccharide main chain with phosphate substitutions.

6.1.4.2 The Capsules of Other Gram-Negative Bacteria

6.1.4.2.1 Gram-Negative Rods. The capsules of the Klebsiellae and those of E. coli group 1 resemble each other in various aspects. Firstly, they are well developed and therefore morphologically well detectable. Secondly, in primary structure and mechanism of biosynthesis, the Klebsiella CPS resemble those of E. coli group 1; in some cases, the same structure of the polysaccharides was found in both species. Thirdly, the O-antigens being coexpressed with CPS often contain simply composed O-specific polysaccharide chains, mostly representing homopolysaccharides. In contrast to *E. coli*, however, no K_{LPS} forms seem to exist. Besides the encapsulated forms of Klebsiella, there are others forming a loose slime instead of a capsule. Serologically and chemically, there are no differences between capsule and slime antigens.

The Vi-antigen of Salmonella Typhi received its name since it was believed that its presence would render the strains virulent (virulence antigen). Meanwhile, this assumption has proved incorrect, yet the term has remained. The Vi antigen is not restricted to strains of Salmonella Typhi, but can also be found in *Citrobacter ballerup* and *E. coli* strains. Its structure is that of a poly-*N*-acetyl-galactosamine uronic acid which to a great extent is *O*-acetylated to O-3. The *O*-acetyl group is important for the receptor function of the Vi-polysaccharide for Vi-phage II. After its binding to the Vi-polysaccharide, the phage hydrolyses the *O*-acetyl bond and thus destroys its own receptor. It is possible that this reaction is important for the stepwise movement of the phage along the polysaccharide chain from the outside to the inside and thus necessary for penetration through the capsule.

The M-antigen (M = mucus) is discussed in this connection though it does not represent a capsule in sensu strictu, but rather a capsule-like slime. This is not anchored in the cell wall and can be found in many Enterobacteriaceae. The M-antigen is often present in addition to the capsule. It cross-reacts serologically with some type-specific K-antigens, such as K 30 of *E. coli* or K8 of *Klebsiella*. Since such slimes are developed only at temperatures below 30 °C, they cannot play any role in the infection process in warm-blooded beings. The antigen is not type-specific, although minor variations in the composition have been found. The M-antigen-forming polysaccharide (structure see Table 6.3) is called colanic acid.

The capsule-forming capacity of other Enterobacteriaceae like *Enterobacter* and *Serratia* should be noted.

Haemophilus capsules are characterised by a rather simple composition and small repeating units. The CPS of *H. influenzae* type B, causing very severe diseases, is a polymer of D-ribofuranosyl-D-ribitol phosphate. In type A CPS, only glucose, ribitol and phosphate have been detected, in that of type C *N*-acetylglucosamine, galactose and phosphate, and in type F a phosphorylated galactosamine disaccharide.

6.1.4.2.2 Gram-Negative Cocci.

The above-mentioned capsules of *N. meningitidis* are of clinical importance, since the CPS of type B is a (2→8)-bound poly-*N*-acetylneuraminic acid, and the CPS of type C a (2→9)-bound one. This substance is an essential component in types W-135 and Y. Some of the CPS possess a hydrophobic diacylglycerol region such as those of *E. coli*. *Neisseria gonorrhoeae* tends to form capsules as well; however, to a lesser extent.

6.1.5 Capsules of Gram-Positive Bacteria

6.1.5.1. Streptococci and Staphylococci

Of particular significance in human medicine are the capsules of Streptococci and of Staphylococci.

Especially the former show a multitude of types. More than 80 types have to date been identified in *Streptococcus pneumoniae* (Pneumococci). Due to the high mortality rate of pneumonia caused by these bacteria, mixtures of CPS from the frequent capsule types are applied as vaccines. The following example underlines the unambiguity of the immunologic recognition: although the CPS 19A and 19F have almost identical structures and do cross-react serologically, the 19F-CPS hardly creates protection given by vaccination against 19A infections. *Streptococcus pyogenes* forms a non-immunogenic capsule of hyaluronic acid.

Group B streptococci frequently cause neonatal bacterial sepsis and meningitis. Among the four capsular types described, type III has been found to be particularly effective in this regard. This type and types Ia and Ib contain in its CPS the terminal region α-NeuNAc-(2→3)-β-Gal as a branch of the polysaccharide chain; this sequence also appears in the human blood group substances M and N.

In the case of Staphylococci, *Staphylococcus aureus* has been found to be the most potent capsule producer; 11 serotypes have been differentiated. These capsules can be divided into two groups on the basis of colony morphology. Strains producing capsules of type 1 and 2 are heavily encapsulated and grow mucoid on solid media. Serotype 3-11 strains form microcapsules and grow non-mucoid. A number of strains produce also teichoic acids besides the capsules, the isolation and purification procedures of which prove to be very complicated, so that misinterpretations cannot be excluded.

6.1.5.2 Mycobacteria

A capsule was first described in 1959 as capsular space between the membrane of the phagosome and the cell wall of internalised mycobacteria. Its presence was confirmed in the mid-1980s by electron microscopy and chemical analyses. Capsule material can be extracted by mild extraction methods and can also be isolated from culture filtrates, and is therefore thought to be loosely attached to the cell wall surface. It comprises proteins, including lipoproteins, and polysaccharides, including an α-$(1{\rightarrow}4)$-linked glucan similar to glycogen, an arabinomannan and a mannan. Several of the capsule proteins have been characterised, especially lipoproteins of different size (19, 30-31 and 38 kDa). Other proteins include enzymes, e.g. a superoxidedismutase (23 kDa) and a β-lactamase (29 kDa).

6.1.6 Biosynthesis of CPS

Whereas there is an increasing body of knowledge on the capsule genes in Gram-negative and Gram-positive bacteria, information on the biochemical pathways that lead to their expression is relatively scarce. Detailed knowledge has mainly been obtained from a few *E. coli* strains, notably of group 2 (Table 6.2), Klebsiella and *Staphylococcus aureus*.

In general, the biosynthesis of CPS represents a drainage of substrates and energy from the general pool of the cell. For the assembly of these polymers, the synthesis of carbohydrate components and their activation as nucleoside diphosphates (only Kdo and NeuNAc are activated as nucleoside monophosphates) is required. Both types of reactions occur under ATP consumption.

The polymerisation mechanism of *E. coli* groups 1 and 4 CPS, as well as that of Klebsiella CPS, is similar to that of the O-specific polysaccharides of LPS from Salmonella O-groups B, D and E, as described in Fig. 2.25. The oligosaccharide repeating unit is built up on undecaprenol diphosphate (undec-PP, at least similar to ACL, see Sect. 2.1.5.3.4) by an initial transfer of a sugar 1-phosphate (Gal-1-P in the present case) from its activated nucleotide diphosphate form, viz. UDP-Gal, to undecaprenol monophosphate. This is followed by transfer of the other sugar units from their respective nucleotide diphosphate forms. The resulting lipid-linked repeating unit is translocated across the cytoplasmic membrane and, at the periplasmic face of the membrane, inserted at the reducing end of the growing polysaccharide chain. During the insertion process, the activating components are liberated and afterwards recycled. In the case of the Klebsiella, the undec-PP liberated during formation of the CPS is dephosphorylated and the undec-P is utilised to start the formation of a new repeating unit. The process is repeated until the complete undec-PP-CPS is synthesised. Because of the transfer of the whole repeating unit, the mechanism of this type of chain growth is called block polymerisation.

In the case of K_{LPS}, the polymerisation reaction is terminated by the transfer of the polysaccharide chain from the carrier lipid to lipid A-core acceptor by a ligase system. Strains forming K_{LPS} lack the chromosomal gene which regulates the length of polysaccharide chains, and this explains the low degree of polymerisation in K_{LPS}. On the other hand, the chain length of the CPS 1 and 4, which in most cases is not uniform, is not influenced by this gene, presumably because ligation to the lipid A core (see Sect. 2.2.5.4) is not essential for their expression on the cell surface. It could be that in the latter case there is an alternative control mechanism.

A different biosynthetic pathway was found for group 2 and presumably also for group 3 CPS. It is described for the K1 and K5 CPS of *E. coli*. Here, the polysaccharide chain is assembled at the inner side of the cytoplasmic membrane by sequential addition of the sugar components (GlcNAc from UDPGlcNAc and GlcA from UDPGlcA in *E. coli* K5; NeuNAc from CMPNeuNAc in *E. coli* K1) to the non-reducing end of the chain. In the case of K5, the catalysis of both steps is carried out by a bifunctional enzyme. The pyrophospate carrier on which this polymerisation occurs is not known in detail. The polymerisation mechanism is known as a single chain mechanism with growth at the non-reducing end. There is no indication of a chain length regulation as the length of both group 2 and 3 CPS is rather heterogeneous.

A polymerisation mechanism similar to that described for *E. coli* type 2 CPS was proposed for the biosynthesis of the capsular hyaluronate of *Staphylococcus aureus* and for CPS of *Streptococcus pyogenes*. They both resemble the *E. coli* K5 antigen, also consisting of an alternating sequence of glucuronic acid and *N*-acetyl glucosamine. Starting also from the UDP derivatives, both components are alternately fixed stepwise directly to the reducing end of the UDP-bound growing chain. The UDP molecules liberated during growth are rephosphorylised to UTP, which is re-utilised.

Finally, the synthesised CPS have to be transported to the bacterial surface. In the case of *E. coli* type 1 and 4 the periplasmic space of the outer membrane must be passed, in the case of type 2 and 3 additionally the cytoplasmic membrane. Due to the high molecular mass of the CPS, the transport mainly through the outer membrane represents a significant problem. It is supposed that this transport takes place at the Bayer´s patches (Sect. 3.1) where both membranes appear to come into close apposition.

Translocation of group 1 CPS requires a protein of the OMA (outer membrane auxillary) family, which presumably contains a β-barrel structure and forms temporarily opened channels. The process is coupled to ATP hydrolysis at least in the later steps of translocation (including channel opening). The machinery necessary for translocation of group 4 CPS is still unknown.

The nascent group 2 CPS is transported across the cytoplasmic membrane by means of an ABC-2 transporter as in the case of LPS of *E. coli* O9. In this transport, a hetero-oligomeric membrane-associated protein complex is involved. This complex is integrated into the membrane, and all its protein components are essential for the formation and stabilisation of this biosynthetic/export complex. With others, it forms a pore through the membrane. The formation of the complex is a prerequisite for the coordination of the polymer initiation, elongation, attachment to the phosphatidic acid (sometimes via Kdo) and export. The process is coupled to ATP hydrolysis.

The translocation of the group 2 CPS across the narrowed (see above) periplasmic space and through the outer membrane needs a multiprotein capsule assembly complex; in this process also porin proteins of the outer membrane are involved. It is unclear, however, how the small porin pores can facilitate the egress of the high-molecular CPS. The lipid-linked intermediates are a prerequisite for the export, however, their exact role in CPS biosynthesis is still unresolved. The terminal phosphatidyl-Kdo may play a role as a common export motif of the CPS.

The missing outer membrane in the case of Gram-positive bacteria renders the CPS transport much simpler, although the thick peptidoglycan layer has to be penetrated (see Sect. 3.2.1.4). In the case of *Staphylococcus aureus*, a channel has been postulated for the penetration of the cytoplasmic membrane and likewise an ABC-transporter system.

The biosynthesis of the capsule peptides mostly takes place according to the mechanisms of the protein biosynthesis.

6.1.7 Capsules in Immunogenicity and Virulence

Bacterial capsules are usually specific bacterial antigens, giving rise to anticapsular antibodies. This is generally the case when CPS are exposed on the cell surface (i.e. when whole encapsulated bacteria are used for immunisation). The immunogenicity of isolated CPS is rather weak, a situation that can be improved by their coupling to carrier proteins. While this is true for adults, the immune response is very low in infants, a serious problem for vaccination. Some CPS seem to be only poorly immunogenic or not immunogenic at all, no matter in which form they are used as immunogens. As has been mentioned above, this is the case with the K1 CPS of *E. coli* and the CPS of *Neisseria meningitidis* group b as well as for the K5 CPS of *E. coli*.

CPS, notably those containing NeuNAc, protect the encapsulated bacteria against unspecific host defence such as phagocytosis. This may be decisive in early stages of an infection, when antibodies are not yet present. When this is combined with an impeded immune response, the outcome of an infection may be serious.

The capsule may also interfere with the defensive activity of complement and/or phagocytosis merely by the thickness of the extracellular layer. The contribution of capsules in resistance to phagocytosis and to complement, especially with Gram-negative bacteria, is, however, not clear. In the case of *Staphylococcus aureus* mucoid-type capsules it has been found that they mask complement factor C3b deposited on the bacterial cell wall. There are many indications that it is not only the presence of a given capsule with its specific CPS structure and charge density, but also the interaction of the CPS with the cell wall LPS of the respective strain. Thus, we are only beginning to understand the molecular basis for the virulence of encapsulated bacteria.

Bibliography

Jann K, Jann B (1997) Capsules of *Escherichia coli*. In: Sussman M (ed) *Escherichia coli –*
Mechanisms of virulence. Cambridge Univ Press, Cambridge, pp 113-143
Jann K, Jann B (eds) (1990) Curr Top Microbiol Immunol 150. (Issue containing 7 reviews)
Whitfield C, Roberts IS (1999) Structure, assembly and regulation of expression of capsules
in *Escherichia coli*. Mol Microbiol 31:1307-1319

6.2 S-Layers

6.2.1 General Remarks

Houwink (1953) was the first scientist who described the presence of a "periodic
macromolecular monolayer" on the cell wall, which was later called the S-layer (S
from surface). S-Layers originate from a very early evolutionary stage. They are
composed of identical protein or glycoprotein molecules, aggregated to crystal-like
planar structures by an entropy-driven process, and thus enclose the entire cell
surface. Two or even more different S-layers may be superimposed. S-layers are
regarded as the most simple biological membranes.

Originally, S-layers were regarded as characteristic of archaea, where they
frequently are the shape-stabilising components of their cell wall. Nowadays, it is
known that they are also wide-spread in both Gram-positive and Gram-negative
bacteria belonging to all phylogenetic groups where, however, they in most cases
have no direct shape-stabilising functions. In archaea they represent a feature
common to most species. They are, however, often lost during prolonged in vitro
cultivation.

S-Layers are differently localised, depending on the bacterial strain. In Gram-
negative archaea, containing no pseudomurein, they are closely associated with the
cytoplasmic membrane. In Gram-positive archaea and bacteria they are bound to
the outer face of pseudomurein or peptidoglycan and in Gram-negative bacteria to
the outer face of the outer membrane.

6.2.2 Composition and Structure

As can be seen in the electron microscope or, much better, in the underwater
atomic-force microscope, S-layers form, depending on their origin, 5-25 nm thick
square, hexagonal or oblique-angled lattices on the cell surface. Centre-to-centre
spacings of the morphological units may vary between 2.5 and 35 nm. They form
pores which in a given S-layer are identical in size (diameters mostly between 2
and 8 nm; for comparison: the permeability of the Gram-positive cell wall reaches
up to about 4 nm) and morphology. The pores may occupy up to 70% of their
surface. The assembly of the S-layer occurs by autaggregation of mostly identical
subunits, joined to each other and to the underlying layers by non-covalent bonds
(electrostatic or hydrophobic, in the case of negatively charged S-layer also via
Me^{2+} bridges; Fig. 6.1). Frequently, they can rather easily be removed from the cell

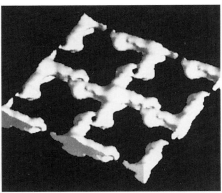

Fig. 6.1. Electronmicrograph of two S-layers: *Comamonas acidovorans* (left) and *Desulfurococcus mobilis* (right) (Engelhardt H, Peters J (1998) Structural research on surface layers: a focus on stability, surface layer homology domains, and surface layer-cell wall interactions. J Struct Biol 124:276-302, with permission)

wall. According to the binding type, chelators, detergents or agents breaking hydrogen bonds may be used for their isolation. In some archae, the S-layers are extremely stable and are isolated by heating in 2% SDS at 100 °C for 30 min.

S-Layer-forming (glyco-)proteins are in most cases weakly acidic (pI = 3-5), whereas those of some lactobacilli are extremely basic (pI = 8-11). Many S-layers appear to be uncharged at physiological pH values, due to the fact that the charged groups are located inside the molecules. The proteins consist largely of acidic and hydrophobic amino acids, although the development of hydrophobic or hydrophilic domains was not detected. The molecular masses of the proteins range between 40 and 200 kDa.

Some S-layer proteins were found to have unusual chemical and physical stability as well as an extreme resistance to proteolysis. For instance, the S-layer proteins of *Thermoproteus tenax* are stable against trichloroacetic acid and also against heating up to 140 °C in the presence of SDS or 6 M guanidine. The chemical basis of this striking stability is largely unknown, but is presumably caused by their spatial structure. They contain a relatively high percentage of β-structures (about 40%), interrupted by aperiodic folding regions and by a smaller percentage of α-helical (about 20%) regions, the latter mainly arranged in the N-terminal part. These secondary structures may produce a hyperstabilization of the protein by exclusively non-covalent interactions. A contribution of the glycan moieties to or an effect of divalent cations on the structural stability are also possible; however, both activities are insufficient to explain the extreme stability. It was found that under the normal growth conditions of the respective organism the quarternary structure of many S-layer proteins is much more stable than their primary structure.

Amino acid composition and structural features of the (glyco-)proteins are not conserved and are of only minor taxonomic significance. In a number of S-layers of (notably Gram-positive) bacteria, but not only there, a widely conserved motif of about 55 amino acids has been found, the S-layer homology (SLH) domain. It can be present in one, but usually in two or three copies. This domain is characterised by two α-helices (H_I and H_{II}) flanking a short β-strand motif (S), which in turn is followed by an intermediate loop region (L_I). The secondary structure pattern H_I-S-L_I-H_{II}-L_{II} is independent of whether one or more SLH domains are present. It was proposed that the SLH domain is a peptidoglycan-binding element. Apart from the SLH domains, the amino acid sequences of S-layer proteins are not very conservative. Nevertheless, this sequence can be similar in the case of S-layers from closely related species.

The glycan moieties are the more characteristic region of the molecules. They are mainly present in archae and Gram-positive bacteria. To elucidate their structure, the isolated S-layer glycoproteins were degraded with proteases. The obtained glycopeptides were purified by preparative chromatographic (gel chromatography, HPLC) or electrophoretic (electrofocusing) procedures. For the analysis of the glycans, the procedures described for LPS-analysis can be used (Sect. 2.2.3.2).

6.2.2.1 Archaea

The three-dimensional structures of the S-layers of the individual archaea are exceptionally various; however, a hexagonal (p6) lattice structure generally predominates.

The glycosidic components of the glycoproteins are distinctly shorter in archaea than in bacteria. They seldom consist of more than ten sugar residues, bound N-glycosidically to an asparagine residue (Asn), or O-glycosidically to a threonine residue of the peptide moiety. Up to now, core structures have not been found.

The halobacteria, as an exception, produce high-molecular mass heteropolysaccharides, strongly negatively charged, due to a high proportion of sulfate residues in the molecule. These substances, in addition, contain relatively large amounts of acidic amino acids (>20%) as well as of uronic acids. The S-layers of *Halobacterium halobium* have been examined in detail (Fig. 6.2). They consist of glycoproteins anchored in the cytoplasmic membrane with their C-terminus. The anchor is formed by a hydrophobic region consisting of 21 amino acids and is located three amino acids away from the C-terminal amino acid. An area rich in threonine residues is in the immediate vicinity of this region. These threonins are O-glycosidically substituted with 15 glucosyl-(1→3)-galactosyl residues (Fig. 6.2B, on the right). An asparaginyl residue at position 2 of the N-terminus of the peptide chain is substituted with an acidic glycosaminoglycan that consists of repeating sulfated pentasaccharide units. It is attached via the hitherto unknown linkage unit GalNAc-1-asparaginyl (Fig. 6.2B, on the left). Finally, ten sulfated oligosaccharides containing glucose, glucuronic acid and iduronic acid, are bound to the peptide chain, with the novel N-glycosyl linkage unit asparaginyl-glucose, at various positions (numbered in Fig. 6.2B, in the middle).

The area rich in threonine residues is supposed to function as spacer, thus creating a region similar to the periplasmic space of the Gram-negative bacteria.

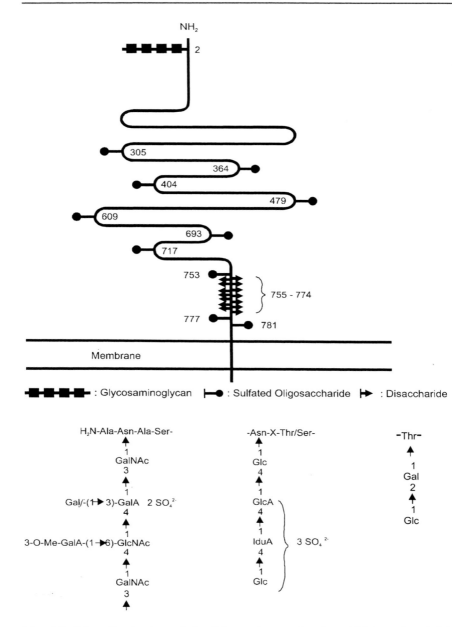

Fig. 6.2. Schematic structure of the S-layer glycoproteins from *Halobacterium halobium* (modified after Kandler O and König H 1993)

6.2.2.2 Bacteria

In bacteria, glycosylated S-layers are mainly found in Bacillaceae (e.g. *Bacillus* or *Clostridium*) and in *Lactobacillus*. Their carbohydrate moieties are long homo- or heteropolysaccharide chains composed of linear or branched repeating units, reminiscent of enterobacterial LPS. They are linked to the peptide moiety via a core-like structure. As in the case of archaea, the linkages are *N*- or *O*-glycosidic. In addition, novel binding types such as Glc→Tyr (or Gal) were found. The same type of polysaccharide may be bound to different amino acids, and a given protein moiety may carry differently composed glycan chains. The structure of *Thermoanaerobacter thermohydrosulfuricus* L111-69 depicted in Fig. 6.3 may serve as an example for the polysaccharide moiety of an S-layer. The 3-*O*-methylation of the terminal rhamnose has to be noticed since it is thought to represent a stop signal.

The glycan of the S-layer of *Bacillus thermoaerophilus* DSM 10155 (Fig. 6.3) consists of disaccharide repeating units composed of rhamnose and and of D-*glycero*-D-*manno*-heptose, which is also a characteristic component of LPS.

S-Layer polysaccharides of bacteria and LPS show distinct similarities. Firstly, the former contain monosaccharides which so far had been regarded as being characteristic for LPS, such as D-*glycero*-D-*manno*-heptose or 3-amino-3,6-dideoxy-D-glucose (quinovosamine). Secondly, they also consist of three regions: (1) a rather variable one consisting of repeating units, (2) a more conserved (core-like) one and (3) a region necessary for the anchoring in the cell wall. Finally, clear parallels can be recognised in the structuring of the cell envelopes; Fig. 6.4 shows the schematic comparison of both.

6.2.3 Biosynthesis

The biosynthesis of the protein moiety occurs corresponding to the general mechanisms of protein biosynthesis; however, details are still to a large extent unknown. By means of sequencing, it was found that most S-layer proteins are produced with an N-terminal secretion signal peptide. The ability of both archaea and bacteria to modify the S-layer proteins posttranslationally is remarkable. These modifications may take place by means of glycosylation and phosphorylation of amino acid residues as well as by the cleavage of C- or N-terminal residues. The synthetic apparatus is submitted to a remarkably stringent control which guarantees that the amount of synthesised S-layer proteins corresponds exactly to that needed. Only small amounts of such proteins are detectable in the growth medium of continuous cultures.

Many bacteria are able to modify the S-layer proteins and thus to create specific changes in the cell surface. These bacteria in their corresponding gene have only a single promotor; it is followed by eight to nine S-layer gene cassettes. The pathogen is able to realign the promotor by a single DNA inversion, which enables expression of different S-layer cassettes. Thus, the antigenic variation required to avoid a host immune response is maintained by gene recombination.

The biosynthesis of the repeating units of the glycan moiety proceeds similarly to that of the enterobacterial LPS by gradual attachment of nucleoside diphosphate-

a)

3-*O*-Me-α-L-Rha-(1→4)-α-D-Man-(1→[3)-α-L-Rha-(1→4)-α-D-Man-(1-α-]$_n$→[3)-α-L-Rha-
(1→3)-α-L-Rha-(1→3)-α-L-Rha-(1→3)-β-D-Gal-(1-α-]-*O*-Tyr

b)

4)-α-L-Rha-(1→3)-β-D-*glycero*-D-*manno*-Hep-(1→

Fig. 6.3 a,b. Complete structures of the polysaccharide moieties of the S-layers from *Thermoanaerobacter thermohydrosulfuricus* L111-69 (**a**) and *Bacillus thermoaerophilus* DSM 10155 (**b**). All linkages are pyranosidic; n = 23-32 in most cases, average 27; [] represents the core structure

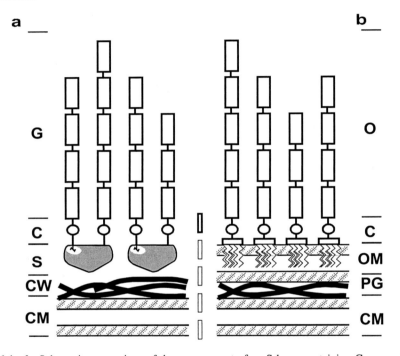

Fig. 6.4 a,b. Schematic comparison of the arrangement of an S-layer containing Gram-positive (**a**) and Gram-negative (**b**) eubacterial cell envelope. (Messner P, Allmaier G, Schäffer C, Wugeditsch T, Lortal S, König H, Niemetz R, Dorner M (1997) Biochemistry of S-layers. FEMS Microbiol Rev 20:25-46, with permission). *G* glycan chain; *O* O-antigen, □ repeating unit; *C* core region; *S* S-layer protein; *OM* outer membrane; *CW* peptidoglycan containing part of the cell wall; *CM* cytoplasmatic membrane

activated monosaccharides. However, there are differences. As an example, the biosynthesis of the hexasaccharide unit of *Thermoanaerobacterium thermosaccharolyticum* is schematically shown in Fig. 6.5:
- There is no lipid carrier in the early stages.
- The addition of glucose and galactose takes place at the reducing end, the following one of rhamnose at the non-reducing end.

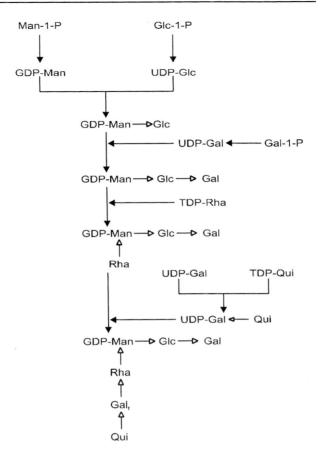

Fig. 6.5. Biosynthesis of the hexasaccharide unit from *Thermoanaerobacterium thermosaccharolyticum* E207-71 (Schäfer et al. 1996). *Qui* chinovosamin; —▷ linkage ; —▶ reaction pathway

- The unit Qui→Gal is incorporated as a disaccharide.

The binding to an ACL-like carrier (as described in Sect. 2.1.3.3), the anchoring in the cytoplasmic membrane, the polymerisation and finally the binding to the peptide moiety are presumably carried out only after completion of the repeating unit.

Secretion is performed via the GSP (Sect. 8.1.2.2), but may also be carried out through specific transport pathways. After having reached the cell surface, the monomers auto-aggregate. Then the lattice pattern determined by their chemical structure (mainly distribution of charged and hydrophobic amino acid residues) is formed, and the likewise predetermined attachment to the respective underlying layer of the cell wall takes place. These processes occur spontaneously in an entropy-driven self-assembly process both in vivo (after removal of the disrupting agent used for their isolation) and in vitro.

A hitherto largely unsolved problem is the incorporation of newly synthesised units into the S-layer. Although this layer is generally proteolysis-resistant, a proteolytic cleavage is a precondition for this process. With Gram-positive bacteria, the incorporation occurs primarily into the cylindrical part of the cell, whereas in the Gram-negative it is random. In all organisms, new S-layer lattices also appear in regions of incipient cell division and the newly formed cell poles.

The genetic background of the biosynthesis and its coordination with that of other cell components are still being investigated.

6.2.4 Fixation of S-Layers in the Cell Wall

S-Layers have been found to be generally linked to the cell wall layer below (Fig. 6.6):
- In Gram-negative archaea with their underlying lipid membranes, S-proteins have a hydrophobic anchor mostly at the C-terminus formed of α-helical coiled coils (Fig. 6.6a).
- In Gram-positive bacteria containing layers with SLH domains, the binding occurs to the peptidoglycan and to a secondary cell wall polymer which in *Bacillus sphaericus* was found to be a teichuronic acid. In the case of layers without SLH domains (e.g. *Bacillus stearothermophilus*) the binding is furnished by electrostatic interactions between the positively charged N-terminus of the S-layer protein and a negatively charged secondary cell wall polymer which is absolutely unknown.
- Gram-positive archaea analogously bind their S-layer to pseudomurein or to methanochondroitin (Fig. 6.6b).
- In the case of the LPS-containing membranes of Gram-negative bacteria, the association takes place via ionic (Ca^{2+}), protein-carbohydrate, or carbohydrate-carbohydrate interactions (Fig. 6.6c). In addition, protein-protein interactions in which hydroxyamino acids partly play a role occur generally between S-layer and cell wall.

6.2.5 Functions of S-Layers

Since S-layers represent in many cases the outermost layer of bacteria and archaea, their functions can be quite diverse and may vary from one cell type to the other. In Gram-negative archaea having the S-layer as the single cell wall component, some of the functions are naturally by far more obvious than in Gram-negative bacteria with their well-developed cell wall. It is supposed, however, that the main function of the layer has not yet been revealed. The fact that the S-layers get lost rather soon during subcultivation in vitro indicates protective functions for the conditions in vivo. In the following, possible functions are presented:
- An essential function is the mechanical and chemical protection of the cell virulence factors since they protect the bacteria against the defence surface and the underlying cell components. S-Layers may thus represent mechanisms of the host. The shielding function of S-layers against low pH values should be noted.

- S-Layers stabilise the cells mechanically. Especially if further cell wall components are lacking, they support the formation and the preservation of the cell shape.
- S-Layers bound to membrane layers via incorporated hydrophobic anchors lead to the stabilisation of the membrane.
- They are able to develop resistance to bacteriophages and phagocytosis, but, on the other hand, they may serve as phage receptors.
- A possible porin or channel function, especially of the SLH domain, is in discussion.
- They have a barrier function restricted by the pore diameters and thus are able to protect the cells against high molecular substances such as lytic enzymes, complement, or biocides.
- They create a kind of periplasmic space in cooperation with the peptidoglycan or the cytoplasmic membrane, at least in Gram-positive bacteria.
- Due to their negative charges, they are able to bind small cations.
- They play a role in cell recognition as well as adhesins that promotes but not mediates bacterial invasion.
- They may represent adhesion sites for cell-associated exo-enzymes such as proteases.

Some properties of the S-layer glycoproteins have been used practically. An important aspect of their application is the ease with which they can be obtained in large quantities. Their self-assembly ability may be utilised for the binding of monolayers of enzymes, antibodies, protein A and others, but also of antigens and haptens for the production of antibody. Their re-crystallisation on liposomes imitates the supramolecular structure of the cell envelopes of Gram-negative archaea. On such particles, functional molecules may be immobilised and used as a basis for solid-phase immunoassays, as bioanalytical sensors or as affinity microparticles. Their possible application as medical diagnostics, as vaccines for drug targeting or delivery, or for gene therapy depends on the ability of S-layer subunits to recrystallise into coherent lattices on functional lipid membranes, including liposomes.

The use of S-layers as ultrafiltration membranes is based on the formation of pores of defined molecular mass cutoffs. Defined chemical modifications may alter their hydrophilicity/hydrophobicity index as well as their ion-exchange properties.

Recently, the application of S-layers in nanostructure technologies has been very useful. After crystallisation on silicon wafers, they can be used as matrices for controlled biomineralization; S-Layers coated with metal oxides can be employed as masks for nanostructuring of semi-conductor surfaces.

Bibliography

Engelhardt H, Peters J (1998) Structural research on surface layers: a focus on stability, surface layer homology domains, and surface layer-cell wall interactions. J Struct Biol 124:276-302

FEMS Microbiol. Rev 20(1997) Issue with six reviews

Houwink AL (1953) A macromolecular monolayer in the cell wall of *Spirillum* spec. Biochim Biophys Acta 10:360-378

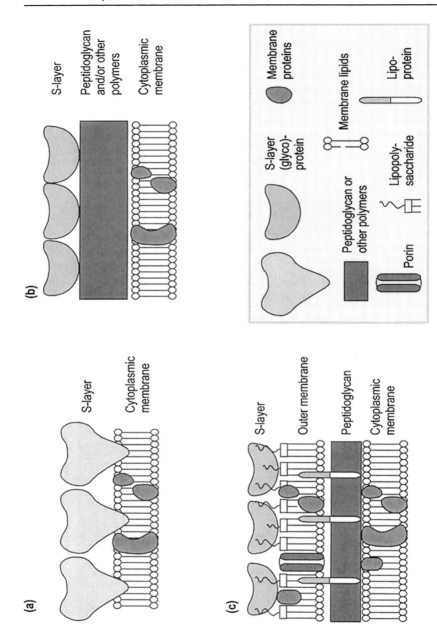

Fig. 6.6 a-c. Schematic illustration of the supramolecular architecture of the three major classes of procaryotic cell envelopes containing S-layers. **a** Gram-negative Archaea; **b** Gram-positive Archaea and Bacteria; **c** Gram-negative Bacteria (Sleytr UB, Beveridge TJ (1999) Bacterial S-layers. Trends Microbiol 7, 253-260, with permission). For details see text

Kandler O, König H (1993) Cell envelopes of archae: structure and chemistry. In: Kates M, Kushner DJ, Matheson AT (eds) The biochemistry of Archae (Archaebacteria). Elsevier, Amsterdam, pp 223-259
Messner P, Allmaier G, Schäffer C, Wugeditsch T, Lortal S, König H, Niemetz R, Dorner M (1997) Biochemistry of S-layers. FEMS Microbiol Rev 20:25-46
Schäfer C, Wugeditsch T, Neuninger C, Messner P (1996) Are S-layer glycoproteins and lipopolysaccharides related? Microb Drug Resist 2:17-23
Sleytr UB, Beveridge TJ (1999) Bacterial S-layers. Trends Microbiol 7:253-260

6.3 Sheaths

A number of filamentous bacteria and archaea living mostly in water or mud create tubular protein sheaths surrounding chains of several cells which are produced merely by growing cells. Their presence is beneficial for the bacteria, especially since they facilitate their attachment to solid surfaces.

The electronmicroscopic detection of sheaths is the only definite proof of their existence. Sheaths may be isolated in the same way as the cell walls of Gram-positive bacteria.

Sheaths form an amorphous polymeric network, such as in *Thiothrix* or in some cyanobacteria, or they are structured as regularly as S-layers as in *Methanospirillum*. The latter are mainly quite stable and, thus, resist the inner pressure of the bacteria. Even if the sheaths closely surround the cells, there is generally no direct linkage between the two. Therefore, the bacteria can escape the sheaths and leave them empty.

The isolated sheath material of *Methanospirillum hungatei* has been examined in detail. It contains a complex spectrum of 18 amino acids (65-72% of the mass) and five neutral sugars (6-8% of the mass); however, it is not known whether or not the sugars are bound to the protein. It has a two-dimensional paracrystalline structure similar to that of S-layers which is composed of a succession of stacked tyres with a diameter of about 2.5 nm. The sheath has small pores allowing the passage of solutes with a radius of up to 0.3 nm.

An important function of sheaths is to oxidise Fe^{2+}- (*Sphaerotilus*) or Mn^{2+}-ions (*Leptothrix*) and, additionally to, encapsulate the cells with an envelope of $Fe(OH)_3$ or MnO_2 or to impregnate the sheaths, respectively. This results in an additional shield for the bacteria. A slime composed of acidic polysaccharides located on the surface of the sheaths seems to be of importance for the accumulation of both ions. In the case of iron the details are largely unclear due to the autoxidability of Fe^{2+}, however, it is known that the Mn^{2+}-oxidation is caused by a protein-like substance present in the sheath.

Bibliography

Mulder EG, Deinema MH (1981) The sheathed bacteria. In: Starr MP, Stolp H, Trüper HG, Balows A, Schlegel HG (eds) The procaryotes, vol I. Springer Berlin Heidelberg New York, pp 425-444

6.4 Filamentous Proteins

Many bacteria contain filamentous protein material on their surface which is designed to function in different ways: as flagella, fimbriae, fibrils and pili. Flagella serve as a means of locomotion. The functions of fimbriae and fibrils are more varied: besides their adhesive function, protection of the cell surface, enlargement of the active surface to improve respiration and uptake of nutrients are discussed. The term pili is not unambiguously defined in literature and is sometimes used synonymously with the term fimbriae, but in most cases is reserved for the definition of the conjugative organs of the bacteria (sex pili).

6.4.1 Flagella

Flagella, screw-shaped proteinaceous cell appendages up to 20 μm long and 10-30 nm thick (depending on the species), are the bacterial organs of locomotion. Their rotation of up to 3000 rpm may propel the bacteria with a speed of up to 20 μm s^{-1}, i.e. a multiple of the body length. The rotation is possible either clockwise or counterclockwise, in this way different kinds of movement are caused. Counterclockwise rotation results in forward motion (smooth swimming), clockwise rotation generates non-productive tumbling. For impulse compensation, the cell bodies counterrotate at an essentially lower speed. Both senses of rotation alternate such that the bacterial path resembles a Brownian motion.

The number and localisation of the flagella on the bacterial surface (e.g. polar or pertrichal) are type-specific; mostly 5-10 flagella are found per cell. The presence of hyperflagellated cells often causes the swarming phenomenon, i.e. the rapid and coordinated population migration across solid surfaces.

Flagella consist morphologically of three parts: the basal body, the flagellar hook and the filament. The basal body fixes the flagella into the cell wall, it represents the motor for the flagellar rotation. The flagellar hook is a slightly crooked cylinder with a somewhat larger diameter than that of the filament. Two proteins (FlgK and FlgL) mediating the binding of the filament are associated with the hook. With more than 95% of the total mass, the filament makes up the major part of a flagellum (Fig. 6.7). It is very rigid and therefore well suited for its function as a propeller.

The isolation of the complete flagella from intact bacteria causes difficulties. On shaking the bacteria, flagella break easily, which occurs in or very close to the hook. Therefore, such isolated material contains only the filaments and about half of the mass of the hooks. To obtain the complete flagella, it is better to use spheroplasts, which are extracted with Triton X-100. The flagella are precipitated from the extract with $(NH_4)_2SO_4$ and purified by means of density gradient centrifugation or ion exchange chromatography. The three regions of flagella can be separated from each other and isolated in a pure state (see below).

The flagellar filaments consist of aggregates of about 20 000 copies of a 50-60-kDa subunit called flagellin assembled in such a way that they form an almost hexagonal cylindric lattice. This arrangement may give rise to up to nine different forms. In the most stable form, the flagellin monomers are arranged as a left-

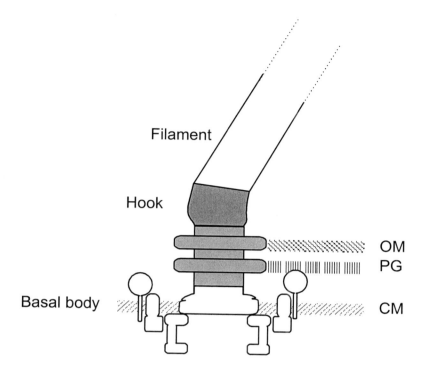

Fig. 6.7. Model of flagellae. *OM*   outer membrane; *PG*   peptidoglycan; *CM*   cytoplasmic membrane

handed helix in such a way that they simultaneously form 11 longitudinal flagellin threads (fibrils) alongside the axis of the cylinder wall. The cylinder is hollow inside. The filaments show polymorphism, which is explained by the fact that the fibrils may appear in a long (extended) and a short (compressed) state. Depending on the ratio of the fibrils in one of both states, further helical arrangements are formed besides that described, even right-handed ones and two stretched ones. During formation of the filaments, environmental conditions, such as pH or ionic strength, may play a role. These arrangements explain the macroscopically visible curved shape of the flagella. Recently, it was found that a second α-helical cylinder is embedded in the one described; both cylinders are interconnected by spokes. Such an arrangement distinctly increases the rigidity of the filaments.

Each of the flagellin subunits forms an outward-directed knob. It has no structural function but is serologically well detectable.

By means of detergents, pH shifting or heating, the filaments can be disaggregated into the monomers, indicating that flagellin aggregation does not involve the formation of covalent bonds. Reaggregation also may occur in vitro, resulting in a filament which cannot be distinguished from the natural one.

Monomeric flagellins have molecular masses of about 50 kDa, and consist of about 380 amino acids. Both data, however, vary distinctly, depending on the bacterial species investigated. Flagellins contain many non-polar, hydrophobic amino acids. Neither cysteine nor tryptophan were detected, proline and histidine are rare constituents, and methionine, tyrosine and phenylalanine are minor components. However, ε-N-methyllysine, also occurring in the actin of muscles, is a characteristic component. Flagellin molecules consist of three domains: both the C-terminal and the N-terminal domains are responsible for the specific aggregation of the subunits. The role of the central domain is still unknown; it can be deleted without affecting assembly.

The flagellar hook consists of helically arranged fibrils and, thus, is structurally related to the filament, but it is built of quite distinct subunits. The proteins forming the hook subunits are more stable towards heating and disaggregating agents than those forming the filament. The helical wavelength of the hook is rather short (approximately 130 nm). Thus, the hook generates less than one-half of a helical turn, which explains its crooked form. In contrast to the filament, the hook has a rather well-defined length (about 55 nm), and consists of about 130 subunits. The hook is much less rigid than the filament. This is necessary for its presumed function as a flexible coupling transmitting the torque, generated by the basal body and vertically coming out of the cell wall, to the filaments arranged parallel to the cell surface. Between the hook and filament are two junction proteins, presumably acting as adapters between the two components; this seems to be necessary because of the pronounced difference in their mechanical properties.

The basal body consists of a central cylindric shaft with a diameter of 13 nm on which several discrete rings (L, P, and MS in *E. coli*) are situated. The shaft is a rod-like structure that transmits torque from the motor to the hook. The rings are located around the shaft at the level of the outer membrane (L-ring, from LPS, in Gram-negative bacteria), of the peptidoglycan (P-ring, 26 nm), and of the cytoplasmic membrane (MS-ring; membrane/supramembrane). An additional ring called C-ring (from cytoplasm) or bell is positioned in the cytoplasm adjacent to the MS-ring. All components are composed of partially diverse subunits.

The rings have different functions. It is supposed that the L- and the P-ring form a kind of bushing through which the shaft passes and thus facilitate rotation. The MS-ring, on the contrary, is tightly linked to the shaft and serves as a mounting plate for the fastening of the proteins necessary for motor rotation and switching (FliG, presumably at the MS-ring; FliM and FliN, both at the C-ring). The proteins actually necessary for the drive (MotA, MotB, possibly as a complex) are ring-like arranged as particles (about 11 studs) around the MS-ring in the cytoplasmic membrane. MotB as a transmembrane protein anchors the stator in the membrane, in which its covalent binding to the peptidoglycan network serves primarily. Besides four transmembrane sections, MotA has an extended cytoplasmic region. This region presumably plays a role for the energy supply of the motor.

In most bacteria, the energy for the rotation is drawn from the transmembrane potential (proton-motive force) of the cytoplasmic membrane; ATP plays no direct role as an energy supplier. The mechanism of the rotation, i.e. the transformation of the protonmotive force into rotation energy, is not clear, but conformational modifications caused by the protonation are discussed.

The initial stage of the biosynthesis of flagella is the assembly of the MS-ring embedded in the cytoplasmic membrane. This is followed by that of the complete C-ring. During the next stages, the shaft of the basal body (in two steps) is attached. Then the subunits of the P- and L-rings are exported to the periplasm and the outer membrane, and via self-assembly processes they form rings around the basal-body shaft. The flagellar hook and, after incorporation of the hook-filament junction proteins FlgK and FlgL, the filament are composed stepwise by distal addition of the subunits. MotA and MotB are synthesised in the cytoplasmic membrane and joined around the MS-ring at a still unknown moment. The building blocks for the flagellar regions which are situated outside the cytoplasmic membrane are synthesised in the cell. Interestingly, their incorporation into the filament occurs at the cell-distal end of the flagellum. As it is unlikely that the monomers are excreted by the cell and captured by the growing ends (there are practically no monomers to be found in the culture supernatant), it is suggested that the monomers are synthesised without a signal sequence and transported to the tip of the growing flagellum via the flagellin-specific pathway through the interior of the flagellum and thereafter integrated. Channels through the flagellar hooks and the filaments have been demonstrated, those through the shaft of the basal bodies are probable. The controlling mechanism of the sequence of the exported subunits is largely unknown.

As can be expected, the synthesis of P- and L-rings takes place by a different mechanism. Their subunits are formed as preproteins, i.e. containing a signal sequence, exported via the Sec-specific pathway and joined together after cleavage of the signal sequence.

An intriguing aspect of the flagella biosynthesis is how the secretion apparatus identifies which protein must be secreted and incorporated at which moment, especially when more than one putative candidate is present. It has been proposed that specific signal sequences within the proteins serve this phenomenon.

In bacteriological routine laboratories, the presence of flagella is in most cases proven in two different ways, i.e. by the motility of the bacteria and serologically by the presence of the so-called H-antigens. Besides the O-antigens, H-antigens are important for the classification of a number of Gram-negative bacteria, such as Salmonellae. Their serological specificity is governed by the structure of the flagellins in their superstructural array. During the disintegration of the filaments into the flagellin subunits some determinants disappear, others are exposed and can then be detected. Therefore, three kinds of antibodies are found: (1) those reacting only with the filament, (2) those reacting only with the flagellin and (3) those reacting with both. Antibodies directed against the flagellar hook or the basal body do not show any cross reaction with the filament-(=H-)antigens. Purified flagellin is weakly immunogenic in rabbits and strongly immunogenic in rats. The complete filament is more efficient. It, however, is also able to induce immunotolerance.

Recently, it has been found that flagella are not the only organs for bacterial movement. Particular polar-arranged fimbriae (pili) may also serve this purpose (see below).

Bibliography

MacNab RM (1996) Flagella and motility. In: Neidhart FC, Curtiss R III, Ingraham IL, Lin ECC, Low KB, Magasanik B, Reznikoff WS, Riley M, Schaechter M, Umbarger ME (eds) *Escherichia coli* and *Salmonella*: cellular and molecular biology, 2nd edn. ASM Press, Washington, pp 123-145

6.4.2 Fimbriae and Fibrils

Fimbriae, also called pili, are hair-like, proteinaceous appendages on the surface bacteria. They are shorter and generally thinner than flagellae. Fimbriae generally extend 1-2 μm from the bacterial surface and possess diameters ranging between 2 and 10 nm. They can be roughly divided into two morphological groups, of which those of the first group are about 7 nm in diameter with a central axis hole of about 1.5 nm in diameter. These fimbriae appear mostly in Gram-negative bacteria. Those of the second group are much thinner, with diameters between 2 and 3 nm; they have also been designated fibrils (fibrillae) and can also be found rather frequently in Gram-positive bacteria. The number of fimbriae is higher than those of flagellae; they may be so numerous that they completely cover the bacterial surface. For this reason, some of the fimbriae were earlier regarded as capsules, for example the K88- and K99-fibrils in *E. coli*. According to their origin, fimbriae may be very diverse both morphologically and in their chemical composition. Frequently, bacteria from a single isolate are equipped with fimbriae of more than one type. The percentage of fimbriated bacteria depends on the growth phase and the cultivation conditions.

Fimbriae provide the bacteria with adhesive properties and are involved in the specific adherence of bacterial cells to host tissue. They are often identified by an altered colony morphology of the bacteria or by their ability to agglutinate certain yeast cells or erythrocytes of diverse origin. The agglutination may be prevented by, e.g. the addition of mannose (type I-fimbriae), α-Gal-(1$\rightarrow$4)-β-Gal (P-fimbriae), or α-NeuNAc (S-fimbriae). The serological or the electron microscopical proof is also very convincing. These methods even detect the presence of several fimbrial types in the same bacterium. A separation of fimbriated and non-fimbriated bacteria can be performed by means of electrophoresis, hydrophobic interaction chromatography or ion exchange chromatography.

In this context it should be noted that bacterial adhesion can also be caused without fimbriae, namely by a-fimbrial adhesins (AFA-1, AFA-3) and non-fimbrial adhesins (NFA-1, NFA-3) that are located immediately at the cell surface, furnishing a fine mesh or coil. They often possess blood-group specificity.

There are a number of methods for typing or characterization of fimbriae. Among them are:
- Phenotypic (morphological) criteria.
- Agglutination specificities of the bacteria (see above) and the inhibition of the agglutination by saccharides.
- Serological methods (antigenic determinants).

- The pattern of their subunit(s) called fimbrins (or pilins) in SDS-PAGE of isolated fimbriae (Sect. 2.2.3.2).
- Determination of the amino acid sequence of the fimbrins and characterisation of the operons coding for their biosynthesis may be used, but often provides a limited amount of information regarding the relatedness of fimbrial operons.

Some of the more important fimbriae have been included into the serological typing scheme, for example in the O:K:H:F-classification of *E. coli.*

For the isolation of fimbriae, they have to be sheared from the bacterial surface under such mild conditions that the cells remain intact (usually with an omnimix). By means of precipitation with $(NH_4)_2SO_4$, isoelectric precipitation, ion-exchange chromatography, and density gradient centrifugation fimbriae can be obtained in good purity.

All fimbriae are mainly composed of major fimbrial subunits which determine their structure. Besides these, minor fimbrial subunits were detected. The latter have different functions, e.g. in the process of biosynthesis or, in some fimbrial systems (type I- or P-fimbriae) representing the adhesins.

The morphology and the structure of the fimbriae are quite diverse according to the type. Some of them, like for example P- or type I-fimbriae of *E. coli,* possess a long (up to 2 μm) rigid proximal shaft of about 7 nm in diameter in which in the case of the P-fimbriae 3.3 PapA-subunits per turn are arranged as a right-handed helix enclosing a central channel of 1.5 nm in diameter (the *pap* [from pyelonephritis associated pili] gene clusters encode for the P-fimbriae; see Fig. 6.8). To this shaft a short, flexible stretch of approximately 2 nm in diameter is distally adjoined consisting of PapE subunits. Even further outside, there is a tip consisting of PapG subunits representing the adhesins (lectins) for its specific binding to oligosaccharidic receptors, i.e. the Gal-$\alpha(1{\rightarrow}4)$Gal disaccharide receptor found in the globoseries of glycolipids of the human kidney. Fimbriae of this type are often embedded in a polysaccharide capsule; only their distal part are reachable for receptor recognition.

In some fimbriae, such as K88 or CFA/I (colonization factor antigen), the adhesive properties are not localised at the tip of the fimbriae, but distributed over the whole fimbrial surface. The structure of these, as well as that of F41 fimbriae, is totally different. They are rather thin (K88 2,1 nm, CFA/I 7 nm), flexible, and have no central channel. Fimbriae of this type are much more flexible and surround the bacteria in a capsule-like manner.

A hydrophobic C-terminal motif is essential for the aggregation of the subunits in the fimbriae. The strength of binding of these subunits to each other depends very much on the nature of the fimbriae. Thus, the facility to disaggregate them differs correspondingly. In some cases, a very diluted buffer devoid of bivalent cations can be already sufficient, in other cases solutions of urea or guanidine have to be applied. The molecular mass of the fimbrins ranges mostly between 14 and 30 kDa.

The adhesive properties of fimbriae have been investigated in detail in the context of seeking criteria for bacterial pathogenicity and virulence. The adhesion is caused by binding of oligosaccharidic components of host mucosae (glycoproteins, glycolipids such as gangliosides) by means of the already described lectins. Due to the high information density of oligosaccharides (Sect. 2.2.1), binding is very specific with regard to both the host and particular organs. Lectins bound to the fimbriae also mediate the adhesion of oral bacteria such as *Streptococcus sanguis* to the glycoproteins of the dental enamel originating from saliva.

The assembly of fimbriae occurs at the cytoplasmic membrane. The operon coding for the genetic information of the biosynthesis may be localised both in the chromosome (CFA/I) and on a plasmid (P, S, type I). The complete subunits contain a signal sequence and are exported by means of the protein secretion machinery (Sec system and a translocation-ATPase, see Sect. 2.3.2.6), at least in the case of type I-fimbriae. The folding into the native form mostly takes place in the periplasm. In contrast to flagella, the subunits of type I-fimbriae are attached to the growing fimbriae in the outermost part of the periplasmic space and the fimbriae are pushed through the outer membrane (Fig. 6.8). PapD represents a molecular chaperone, whose function is to bind the proteins and escort them from the cytoplasmic membrane to an outer membrane assembly site and allow correct folding and assembly into the growing fimbriae. Usher proteins (PapC in the case of P-fimbriae) embedded in this membrane are important for the process (chaperone/usher-dependent pathway). They are arranged in O-shaped rings forming central pores with diameters of 2-3 nm in the outer membrane that allow the passage of the tip region (PapE subunits). As the fimbrial rod (PapA subunits) is too large for passage through the pores, its transport occurs in an unravelled state representing a fibre of 2-3 nm (not shown in Fig. 6.8). Finally, the intact fimbrial body is formed and anchored to the outer membrane. PapF is the subunit that joins PapG to the rest of the tip fibrillum; PapK seems to play a role in regulating the length of the tip fibrillum.

The type IV-pili are polar-arranged fimbriae of 5-6 nm in diameter and 1-2 μm in length. They are, besides adherence, for instance at *Vibrio cholerae* or invasion of human intestinal cells at Salmonella Typhi, mainly associated with bacterial motility without using of flagellae. The movements are known as social gliding (smooth motion in an interfacial plane directed along the long axis of the cell) in *Myxococcus xanthus* or as twitching motility (non-flagellar movement) in organisms such as *Pseudomonas aeruginosa* and *Neisseria gonorrhoeae*. Classical type IV-pili appear to be composed of one major component, the type IV-pilin. However, several minor pilins have been found at least in *Pseudomonas aeruginosa*, the exact role of which is still unknown. The spatial structures of type IV-pili of different origin and their composition from pilins are very similar to each other and do not differ basically from those of other fimbriae. The type IV-fibre forms a three-layered helical structure of coiled α-helices surrounded by a β-sheet. The driving mechanism has not yet been entirely elucidated; it is possible that only retraction or a dynamic balance between retraction and extension provides a force for the movement.

The biosynthesis of type IV-pili starts with that of the precursor proteins of the subunits. They are transitorily anchored in the cytoplasmic membrane by an N-terminal hydrophobic domain. Assembly of type IV-pili occurs by expelling the subunits from this membrane. According to this mechanism, the growing type IV-pilus remains anchored in the cytoplasmic membrane (in contrast to type I-fimbriae). A secretin component forms a pore through the outer membrane large

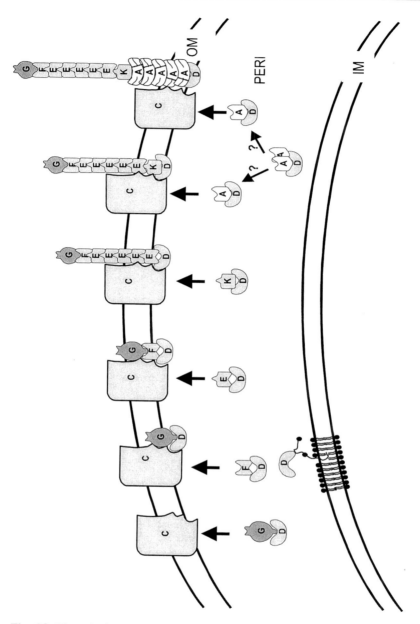

Fig. 6.8. Biosynthesis mechanism of P-fimbriae (Dodson KW, Jacob-Dubuisson F, Striker RT, Hultgren SJ (1997) Assembly of adhesive virulence-associated pili in Gram-negative bacteria. In: Sussman M (ed) *Escherichia coli*. Mechanism of virulence. Cambridge University Press, Cambridge, pp 213-236, with permission). *A, C, D, E, F, G, K* Gene products of the *pap*-region; see text; *OM* outer membrane; *IM* cytoplasmic membrane; *PERI* periplasmic space

enough to enable the complete pilus (contrast to type I-fimbriae) to be secreted into the outer medium. An adhesin is incorporated at the tip of the filament during the first steps of the synthesis.

The biosynthesis of the fimbriae of Gram-positive bacteria has not yet been investigated in detail, but seems to differ from that of the Gram-negative ones.

Fimbriae are immunogenic: between fimbriae of one group or one subtype there are frequently serological cross-reactions even if the bacteria producing them belong to different genera. This causes the risk of false results in the serological sorting of the strains.

Bibliography

Dodson KW, Jacob-Dubuisson F, Striker RT, Hultgren SJ (1997) Assembly of adhesive virulence-associated pili in Gram-negative bacteria. In: Sussman M (ed) *Escherichia coli*. Mechanism of virulence. Cambridge University Press, Cambridge, pp 213-236
Wall, D Kaiser D (1999) Type IV pili and cell motility. Mol Microbiol 32:1-10

6.4.3 Sex Pili

Sex pili predominantly serve for transfer of plasmids and for binding of sex-specific phages. The plasmid transfer starts with the binding of the tip of the donor cell pilus to a receptor at the surface of the recipient cell. By gradual degradation (depolymerisation) of the pili from the inside, both cells approach each other, until a stable direct cell-cell contact takes place. The plasmid-DNA is then linearised and the relatively thin strand obtained is pushed through both cell walls; helper proteins play a role in this process.

Per donor cell, one to three sex pili can be found. They may be roughly divided into three morphological groups: (1) thin, flexible, (2) thick, flexible, (3) thick, rigid. A further subdivision may be carried out serologically or by means of phages. The classification system mostly accepted today is deduced from the incompatibility groups of the plasmids determining the sex pili. The length of the sex pili is diverse; in the case of F pili (group 2) it amounts to $1 - 2$ µm.

The thin, flexible sex pili have a diameter of 6 nm and are mostly produced together with other pili. The diameter of the thick flexible sex pili amounts to 9-13 nm; they contain an inner channel. The rigid sex pili are essentially shorter than the flexible ones, their diameter amounts to 10-11 nm.

Like flagella and fimbriae, sex pili are composed of subunits (pilins). The F pilins have been examined best; they are polypeptides consisting of 70 amino acids. The molecules contain α-helical regions up to a rather high percentage of 65-70. In the pilins, the helices are arranged as three- or four-helix bundles. The helices lie perpendicular to the pilus axis and are inter-connected by loops. There is no evidence for a minor protein at the tip of the pilus; however, the configuration of the pilin subunits at the tip may be different from that at the shaft (see below).

The subunits in the F pili are also helically arranged (25 in two turns with an incline of 32 nm) and form a cylinder of 8 nm in diameter with an axial channel of

2 nm in diameter. The sex pili narrow distally into a tip. It is supposed that this is caused by a conical arrangement of identical subunits.

Sex pili are stable towards guanidine hydrochloride or deoxycholate, whereas they are disaggregated by SDS or sarkosyl.

The biosynthesis of F pili is determined by the F-plasmid transfer region (*tra*-region) which resembles that one of the fimbriae. The primary synthesis product, namely propilin, consists of 121 amino acids. The mature pilin is produced by cleaving off the signal sequence which is formed by first 51 amino acid residues. The pilins are produced in excess and deposited as a pool in the cytoplasmic membrane. The formation of the sex pili occurs by an energy-dependent, proximal assembly of the pilins and a simultaneous pushing through the outer membrane, possibly by using the fusion points between cytoplasmic and outer membrane. The depolymerisation of the sex pili described above, aiming at a closer contact of the cells, occurs also from the inside; the liberated pilins are deposited in the pool of the cytoplasmic membrane and reused.

Bibliography

Firth N, Ippen-Ihler K, Skurray RA (1996) Structure and function of the F factor and mechanism of conjugation, In: Neidhart FC, Curtiss R III, Ingraham IL, Lin ECC, Low KB, Magasanik B, Reznikoff WS, Riley M, Schaechter M, Umbarger ME (eds) *Escherichia coli* and *Salmonella*: cellular and molecular biology, 2nd edn. ASM Press, Washington, pp 2377-2401

Frost LS (1993) Conjugative pili and pilus-specific phages. In: Clewell DB (ed) Bacterial conjugation. Plenum Press, New York, pp 189-221

7 Cell Wall Models

The composition and structure of cell walls are not invariable even under in vitro conditions. They do vary in dependence on cell age, composition of the cultivation medium, pH, redoxpotential, and other parameters. Under in vivo conditions such relationships are even more expressed, especially with regard to colonisation sites in the host and to the phase of progressive infection. Bacterial response to variations in the environment is partially regulated by switch-on/switch-off processes which serve the adaptation of the microorganisms to changing environmental conditions during the course of infection.

Most data dealing with composition and structure of cell walls are obtained from investigations of in vitro-cultivated bacteria; however, it has repeatedly been found that these may distinctly differ from those obtained from bacteria grown in vivo. Therefore one should be rather cautious with the assessment of experimental results and should not draw too far-reaching conclusions.

On the basis of the chemical composition and structure of the cell wall components as well as of the morphological structures observed in the electron microscope after different preparation techniques, a number of models of Gram-negative and Gram-positive cell walls have been developed in the past. To the same extent as experimental techniques and evaluation procedures advanced, providing more and more comprehensive' information, the models became more detailed and closer to reality. However, it has to be realised that each model can represent only an average of several possible structures or only one of these.

In most bacterial cell walls one common electron-dense rigid structure of varying thickness can be seen: the peptidoglycan (murein). Even some archaea contain a comparable component, i.e. pseudomurein. As described in detail in Section 3.2.1, according to the at present most accepted model, the basic structure of peptidoglycan consists of glycan strands arranged parallel to the cell surface which are cross-linked by peptide bridges and thus form a stable net-like layer. Gram-negative cell walls mainly contain only one such layer (thickness 2-2.5 nm), and the Gram-positive ones several of them arranged in layers and linked to each other (thickness 30-40 nm). It should be noted again that according to Koch (1998; Fig. 3.5) a zig-zag arrangement of the peptide-linked glycan strands (forming hexagons) is much more probable, since due to the internal cell pressure linear strands should not exist in the cross-linked peptidoglycan.

According to Labischinski et al. (1979), the peptidoglycan layers of Gram-positive bacteria are arranged parallel to the cell surface at a distance of 4.2 nm from each other, however, the polysaccharide chains of adjacent layers mostly do not point in the same direction (Fig. 7.1). The resulting diverse spatial orientations of the single peptidoglycan layers additionally stabilise the cell wall.

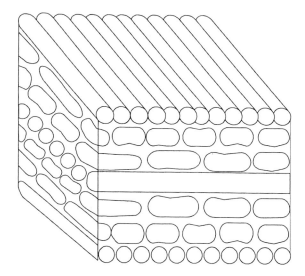

Fig. 7.1. Schematic presentation of the possible rotation of adjacent peptidoglycan layers against each other (Labischinski H, Barnickel G, Bradaczek H, Giesbrecht P (1979) On the secondary and tertiary structure of murein. Low and medium-angle X-ray evidence against chitin-based conformation of bacterial peptidoglycan. Eur J Biochem 95:147-155, with permission)

A recently proposed and fundamentally new model of the peptidoglycan (Dmitriev et al. 1999) rejects the multilayered structure and assumes only one layer in which the glycan strands run perpendicular to the cell wall plane (Fig. 7.2). The disaccharide units of the strands are cross-linked by peptide bridges running in this plane with four adjacent strands. In the case of Gram-negative bacteria the thickness of the layer corresponds to the height of one turn of the polysaccharide helix (four disaccharide units; 3.92 nm), which is necessary for complete cross-linking. Remaining saccharide moieties are assumed to be either far less or not at all cross-linked. The glycan strands in the Gram-positive cell wall comprise 100–200 disaccharide units and are cross-linked throughout the entire length, which is in agreement with the thickness of the (streptococcal) cell wall of 30-40 nm. In this way, a thick matrix with helix-shaped channels is formed that harbour either teichoic acids or lipoteichoic acids piercing outwards through the wall. According to this model, new complete cell wall structures (including pili and fimbriae) of both Gram-negative and Gram-positive bacteria are produced in toto by the cytoplasmic membrane under the protection of the old cell wall, pushed outwards after stepwise lysis of the old cell wall in this region, and stretched by means of the internal cell pressure. By this assumption the problem of penetration of macromolecules across the peptidoglycan layer could be solved.

Gram-Negative Bacteria

Electron micrographs of Gram-negative bacteria show three or four electron-dense regions in the cell envelope. The innermost is the cytoplasmic membrane which does not belong to the cell wall, the next region but one is the outer membrane. The periplasmic space containing the peptidoglycan layer is located between both membranes (Fig. 7.3). Sometimes a fourth, outermost region can be seen, which is presumably formed by the S-layer.

The periplasm was originally regarded as a space located between both membranes, present in a liquid state, and interlaced by a thin peptidoglycan layer.

New preparation techniques for electron microscopy led to the opinion that the periplasm represents a gel in which the peptidoglycan is embedded as a loosely organised three-dimensional network.

The structure models of the outer membrane were developed on the basis of the "fluid mosaic model of biological membranes" (Singer), although the outer membrane is only partially comparable to a typical biological membrane bilayer, especially due to its more pronounced asymmetry and the significanctly lower fluidity of its outer leaflet.

In a model published in 1974, Leive assumed the inner monolayer of the outer membrane to consist of phospholipids and proteins, and the outer one of LPS, proteins and only a few, if at all, phospholipids (Fig. 7.4). Thus, the outer membrane was proposed to be highly asymmetrical. Part of the LPS is hydrophobi-

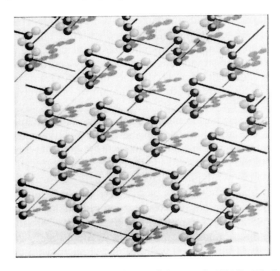

Fig. 7.2. Model of the peptidoglycan layer after Dmitriev et al. (1999). *Black balls* MurNAc; *white balls* GlcNAc; *straight lines* peptide bridges in the upper turn of the helix. For reasons of clarity, only the short saccharide chains of the Gram-negative bacteria are outlined; those of the Gram-positive cell wall are much longer (100-200 disaccharides)

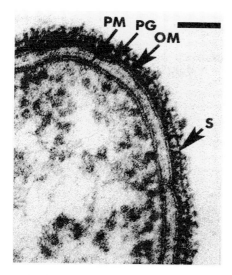

Fig. 7.3. Electron micrograph of the Gram-negative cell envelope (Sleytr and Beveridge 1999, with permission). *PM* Periplasmic membrane; *PG* peptidoglycane; *OM* outer membrane; *S* S-layer

cally bound, the rest via Mg^{2+} and Ca^{2+} bridges. The latter fraction is extractable by EDTA. Both forms of LPS are in an equilibrium. The polysaccharide chains of the LPS rise from the cell surface into the surrounding medium. The proteins are mainly integral ones, to a great extent present as transmembrane proteins. They are also mainly hydrophobically bound. The membrane is linked to the peptidoglycan via anchor molecules (for example Braun´s lipoprotein). In the peptidoglycan layer, larger pores are postulated as being of importance for the transport of substances through the cell wall. The Leive model at that time was of significance because it explained the extractibility of a part of LPS by means of EDTA and postulated a high membrane asymmetry.

The cell wall model of the Nikaido group (Smit et al. 1975) does not include the peptidoglycan layer, but deals with the outer membrane. On the basis that deep rough mutants created by point mutation (see Sect. 2.2.4.2) have less 33/36 K-proteins in their outer membrane, it is concluded that such a mutation may not only modify the LPS structure, but unexpectedly also the protein content. There are two explanations for this finding:

1. The protein biosynthesis is continued, but its incorporation is interrupted due to the lacking attachment site in the modified LPS.
2. Due to the lacking attachment site, the protein biosynthesis is interrupted via a feedback mechanism.

Examinations of different deep R mutants revealed the attachment sites for proteins in distinct regions of the LPS molecules. Truncation of LPS chains caused by the transition of the bacteria from the S to Ra form does not affect composition and structure of the outer membrane. However, the transition to the deeper rough

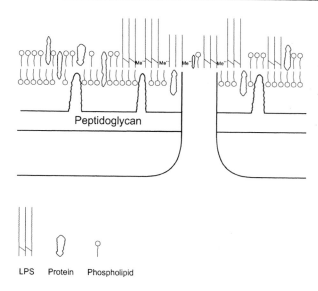

Peptidoglycan

LPS   Protein   Phospholipid

Fig. 7.4. Model of the Gram-negative cell wall (after Leive 1974)

forms, especially to Rd and Re, causes a decrease in the protein content, i.e. the attachment sites must be located in the corresponding LPS region.

In Re mutants, phospholipids are detectable also on the cell surface. In accordance with the Leive model, it can be concluded that S strains and Ra to Rc mutants contain exclusively LPS and (mainly transmembrane) proteins in the outer leaflet of their outer membrane (Fig. 7.5), whereas the inner leaflet consists of phospholipids and proteins. In deeper rough forms, the outer monolayer contains phospholipids in addition to LPS, presumably to compensate the lower protein content, caused by the lacking attachment sites. Membranes of such mutants contain regions consisting of real phospholipid bilayers which cannot be found in membranes containing more complete LPS.

DiRienzo et al. (1978; Fig. 7.6) published a very illustrative model of the Gram-negative cell envelope. It is based on preceding models, however, develops new conceptions of the protein pores present in the outer membrane. They were thought to consist of trimers of the so-called matrix proteins and are stabilised by lipoprotein molecules; the trimers are arranged in such a manner that they form one membrane-spanning channel at its axis. It was further postulated that the cell surface is not homogeneously covered by the O-specific polysaccharide chains of the LPS, but rather contains isle structures. In addition, a stabilising function of the complex of Braun's lipoprotein with one of the major proteins (later named OmpA; see Sect. 2.3.3) was assumed. In this model, the adhesion zones between outer and cytoplasmic membranes (Bayer's patches) are not shown.

In the cell wall model of Hancock et al. (1994; Fig.7.7) the progress that was made during the 15 years since DiRienzo had published his model becomes evident. Above all, this concerns the fact that the proportions and the spatial

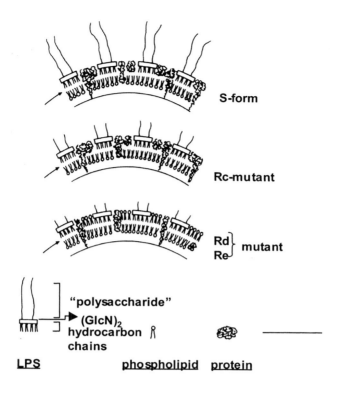

S-form

Rc-mutant

Rd⎫
Re⎭ mutant

"polysaccharide"
(GlcN)$_2$
hydrocarbon R
chains

LPS   **phospholipid   protein**

Fig. 7.5. Model of the Gram-negative cell wall (Smit, Kamio and Nikaido 1975)

arrangements of the individual cell wall components were exactly determined down to molecular dimensions. The length of the polysaccharide chains of the LPS presented in the figure, however, is not constant as indicated. In reality, it is rather heterogeneous and many of the chains are essentially longer (up to 40 repeating units instead of the two shown). For clarity, only the amino acid backbones of the cross-bridging peptide chains of the peptidoglycan and of all the protein components are shown. The pore-forming OmpF molecules depicted in Fig. 7.7 are supposed to be present as trimers forming (unlike the trimers in the DiRienzo-model) three channels indicated in the figure by arrows.

All the models are static by nature. It, however, has already been mentioned (for example in Sect. 2.1.3.1) that especially the hydrocarbon chains in the lipid bilayers adopt a number of spatial arrangements, mainly at higher temperatures. Thus, the gaps in the lipid bilayer shown in the last figure usually exist temporarily.

As revealed by X-ray diffraction and Fourier-transformed infrared spectroscopy, the LPS layer furnishes a highly ordered quasicrystalline structure with very low fluidity, whereas the inner leaflet containing phospholipids is fluid.

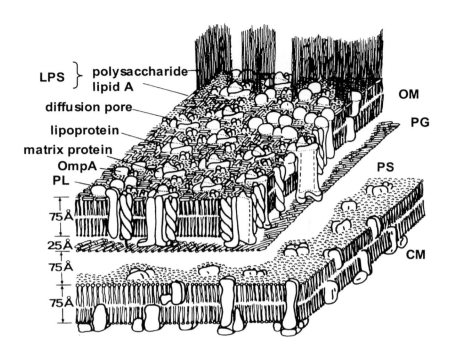

LPS } polysaccharide
        lipid A

diffusion pore

lipoprotein

matrix protein

OmpA

PL

75Å

25Å

75Å

75Å

OM

PG

PS

CM

Fig. 7.6. Model of the Gram-negative cell envelope (DiRienzo et al. 1978)

The cell wall components are linked to each other by at least non-covalent bonds, in particular those of the outer membrane bilayer. The fixation of the membrane to the peptidoglycan layer takes place via Braun's lipoprotein. One terminus of these molecules is hydrophobically bound in the inner monolayer of the outer membrane via fatty acid residues and fixation to the peptidoglycan occurs via a peptide linkage between the ε-amino group of its C-terminal lysine and the carboxyl group of every 10th-12th *m*-diaminopimelic acid. The molecules are mostly arranged as trimers to which an OmpA-molecule is additionally bound, probably as a monomer (not shown in the figure). OmpC and OmpF are also so strongly bound to peptidoglycan via secondary valencies which are strong enough to prevent their removal even in a 2% SDS solution at 60°C.

Figure 6.7 shows a model of the fixation of the flagellae into the outer membrane, Fig. 6.6 that of the fixation of the S layers at the outer membrane.

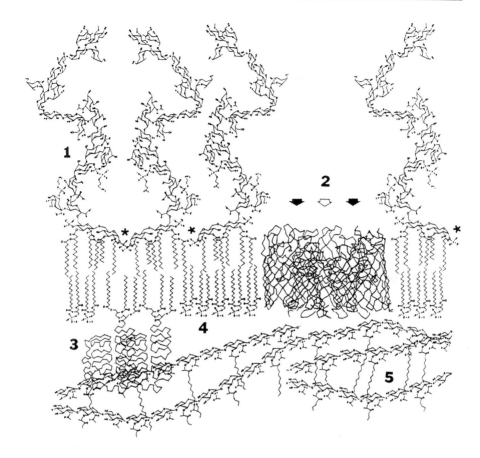

Fig. 7.7. Model of the Gram-negative cell envelope after Hancock et al. (1994). *1* LPS; *2* trimeric OmpF-porin (⬇ channels in the front, ⇩ channel in the back); *3* BRAUN′s lipoprotein; *4* phospholipid; *5* peptidoglycan; * bivalent cation

7.2 Gram-Positive Bacteria

7.2.1 "Typical" Gram-Positive Bacteria

The Gram-positive cell walls in general are much thicker and far less structured than the Gram-negative ones. The electron micrographs of the cell envelopes of streptococci, staphylococci or others (Fig. 7.8) mostly show two or three electron-dense regions. The innermost one represents the cytoplasmic membrane, more outside is located the actual cell wall and finally, if present at all, a surface layer is observed.

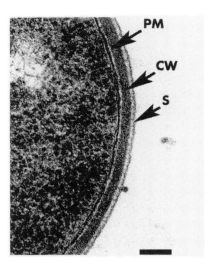

Fig. 7.8. Electron micrograph of the Gram-positive cell envelope. (Sleytr UB, Beveridge TJ (1999) Bacterial S-layers. Trends Microbiol 7,253-260, with permission). *PM* periplasmic membrane; *CW* cell wall; *S* S-layer

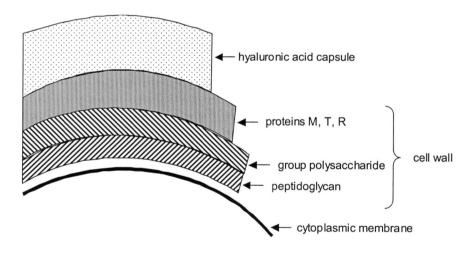

Fig. 7.9. Early model of the Gram-positive cell wall. (Krause 1963)

The peptidoglycan represents a main component of the Gram-positive cell walls. It is much thicker than that of the Gram-negative bacteria because the sacculus consists of more than one layer. From the multilayer version of structure an enlargement model was developed which is much simpler than that of the Gram-negative bacteria. As discussed in Section 3.2.2.5, the bacteria during incorporation of new peptidoglycan strands need not take precautions to keep the cell stable against its own pressure. The preexisting multiple murein layers are sufficient for such a stabilisation.

Besides the peptidoglycan, Gram-positive cell walls contain substantial amounts of polysaccharides and proteins. In "layer type" models as the highly schematic one developed by Krause (1963) for the streptococcal cell wall (Fig. 7.9) each of the individual substance groups was assumed to form an individual layer. The peptidoglycan as the innermost cell wall layer was supposed to be located on top of the cytoplasmic membrane, but separated from it by a narrow electron-transparent region of unknown character. Further outwards, a layer containing the group-specific polysaccharides (teichoic acids) was postulated (electron microscopically not visible), followed by a layer consisting of the proteins (proteins M, T, R). The outermost layer was assumed to be formed by hyaluronic acid, representing a capsule. This model was justified by the observation that individual components of the cell wall may cover others and thus prevent their proof. In the case of the similarly structured cell wall of *Staphylococcus aureus*, for example, it is well known that their protein A may cover some phage receptors and thus block them.

A range of findings, however, could not be explained by the layer models. Besides proteins, also polysaccharide antigens could be simultaneously detected on the outer cell surface, and even the peptidoglycan was found to act as a phage receptor. In *Staphylococcus aureus*, for instance, teichoic acids contribute to the charge of the cell surface but cease to do it below pH 5 to 6. This was believed to be the result of a pH-dependent rearrangement in the cell wall, thus opposing a layer model. On the other hand, the same polysaccharide antigens could be detected at the inner cell wall surface. Therefore more recent models, although also based on electron microscopical results, postulate limited mutual penetrations of layer-forming components, mainly with regard to the presence of polysaccharides in the peptidoglycan and in the protein layer.

However, these models were also not tenable. For instance, they did not explain why in *Staphylococcus aureus* both protein A and teichoic acids, making up to 50% of the cell wall mass, are covalently bound in the peptidoglycan layer and are simultaneously detectable at the cell surface. Nor were the models consistent with the results of immunoelectron microscopic investigations (Sect. 4.2.2). Using this method, all main components of the cell wall could be detected at its outer and some of them likewise at its inner surface. Finally, a basic objection to the layer type models came from the observation that after extraction of the different polymers from the cell wall, parts of them could still be detected on the remaining peptidoglycan.

The mutual penetration of individual substance groups was found to be much more pronounced than anticipated. The teich(ur)onic acids, for instance, are predominantly bound in the peptidoglycan layer by covalent linkages, the lipoteichoic acids completely span the peptidoglycan and are anchored in the underlying cytoplasmic membrane by hydrophobic forces. Actually, most of the proteins were found to be located close to the cell wall surface as many of them can easily be solubilised by proteolytic enzymes and can also be detected by agglutination with specific antisera.

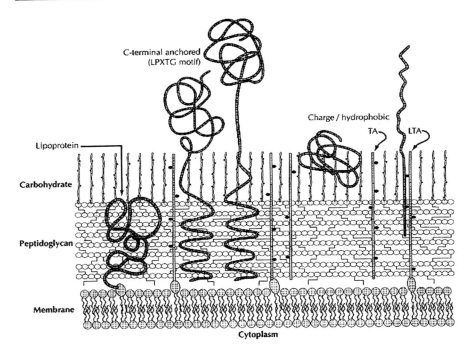

Fig. 7.10. Model of the Gram-positive cell wall. (Fischetti VA (2000) Surface proteins of Gram-positive bacteria. In: Fischetti VA, Novick RP, Ferretti JJ, Portnoy DA, Rood JJ (eds) Gram-positive pathogens. ASM Press, Washington, pp. 11-24, with permission). *TA* Teichoic acid; *LTA* lipoteichoic acid

Therefore, models assuming a mosaic-like cell wall structure have been proposed. According to these models the Gram-positive cell walls consists of only one thick layer which is not homogeneously composed (Fig. 7.10). All components more or less penetrate each other, however, proteins and polysaccharides are more predominantly present in the outer part and the peptidoglycan predominantly located in the inner region of the cell wall. This postulation is in agreement with the assumption that during biosyntheticenlargement of the peptidoglycan, the uppermost layers are stepwise enzymatically degraded and thus present in reduced amounts. The long glycan strands of teichoic and teichuronic acids, or especially the membrane-bound lipoteichoic acids, are assumed to more or less span the whole layer. Proteins like staphylococcal protein A are covalently bound to the peptidoglycan layer but simultaneously detectable on the surface. Other proteins like proteins M and T of group A streptococci, likewise being fixed in the peptidoglycan layer, may form fibrils (filaments) that project outwards from the cell surface. In the cell wall, they are non-covalently attached to the cell wall, for instance to lipoteichoic acids. For some proteins, their proline-rich regions at the C-terminus (Sect. 4.2.1) are fixed by a strong cross-linking to newly incorporated peptidoglycan.

It could not be decided whether or not there is a complete mutual penetration of the individual components in molecular dimensions or if isles of each of the components exist.

This model, however, is not in complete accordance with the electron micrographs (Fig. 7.8) as the outermost layer which may frequently be detected there is missing. This layer has been regarded as an S layer; Fig. 6.6 shows a model of the localisation of this layer on the cell wall.

7.2.2 "Atypical" Gram-Positive Bacteria

Mycolata, like e.g. Mycobacteria, Corynebacteria and Nocardia belong to the Gram-positive bacteria; however, their cell walls are quite differently constructed. In composition and structure they recall much more the Gram-negative cell wall. Mainly a membrane-like region can be found which is superimposed to the peptidoglycan layer and linked to it; however, the similarities in composition and structure between this region and the outer membrane of the Gram-negative bacteria are restricted.

Figure 7.11 shows the electron-microscopic picture of an ultrathin section of a mycobacterial cell envelope. The innermost layer showing a moderate electron density (CM) is formed by the cytoplasmic membrane, the superimposed electron-dense layer (EDL) by the peptidoglycan. The electron-transparent layer (ETL) with a thickness of 9-10 nm presumably represents the mentioned membrane-like region, it is formed by hydrophobic components. The nature of the superimposed electron-dense outer layer (OL) is still unclear; it seems to contain negative charges.

Figure 7.12 shows a model compatible with the electron microscopic pictures and the analytical-chemical results. The main component of the cell wall is represented by the mycolyl-arabinogalactan peptidoglycan complex. Anchoring of arabinogalactan to peptidoglycan occurs via the described Rha-GlcNAc-P bridge (Sect. 4.3.1.2) and the mycolic acid residues being ester-linked to the terminal arabinoses are arranged parallel to each other, forming the hydrophobic region together with other amphiphiles.

Because of its membrane-like structure, the hydrophobic region represents a very effective penetration barrier against hydrophilic substances. Its effectiveness depends on the arrangement, the length and the packing density in particular of the mycolic acid molecules. The innermost zone of this region is formed by meromycolate and α-chains, densely packed, and arranged in a quasicrystalline form. The packing of the middle zone located above the innermost one, mainly consisting only of meromycolate chains, is distinctly looser due to the absence of the α-chains and to the bulky substituents (Sect. 4.3.1.1). Thus, this zone is rather fluid. In the outer zone the packing density is again higher due to the embedded extractable lipids (Sect. 4.3.3), the glycolipids with their hydrophilic components directed outwards. Thus, this zone is less fluid than the middle part, but more fluid than the innermost one. The outward-directed hydrophilic moieties of the glycolipids are immunologically active; they induce formation of specific antibodies.

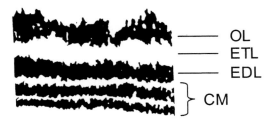

Fig. 7.11. Schematic presentation of the electron-microscopic appearance of the mycobacterial cell envelope in thin sections. (Brennan PJ, Nikaido H (1995) The envelope of Mycobacteria. with permission, from the Annul Review of Biochemistry Volume 64 ©1995 by Annual Reviews www.AnnualReviews.org). *CM* cytoplasmic membrane; *EDL* electron dense layer; *ETL* electron transparent layer; *OL* outer layer

The zone containing α-chains is topologically comparable to the inner leaflet of the outer membrane of the Gram-negative cell wall, the zone containing the extractable lipids to the outer leaflet. However, differences exist. As already mentioned, the inner zone of the membrane-like region is less fluid than the outer one. This is in contrast with the outer membrane, where the inner monolayer is more fluid than the outer one. A second difference results from the fact that both monolayers of the outer membrane are connected by mainly hydrophobic forces, while in the case of the mycolata, the long acyl chains of the meromycolate moieties covalently connect inner and outer zone. Unexpectedly, both zones can be separated by means of the freeze-fracture technique (Sect. 2.4.3). To explain this, a localization of the mycolic acid exclusively in the inner, therefore very thick, zone (now representing a monolayer) was stated, whereas the outer monolayer was supposed to be formed merely by the extractable lipids.

The connecting piece between the mycolic acid region and the peptidoglycan is represented by the arabinogalactan. Its long and flexible saccharide chains allow the mycolic acid moiety some mobility which is beneficial to its embedding into the "membrane" plane.

The localisation of a further component of the cell wall, namely the lipoarabinomannan (Sect. 4.3.2), has not yet been clearly elucidated. It is possibly fixed via its phospholipid moiety in the cytoplasmic membrane and spans all cell wall layers. However, there are also reasons arguing against a localisation of the polysaccharide chains in the hydrophobic membrane-like region, such as the hydrophilicity and the bulky shape of its arabinan moiety, thus favouring an anchoring of phospholipid moiety in this region.

A completely new model of the mycobacterial cell wall has been very recently proposed by Dmitriev et al (2000). It postulates a vertical arrangement of the constituents instead of horizontal layers and explains thus the formidable impermeability of this wall.

The mycolic acid region is poor in proteins; only a small amount of a pore-forming protein with weak activity has been detected, causing the low penetrability towards hydrophilic substances (2-3 orders of magnitude lower than in *E. coli*).

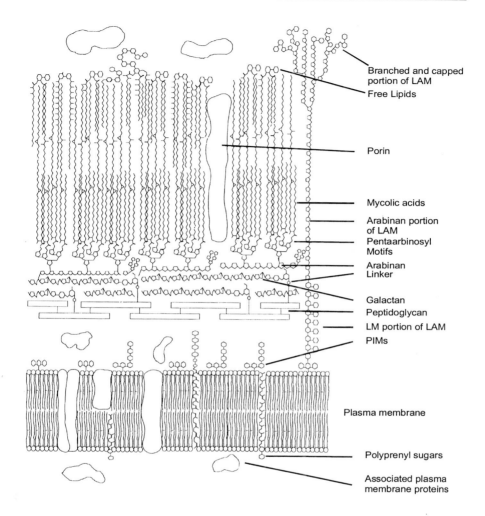

Branched and capped portion of LAM
Free Lipids
Porin
Mycolic acids
Arabinan portion of LAM
Pentaarbinosyl Motifs
Arabinan Linker
Galactan
Peptidoglycan
LM portion of LAM
PIMs
Plasma membrane
Polyprenyl sugars
Associated plasma membrane proteins

Fig. 7.12. Proposed structure of the mycobacterial cell wall. (Lee RE, Brennan PJ, Besra GS (1996) *Mycobacterium tuberculosis* cell envelope. Cur Top Microbiol Immunol *215*, 1-27, Springer-Verlag)

However, owing to the crystal-like structure of the innermost zone, the penetrability towards hydrophobic substances is also reduced (approximately 1 order of magnitude lower than in *E. coli*). These are explanations for the low growth rate and the high natural resistance of mycobacteria to the defence mechanisms of the host as well as to antibiotics.

Bibliography

Brennan PJ, Nikaido H (1995) The envelope of Mycobacteria. Annu Rev Biochem 64:29-63

Daffé M, Draper P (1998) The envelope layers of mycobacteria with reference to their pathogenicity. Adv Microb Physiol 39:131-203

DiRienzo JM, Nakamura K, Inouye M (1978) The outer membrane proteins of Gram-negative bacteria: biosynthesis, assembly, and functions. Annu Rev Biochem 47:481-532

Dmitriev BA, Ehlers S, Rietschel ET (1999) Layered murein revisited: a fundamentally new concept of bacterial cell wall structure, biogenesis and function. Med Microb Immunol 187:173-181

Dmitriev BA, Ehlers S, Rietschel ET, Brennan PJ (2000) Molecular mechanisms of mybacterial cell wall: from horizontal layers to vertical scaffolds. Int J Med Microb 290:251-258

Fischetti VA (2000) Surface proteins of Gram-positive bacteria. In: Fischetti VA, Novick RP, Ferretti JJ, Portnoy DA, Rood JJ (eds) Gram-positive pathogens. ASM Press, Washington, pp. 11-24

Hancock REW, Karunaratne DN, Bernegger-Egli C (1994) Molecular organization and structural role of outer membrane macromolecules. In: Ghuysen JM, Hakenbeck R (eds) Bacterial cell wall. New comprehensive biochemistry, vol 27. Elsevier, Amsterdam, pp 263-279

Koch AL (1998) Orientation of the peptidoglycan chains in the sacculus of *Escherichia coli*. Res Microbiol 149:689-701

Krause RM (1963) Antigenic and biochemical composition oh hemolytic streptococcal cell walls. Bact Rev 27:369-380

Labischinski H, Maidhof H (1994) Bacterial peptidoglycan: overview and evolving concepts. In: Ghuysen JM, Hakenbeck R (eds) Bacterial cell wall. New comprehensive biochemistry, vol 27. Elsevier, Amsterdam, pp 23-38

Labischinski H, Barnickel G, Bradaczek H, Giesbrecht P (1979) On the secondary and tertiary structure of murein. Low and medium-angle X-ray evidence against chitin-based conformation of bacterial peptidoglycan. Eur J Biochem 95:147-155

Lee RE, Brennan PJ, Besra GS (1996) *Mycobacterium tuberculosis* cell envelope. Cur Top Microbiol Immunol 215:1-27

Leive L (1974) The barrier function of the Gram-negative envelope. Ann NY Acad Sci 235:109-124

Nikaido H (1996) Outer membrane. In: Neidhart FC, Curtiss R III, Ingraham IL, Lin ECC, Low KB, Magasanik B, Reznikoff WS, Riley M, Schaechter M, Umbarger ME (eds) *Escherichia coli* and *Salmonella*: cellular and molecular biology, 2nd edn. ASM Press, Washington, pp 29-47

Sleytr UB, Beveridge TJ (1999) Bacterial S-layers. Trends Microbiol 7:253-260

Smit J, Kamio Y, Nikaido H (1975) Outer membrane of Salmonella typhimurium: Chemical analysis and freeze-fracture studies with lipopolysaccharide mutants. J Bacteriol 124:942-958

Vaara M (1999) Lipopolysaccharide and the permeability of the bacterial outer membrane. In: Brade H, Morrison DC, Opal S, Vogel S (eds) Endotoxin in health and disease. M Dekker, NewYork, pp 31-38

8 Cell Wall Functions

The cell wall has to fulfil numerous functions, some of which are apparently contradictory, like e.g. the demarcation of the cell interior on the one hand and the mediation of interactions with the environment on the other hand. These functions can be executed neither with only one cell wall layer, especially in a more complex environment, nor with a cell wall architecture which is constant in all growth phases.

8.1 The Cell Wall as a Transport Organ

8.1.1 General Remarks

The most simple form of substance transport across a semipermeable wall or membrane is diffusion. It takes place as a process equalising solutions of higher with those of lower concentrations, in the direction of the concentration gradient. The steeper the gradient, the higher is the diffusion rate. The geometry of the molecules is also of importance. Larger and/or more bulky molecules usually diffuse more slowly than smaller and/or globular ones. In each case exists an upper limit in the molecular size above which penetration of a wall is impossible; it depends on the nature of wall or membrane.

In the case of smaller concentration differences or of larger molecules, the rate of free diffusion may be so low that an adequate supply of the cell is not guaranteed. For example, only about 1% of the glucose required for normal growth of Gram-negative cells can be provided in this way. For this reason, cell walls have a second transport system at their disposal, the facilitated diffusion. It differs from simple diffusion by the following criteria:

- It has a saturation kinetics (Fig. 8.1). In the case of slight concentration differences, its rate is clearly higher than that of free diffusion, in the case of large differences, it tends towards an asymptotic value.
- It is relatively specific to the substance to be transported.
- It may be specifically inhibited: competitively by means structure analogues, non-competitively by blocking specific functional groups.
- Its temperature coefficient is different from that of free diffusion.

Facilitated diffusion can be understood by the assumption that the substance to be transported is linked to a specific transporter molecule. Transport can be carried

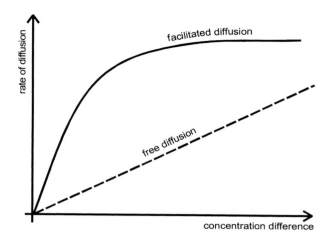

Fig. 8.1. Kinetics of free and facilitated diffusion

out both in the direction of and even against a concentration gradient (passive and active transport, respectively).

In the case of passive transport, the energy is taken from the concentration gradient, and the process can take place only until the concentration difference has disappeared. The meaning of the substrate binding to the transporter is to elevate its concentration on the cell surface and to facilitate the diffusion process.

Active transport, on the contrary, needs additional energy which may be taken from an electrochemical potential or be obtained by coupling to an exergonic process such as ATP hydrolysis.

Diffusion through bacterial cell walls differs very much depending on the nature of the bacteria. The "typical" Gram-positive cell wall is rather penetrable since it contains only the peptidoglycan and sometimes the S layer, which may have a blocking or reducing effect on the penetration of solutes. In most cases they hardly represent a barrier for hydrophilic molecules of up to approximately 50 kDa. Since the peptidoglycans of Gram-positive bacteria (*Bacillus subtilis*) are essentially thicker than those of Gram-negative ones (*E. coli*), differences in the permeability of both should be anticipated. However, this is not the case.

The Gram-negative cell walls and those of "untypical" Gram-positive bacteria like mycolata represent a much more effective penetration barrier. They contain membranes or membrane-like structures, the hydrophobic interior of which is able to block penetration of hydrophilic molecules even of lower molecular mass such as monosaccharides, amino acids, nucleosides and hydroxonium, alkali or alkaline earth ions. However, hydrophobic substances (also gases like O_2, N_2, CH_4) are able to penetrate phospholipid bilayer membranes; the faster, the better the substance is soluble in the hydrophobic membrane interior (hydrophobic transport pathway). In the case of Gram-negative cell walls, this pathway is of minor importance, which is mainly due to the presence of LPS in the outer leaflet, the hydrophilic polysaccharide chains of which form a layer which hinders hydrophobic substances from reaching the membrane surface. This is proven by the observation that the outer membrane of deep-rough mutants is penetrated much more easily by hydrophobic molecules since they contain LPS with far shorter polysaccharide chains.

In order to enable the diffusion of hydrophilic molecules of low molecular mass, which is essential for the bacteria, their membranes or membrane-like structures contains water-filled channels embedded into the lipid bilayer (hydrophilic transport pathway).

A third possible diffusion pathway is generated by the substrate itself (self-promoted pathway). However, predominantly it does not serve the vital processes of the bacteria, but is generated, for example, by external substances of mainly polycationic character (aminoglycoside antibiotics, basic peptides) and results in damage to the bacteria (see Sect. 8.6).

Contrary to earlier opinions, it is now presumed that not only free diffusion but also passive and active transport through the outer membrane of Gram-negative cell walls of hydrophilic substances are carried out by transmembrane proteins which form water-filled channels extending through the entire membrane. There are, however, differences in the mechanism of action in dependence on the kind of transport.

Systems carrying out active transport with particular effectiveness are called permeases or transporters. Such systems have been found mainly in microorganisms, since they must be able to grow fast even in diluted nutrient solutions. Permeases may be both constitutive and inducible and often consist of several components (compare to the classical β-galactoside-permease of Monod).

Membrane transport depends on the metabolic requirements of the cell. Therefore there are control mechanisms which are able to increase or decrease the particular transport rates according to demand.

8.1.2 Transport Across the Bacterial Cell Wall

The much more expressed permeability of the "typical" Gram-positive cell wall compared with the Gram-negative and the "untypical" Gram-positive ones results in both advantages and disadvantages for the bacteria. A better permeability clearly enlarges the range of substances coming into question as nutrients and also facilitates the excretion of higher-molecular substances into the surrounding medium with the intention of improving the conditions of bacterial growth and multiplication (enzymes, toxins). One disadvantage is the possibility of facilitated penetration of substances damaging the vital processes of the bacteria. Thus, in most cases Gram-positive bacteria are distinctly more sensitive towards antibiotics than Gram-negative ones.

8.1.2.1 Transport from the Exterior to the Interior

A number of methods have been developed to determine the relationship molecular size/permeation velocity and that of the upper molecular mass limit of permeability. For this purpose, intact bacteria, isolated cell walls and liposomes or planar model membranes put together from the individual cell wall components can be used (Seydel et al. 1989; Wiese and Seydel 2000).

By using intact Gram-negative bacteria which contain a β-lactamase in their periplasmic space, the hydrolysis rate by the bacteria of β-lactam antibiotics added to the medium is measured. The data are corrected on the basis of the hydrolysis rate which was determined using liberated β-lactamase; the difference is an expression in the diffusion rate.

In a method similar to gel filtration, isolated cell walls may be used as the stationary phase and molecules of defined size as the mobile phase. The pore size can be deduced from the ratio elution volume/total volume.

In the liposome swelling assay, liposomes corresponding to the intact outer membrane in composition and structure are formed. They are filled with the solution of a non-penetrating polysaccharide and then brought into an isotonic solution of the test solute. The rate of influx of this solute through the channels is determined from the initial rate of liposome swelling, which can easily be determined quantitatively by turbidity determinations. The method has the advantage that the composition of the liposome membrane may be varied, thus allowing the investigation of the influence of the individual membrane components (for example of the single pore proteins) on permeability.

The results of the individual methods correspond to each other. The upper exclusion limit of free penetration of Gram-negative cell walls such as of *E. coli* for oligosaccharides and peptides amounts to a molecular mass of about 600 Da, whereas in the case of globular proteins about 50 kDa are indicated for Gram-positive cell walls. In the case of high-molecular substances, the molecular geometry represents an essential factor and therefore it is more obvious to indicate the maximum radius of the molecules instead of the molecular mass. In Gram-positive bacteria, the radius amounts to approximately 4 nm, in the Gram-negative to not more than one tenth of this value, however, with distinct differences in dependence of the nature of the bacteria. Larger proteins may be able to penetrate the cell wall after unfolding, thus creating long peptides with correspondingly reduced diameters. This happens, for example, in the case of colicin A (Lazdunski et al. 1998).

The molecular mass limit of about 600 Da in Gram-negative bacteria is valid only for hydrophilic molecules. The diffusion of these molecules shows neither a greater temperature dependence nor a dependence on the completeness of the LPS present in the cell wall. However, this limit is not proper for hydrophobic molecules. Their diffusion is relatively insignificant in S-type bacteria and shows a strong temperature dependence. However, it becomes significant in bacteria containing incomplete LPS (thus in R forms, mainly Rd and Re). The diffusion of molecules of the first type occurs via water-filled pores and that of the second type across the lipid bilayer (hydrophobic pathway).

The permeability of the Gram-negative cell wall limited by the outer membrane can be altered. Modifications in its composition or structure, as, for example, by removal of a part of LPS by means of EDTA, make the membrane clearly more permeable and lead to a drastic increase in sensitivity for instance towards antibiotics.

For transport of hydrophilic molecules across the outer membrane via channels formed by specific membrane proteins basically three types are distinguished:
1. Simple diffusion through open, water-filled channels.
2. Facilitated diffusion through channels in which stereo-specific binding sites for the substance to be transported are located, thus causing an increase of the concentration.
3. In the case of active transport specific binding of the substrate with a distinctly higher affinity and coupling to an energy-supplying process.

The molecular mass limit of 600 Da is not valid for transport pathways 2 and 3. In addition, biosynthesis of both pathways is in most cases inducible.

Ad 1. In general, the channels are formed by transmembrane proteins (porins). In the case of *E. coli* K12, these are mainly OmpF, OmpC and PhoE. Although all three proteins just form simple channels, OmpC and OmpF facilitate the diffusion of neutral and cationic substances, while PhoE is responsible for that of phosphate and other anionic molecules. OmpC channels are more narrow than OmpF channels. Thus, the bacteria have the possibility to reduce the uptake of noxious substances (bile acids, for example) by regulating the biosynthesis of the three proteins. A certain channel function of OmpA has already been mentioned which presumably comes more to the fore after the loss of the other channels.

For a long time, it was assumed that all these channels are mostly present in their open form. Delcour (1997), on the contrary, detected that they change between a short-lived open and a more durable closed conformation, and that this process may be regulated. Thus a further possibility is given to prevent the influx of noxious substances.

After diffusion through the pores, many of the required substrates are bound by periplasmic binding proteins and thus removed from the diffusion equilibrium.

Ad 2. Like the porins, this kind of transport proteins forms transmembrane channels as well. In their interior, they contain a ligand-binding site. The substrate concentration in the channel is elevated by the binding and thus the diffusion equilibrium is changed. Such transport shows a saturation kinetics contrary to that through unspecific channels. The details are described for LamB (see also Sect. 2.3.2.3.2) which has been examined best to the present.

As a means to investigate the transport mechanism of LamB, the protein is incorporated in a planar lipid bilayer and afterwards the electrical conductivity of the pore is measured. Addition of maltose or maltooligosaccharides considerably decreases the conductivity by blocking the pore. The binding constant of the substrate to the ligand-binding site may be calculated from the numerical value of the decrease in conductivity. This binding constant gradually increases in the order from D-glucose to maltopentaose up to 1800-fold and then remains constant for maltohexaose and maltoheptaose. This means that the binding site in the channel has the size of five glucose units, i.e. it is about 2.5 nm long.

In addition, the velocity constants for the binding/outside and for the release/inside have been determined. The former has comparable values for glucose and the malto-oligosaccharides. Therefore, the distinctly different transport rates of sucrose and maltose must be explained by the essentially different velocity constants of the release/inside (Benz 1994). The velocity of the release decreases drastically in the order glucose → maltoheptaose, probably because the affinity of the sugar towards the binding site increases with the number of glucose residues.

After penetration through the outer membrane, the malto-oligosaccharides are transferred from LamB to a specific periplasmic maltose binding protein (MalE) which in turn mediates the binding to specific permease subunits (MalF, MalG) in the cytoplasmic membrane.

Besides the transport of malto-oligosaccharides, the LamB channels also permit the unspecific diffusion of low-molecular substances, but not the penetration of molecules of sizes comparable with that of maltoheptaose.

Ad 3. As an example for the active transport, the uptake of iron is discussed. Although iron belongs to the most common elements in the earth crust, it is extremely difficult for the cells to obtain it. This apparent contradiction can be explained by the fact that iron in the presence of oxygen is stable only as Fe^{3+} and that Fe^{3+} at pH values around 7 exclusively exists in the form of rather insoluble hydroxides and oxyhydroxides. The concentration of iron ions available under these conditions lies more than 10 orders of magnitude below the growth-promoting concentration. Thus, iron has to be transformed into a water-soluble state in which it can be taken up by the bacteria. This occurs by means of chelators (siderophores). The stability constant of the formed iron chelates must be far below the solubility product of the hydroxides and oxyhydroxides. In macroorganisms, host and microbe compete for the iron mainly by synthesis of more and more effective siderophores and of enzymes destroying the iron chelates of the competitor, thus releasing the metal ion.

Basically, all bacterial iron-providing systems consist of three components: (1) a low-molecular siderophore, (2) a membrane receptor for the iron-loaded siderophore which binds the chelate and subsequently transports it across the membrane and (3) a system inside the cell responsible for the release of iron from the siderophore.

Earhardt (1996) differentiates three groups of high-affinity siderophores in the case of *E. coli*:
- Real siderophores of the catecholate or hydroxamate type produced and excreted by the bacteria themselves; among them are the catecholate-type siderophore enterobactin (stability constant of 10^{52}) and the hydroxamate-type siderophore aerobactin.
- Siderophores produced and excreted by other organisms, but taken up from *E. coli*; such as desferri-ferrichrome, coprogen (both of fungal origin) and rhodotorulic acid.
- Siderophore-like substances like citrate, dihydroxybenzoic acid or dihydroxybenzoyl serine.

Iron supply systems of rather low affinity have additionally been described. However, they can be efficient only if iron ions are present in sufficient concentrations.

Due to their relatively high molecular mass and their bulky structure, the velocity of free diffusion of many iron-siderophore chelates through the porin channels is insufficient for a satisfactory supply with iron of the bacterial cells.

Therefore the bacteria have developed several distinct specific transport systems (six in the case of *E. coli* K-12). The transport across the cell wall occurs by means of receptor proteins with molecular masses of 74-83 kDa, the biosynthesis of which is iron-regulated.

The mechanisms of the iron uptake are comparable in all transport systems. The iron-siderophore chelate is specifically bound by a loop superimposed to the channel of the outer membrane receptor-/transport protein (Sect. 2.3.2.3.3) and then transported to the interior. The high specificity of the receptor for the Fe^{3+}-siderophore chelates ("siderophore receptor" in Fig. 8.2) causes an "extraction" of these chelates from the medium and a concentration at the bacterial cell surface. Table 2.13 gives an overview of some iron transport proteins in the outer membrane of *E. coli*.

In the periplasmic space, the iron-siderophore chelates are transmitted to specific periplasmic binding proteins (PBP) by the transport protein, in the case of *E. coli* iron citrate to FecB, iron enterocheline to FepB and the three structurally dissimilar ferric siderophores ferrichrome, iron coprogen or iron aerobactin to FhuD each. The iron-loaded binding proteins are recognised by the specific transport systems (permeases in Fig. 8.2) which are localised in the cytoplasmic membrane. After transport across this membrane, siderophore and iron are separated from each other. The detailed mechanisms of this procedure are not yet completely understood, they differ from siderophore to siderophore. Reduction of Fe^{3+} to the much better soluble and worse chelate-forming Fe^{2+} as well as the enzymatic modification of the siderophores are possible pathways.

In Gram-positive bacteria, the Fe^{3+} siderophore chelates transport into the interior of the cell is much easier. They diffuse through the cell wall and gain direct access to the appropriate transport system in the cytoplasmic membrane.

In this connection, a phenomenon called receptor misuse is worth mentioning. Receptor proteins initiate the transport of substances only when they are sterically adapted. They cannot decide whether the substance to be transported is useful or harmful for the bacteria. This fact can be used either by utilising antibiotics which are structurally similar to specific siderophores (e.g. ferrichrome and the antibiotic albomycin are both transported by the FhuA protein) or by attaching antibiotics via a linker to carriers (peptides, siderophores) for which transporters are present in the membrane. The effectiveness of the drugs can thus be drastically increased.

Transport needs energy. In the case of passive transport this is taken from the concentration difference outwards/inwards of the substance to be transported. In the case of "normal" biological membranes, the energy for the active transport is often taken from ATP hydrolysis or from the electrochemical potential (proton gradient); couplings to oxidation processes (e.g. lactate $\rightarrow$ pyruvate) have also been described. All these mechanisms are not feasible in the outer membrane of Gram-negative bacteria. This membrane does not contain ATP and, due to the possible free diffusion of protons through the porin pores, no electrochemical potential can be built up. The energy for an active transport has thus to be supplied by energy-delivering processes of the cytoplasmic membrane via transmitter molecules. In this energy transmission, TonB protein plays an essential role. This

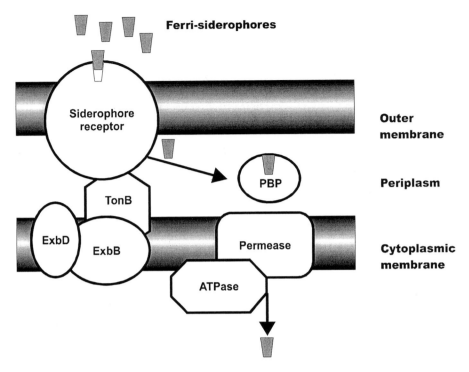

Fig. 8.2. Diagrammatic representation of ferri-siderophore mediated uptake of iron across the Gram-negative cell envelope (see text). (Griffith E, Williams P (1999) The iron-uptake systems of pathogenic bacteria, fungi and protozoa. In: Bullen JJ, Griffith E (eds) Iron and infection. 2nd edn. ©John Wiley Ltd., New York, pp 87-172, reproduced with permission)

molecule is anchored in the cytoplasmic membrane and extends across the periplasmic space to an immediate vicinity of the outer membrane. The energy of the cytoplasmic membrane supplied by supporter proteins (mainly ExbB, also ExbD) creates via the TonB protein a conformational modification in the receptor/transport protein of the outer membrane which permits the specific transport of the substrate into the periplasmic space. In some cases, an outwards-to-inwards variant of the ABC transporter described below is employed. Figure 8.2 shows a summarising diagrammatic representation of ferri-siderophore-mediated uptake of iron across the Gram-negative cell envelope as a special case of the discussed mechanism.

Some substances harming the bacteria develop their effects by deenergising the transport systems. Among them are some colicins (A, B, E1, Ia, Ib, K, N). Having penetrated the outer membrane, they form channels in the cytoplasmic membrane through which protons may diffuse, thus reducing the electrochemical potential. The reenergising mediated by hydrolysis of the ATP present is ineffective because of the permanent draining of energy. It leads only to the exhaustion of the ATP-stores and thus to a breakdown of the anabolism.

The substances which reach the periplasmic space after penetration of the outer membrane are prepared for the transport across the cytoplasmic membrane. Large molecules may be degraded enzymatically. Smaller molecules are bound to binding proteins, by means of which they are linked to the corresponding active transport systems and then cross the cytoplasmic membrane. In parallel, a signal transduction to ATP hydrolysis is carried out with the aim of energy supply. Binding proteins have been detected for quite a number of substrates such as amino acids, sugar(-phosphates), vitamins and inorganic ions. Their specificity may be different: relative unspecificity for monosaccharides, rather high specificity for amino acids.

8.1.2.2 Transport from the Interior to the Exterior

Besides low-molecular end-products of bacterial metabolism which may easily diffuse from inwards to outwards, bacteria also have to transport high molecular substances across the cell wall, predominantly proteins (invasins, toxins, bacteriocins) and polysaccharides.

Bacterial proteins are mainly produced in the cytoplasm from where they have to be transported to the respective destination sites (cytoplasm, cytoplasmic membrane, periplasmic space, outer membrane, or secretion in Gram-negative bacteria; cytoplasm, cytoplasmic membrane, cell wall inclusive periplasm-like space, or secretion in Gram-positive ones; see also Sect. 2.3.2.6). The presorting for this is carried out by a universal mechanism using particular signal peptides. In the first stage, the transport across the cytoplasmic membrane, there are basically no differences between Gram-positive and Gram-negative bacteria. It is mostly brought about by a universal mechanism, the signal peptide-dependent general export pathway (GEP) effected by the Sec-proteins (see Sect. 2.3.2.6). For some proteins, however, Sec-independent transport pathways have been described.

After translocation of the transportant across the cytoplasmic membrane the further cell wall transport becomes different for both Gram-positive and Gram-negative bacteria. For example, in most cases of Gram-positive bacteria an additional signal sequence is necessary for binding of the protein in the cell wall, while in Gram-negative bacteria such a sequence is required for its secretion.

The secretion of macromolecules from Gram-positive bacteria can be carried out more easily due to the lack of an outer membrane, even though peptidoglycan and S layer have barrier functions. After the transport of the extracellular proteins across the cytoplasmic membrane, the signal sequence is split off. If the mature peptide fits through the peptidoglycan meshes, it can pass the cell wall structure by means of simple diffusion. For the secretion process, a special secretion component is necessary. It is presumably represented by a chaperone folding the proteins into a conformation compatible with secretion. The mechanism of the passage of larger molecules across the peptidoglycan is not yet completely clear; in single cases, it may be facilitated by partial autolytic cleaving of some inner-peptidoglycan linkages thus enlarging meshes.

In the case of Gram-negative bacteria, a number of specific pathways exists for the transport across the outer membrane, the most widespread one of which is the

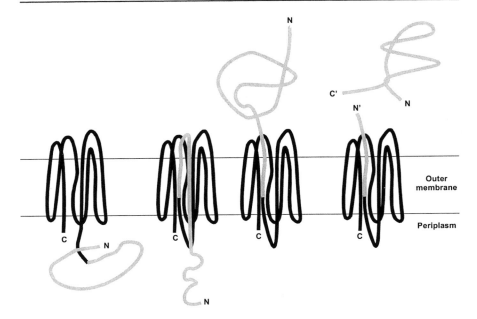

Fig. 8.3. Model for translocation of a protease across the outer membrane. See text. For clarity, the protease is considered to be composed of only two segments, an integral outer membrane β-barrel domain (*black line*) and an amino-proximal region containing the catalytic domain and an intermediate segment (*shaded line*). (Pugsley AP (1993) The complete general secretory pathway in Gram-negative bacteria. Microbiol Rev *57*, 50-108, with permission)

general secretory pathway (GSP). Three pathways are Sec-dependent, i.e. they occur via the GEP and two further ones are Sec-independent.

The GSP represents a two-step process. The first step is the GEP, which allows the proteins to cross first the cytoplasmic membrane and then the outer membrane, via specific terminal branches. For this purpose, the protein is anchored at the periplasmic surface of the cytoplasmic membrane by means of its hydrophobic region located at the *N*-terminus. Then the signal sequence is cleaved and the protein is folded into the secretion form. This folding gives the signal for the passage across the outer membrane. The machinery for this process consists of 12-14 different specific proteins of which, however, mostly only 1-2 are localised in the outer membrane. Diverse exoproteins of the same strain may be secreted using this pathway, and it also serves the synthesis of type IV-pili and the secretion of several filamentous phages.

However, in many cases (mainly with hydrophobic components) the secretion process is poorly understood. The already mentioned Bayer´s patches (Sect. 3.1) as well as specific transporters are in discussion as functioning as possible pathways. One pathway, used among others by a pore-developing *Serratia marcescens* hemolysin (ShlA), needs the assistance of a second large protein, ShlB, which is incorporated in the outer membrane with a β-barrel domain. ShlB mediates the transport of the protein brought into the periplasmic space via GEP across the outer membrane and simultaneously activates the hemolysin.

Some extracellular proteases have an autocatalytic transport pathway (Fig. 8.3). The precursor molecule transported into the periplasmic space via GEP consists of five domains including the signal sequence. The hydrophobic β-domain has a β-barrel structure and, after cleavage of the signal sequence, integrates itself like other OMPs into the outer membrane. With the help of the β-domain the remaining three domains are pushed outwards across the membrane; a channel formation is not excluded in this case. Afterwards, folding of the protease into the active form takes place, then the autocatalytic cleavage from the β-domain and subsequently, autocatalytically as well, the release of the protease domain.

The most important Sec-independent transport pathways are the so-called type I secretion systems, also named ABC secretion systems (ATP-binding cassette). They were discovered during the investigation of the secretion of related exotoxins (E. coli hemolysin for instance) and have been found in both Gram-positive and Gram-negative bacteria. ABC secretion systems are involved in the secretion of a wide range of compounds across bacterial membranes, for example (hydrolytic) enzymes, peptides such as pilins and S-layer components, or polysaccharides such as LPS, CPS or teichoic acids. GEP plays no role in these pathways. An ABC-secretion system of Gram-negative bacteria consisting of one ABC and two supporter proteins extends over both membranes and the periplasmic space. The ABC protein is responsible for coupling the energy of ATP hydrolysis to many physiological, usually transport-related, processes. It contains at least one cytoplasmic domain with a typical ATP-binding site besides one or several integral membrane domains (MD). One of the supporter proteins, a member of the so-called membrane fusion protein (MFP) family, has a hydrophobic segment for anchoring in the cytoplasmic membrane, a hydrophilic segment for spanning the periplasmic space, and a β-structured C-terminus for anchoring in the outer membrane. The second helper protein is an outer membrane protein, in the case of E. coli-hemolysin secretion TolC (Fig. 8.4).

Two types of ABC-dependent transporter complexes can be distinguished. The first is involved in the secretion of proteins and related compounds, the second (ABC-2) in the secretion of polysaccharides and related compounds. They differ in the arrangement of the MD-ABC complex and in the nature of the protein spanning the periplasmic space.

The proteins to be transported via this pathway have no signal sequences, the regulation of their secretion occurs by means of a secretion signal at the C-terminus. The secretion across both membranes presumably takes place in one step.

Besides the described unidirectional pathway, there are ABC systems for the transport in both directions; both systems differ from each other distinctly.

ABC-Dependent transporter complexes have also been described for Gram-positive bacteria. However, they are composed much more simply. In some cases they require only the ABC protein and an additional cytoplasmic membrane protein.

Furthermore, the Yop-secretory pathway is worth mentioning (Yersinia outer proteins, which are plasmid-determined, mostly secreted, and jointly responsible for virulence), which is also used by the Ipa (invasion plasmid antigen) of Shigella and recalls GSP. It is Sec-independent, however, and the proteins to be transported contain no signal sequences. The secretion signal is localised within 50-100 N-terminal amino acid residues. The secretion requires a machinery of about 2ß proteins, one of which is an ATPase and at least another one is a chaperon.

Finally, secretion via outer membrane vesicles (see Sect. 2.4.3) could represent an additional transport mechanism. In this case the diffusion of the mature substances through the outer membrane becomes unnecessary.

Bibliography

Benz R (1994) Uptake of solutes through bacterial outer membranes. In: Ghuysen J-M, Hakenbeck R (eds) Bacterial cell wall. New comprehensive biochemistry, vol 27. Elsevier, Amsterdam, pp 397-423

Binet R, Létoffé S, Ghigo JM, Delepaire P, Wandersman C (1997) Protein secretion by Gram-negative bacterial ABC exporters - a review. Gene 192:7-11

Danese PN, Silhavy TJ (1998) Targeting and assembly of periplasmic and outer-membrane proteins in Escherichia coli. Annu Rev Genet 32:59-94

Delcour AH (1997) Function and modulation of bacterial porins: insight from electrophysiology. FEMS Microbiol Lett 151:115-123

Earhardt CF (1996) Uptake and metabolism of iron and molybdenum. In: Neidhart FC, Curtiss R III, Ingraham IL, Lin ECC, Low KB, Magasanik B, Reznikoff WS, Riley M, Schaechter M, Umbarger ME (eds) Escherichia coli and Salmonella: cellular and molecular biology, 2nd edn. ASM Press, Washington, pp 1075-1090

Griffith E, Williams P (1999) The iron-uptake systems of pathogenic bacteria, fungi and protozoa. In: Bullen JJ, Griffith E (eds) Iron and infection. 2nd edn. John Wiley, New York, pp 87-172

Lazdunski CJ, Bouveret E, Rigal A, Journet L, Lloubès R, Bénédetti H (1998) Colicin import into Escherichia coli cells. J Bacteriol 180:4993-5002

Postle K (1993) TonB protein and energy transduction between membranes. J Bioenerget Biomembr 25:591-601

Pugsley AP (1993) The complete general secretory pathway in Gram-negative bacteria. Microbiol Rev 57:50-108

Pugsley AP, Francetic O, Possot OM, Sauvonnet N, Hardie KR (1997) Recent progress and future directions in studies of the main terminal branch of the general secretory pathway in Gram-negative bacteria - a review. Gene 192:13-19

Seydel U, Schröder G, Brandenburg K (1989) Reconstitution of the lipid matrix of the outer membrane of Gram-negative bacteria as asymmetric planar bilayer. J Membr Biol 109:95-103

Van Voorst F, de Kruiff B (2000) Role of lipids in the translocation of proteins across membranes. Biochem J 347:601-612

Wandersman C (1996) Secretion across the bacterial outer membrane. In: Neidhart FC, Curtiss R III, Ingraham IL, Lin ECC, Low KB, Magasanik B, Reznikoff WS, Riley M, Schaechter M, Umbarger ME (eds) Escherichia coli and Salmonella: cellular and molecular biology, 2nd edn. ASM Press, Washington, pp 955-967

Wandersman C (1998) Protein and peptide secretion by ABC transporter. Res Microbiol 149:163-170

Wiese A, Seydel U (2000) Electrophysiological measurements on reconstituted membranes. In: Holst O (ed) Methods in molecular biology, vol 145: Bacterial toxins: methods and protocols. Humana Press, Totowa, pp 355-370

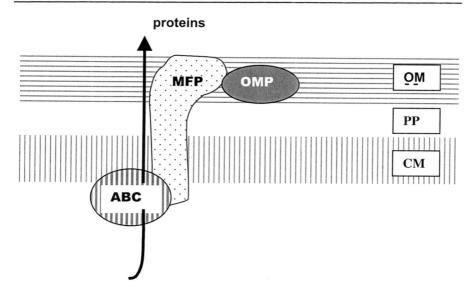

Fig. 8.4. Schematic presentation of an ABC transporter from Gram-negative bacteria. *ABC* ATP-binding casette; *CM*   cytoplasmic membrane; *MFP* membrane fusion protein; *OM*   outer membrane; *OMP*   outer membrane protein; *PP*   periplasmic space. (Wandersman C (1998) Protein and peptide secretion by ABC transporter. Res Microbiol *149*,163-170, with permission)

Young J, Holland IB (1999) ABC transporters: bacterial exporters-revisited five years on. Biochim Biophys Acta 1461:177-200

8.1.3 The Transport (Transduction) of Signals

In order to generate an optimal adaptation to their environment, bacteria have to react sensitively to it and to its changes. This occurs by means of recognition, transformation and emission of signals.

Three types of signal transduction can be differentiated:
1. Reaction to the nature of the environment, especially to attractants and repellents, for example by positive or negative chemotaxis with the help of flagellae. In this case, it is mostly not the concentration itself but rather a concentration gradient in the enormous range of $<10^{-10}$ - $>10^{-3}$ which represents the driving force.
2. Conversation of the bacteria among each other (mainly in colonies and biofilms).
3. Conversation of the bacteria with the host.

The basic function of signal transduction is to transform environmental stimuli into cellular signals. This transformation is mediated by sensor proteins and induces changes of the activity of regulator proteins. Sensory and regulatory functions may be combined in a single protein, but rather often large signal transduction chains are involved until the signal reaches the final effector protein.

The nature of the environmental stimuli transduced in these ways is very broad: simple physical (temperature, pH) or chemical (monosaccharides, amino acids) stimuli, but on the other hand also very specific signals. Many of the signalling molecules made by bacteria belong to the large group of secondary metabolites.

Initial recognition of the signalling molecule requires its binding to a specific receptor. Some of these molecules bind to a soluble receptor, or binding protein, located in the periplasmic space. The ligand-loaded binding proteins can interact with, or activate, a particular receptor bound in the cytoplasmic membrane. Other signalling molecules bind directly to a specific membrane receptor. In each case, the membrane-bound protein carries the signal to the cytoplasm and conveys the proper instructions to the rest of the system.

For the regulation of virulence and other adaptional and regulatory processes, sensor and regulator proteins very often represent a two-component system. The family of two-component systems is typically composed of a sensor kinase at the periplasmic side of the cytoplasmic membrane extending into the periplasma and a regulator protein in the cytoplasm. The sensor kinase catalyses an autokinase reaction, or transfer of phosphate from ATP to a histidine residue (histidine-protein kinase, HK). The HK-bound phosphate is then transferred to a specific aspartate residue on the cognate response regulator protein (RR). This reaction influences the binding of RR to its target, changes its oligomeric state, or activates its function. It is of interest how the cognate HK and RR recognise each other so efficiently and specifically.

The exchange of signals may not be only beneficial. Signals can be sent out to lure bacteria (for instance *E. coli* by *Myxococcus xanthus*) with the aim of killing and consuming them. Bacteriocins are another example of cell-cell signals resulting in lethal interspecies interaction.

To induce a reaction, the signal concentrations have to surpass a threshold. In the case of conversation of bacteria among each other, this may occur in two different ways. In the first, the signals (so-called autoinducers which may be peptides in Gram-positive bacteria and acylated homoserine lactones in Gram-negative bacteria) are continuously emitted by the singly growing bacteria; they cannot be effective due to their low concentration. Only at the emergence of higher bacterial densities, e.g. at desiccation, the concentration may reach its threshold value. This may take place during induction of the formation of spores or of biofilms. In the second way, signals are emitted only in the case of danger. If enough emitting cells are present, the threshold value is automatically reached. This pathway is observed in the case of SOS signals. The threshold values are mostly low in both cases; one nanomol or less are often sufficient.

The chemical conversation with higher organisms may be favourable for both partners, for example if it regulates symbiotic activities or limits the bacterial concentration in a host. The signals of pathogens, however, often induce less positive effects. They generate reactions of host-harming characters, disabling its defence mechanisms and enabling the pathogen to gain a foothold. A number of examples are given in Section 8.3.

The role of the cell wall in signal transduction may be either a passive or an active one. Frequently, the signal molecules are so small that they may easily penetrate even the Gram-negative cell walls via the porin pores, and the higher permeable Gram-positive cell wall does not represent any barrier for them. However, there are cases in which bacterial (cell wall) components may intervene in the signal transduction as a stimulus. The specific binding proteins located in the periplasmic space of Gram-negative bacteria or the periplasm-like space of the Gram-positive ones which represent components of sensor systems have already been mentioned. The location in this bacterial region is optimal for the binding proteins to form an excellent relay station for binding the signalling molecules and then produce cellular signals which are conducted inwards. Different cell envelope receptors may trigger the same intracellular events.

As an example, the transport of iron is mentioned. A soluble primary receptor (siderophore) is secreted by the cells; after binding of the ligand (Fe^{3+}) this receptor interacts with a particular cell wall-bound receptor and activates it, thus giving the signal for the induction of consecutive reactions.

Bibliography

Aizawa S-I, Haarwood CS, Kadner RJ (2000) Signaling components in bacterial locomotion and sensory reception. J Bacteriol 182:1459-1471
Dunny GM, Leonard BAB (1997) Cell-cell communication in gram-positive bacteria. Annu Rev Microbiol 51:527-564
Gross R (1993) Signal transduction and virulence regulation in human and animal pathogens. FEMS Microbiol Rev 104:301-326
Losick R, Kaiser D (1997) Why and how bacteria communicate. Sci Am 276:52-57

8.2 Protective Function of the Bacterial Cell Wall

As single-celled organisms, bacteria have to prevail individually in their struggle with the environment. Even in apparently inert surroundings, such as clean water, they are not out of danger since due to the large difference between the osmotic pressures outside and inside the cells, they are exposed to the risk of bursting. In ecosystems where single-cell organisms and simple multi-cell ones compete with each other, the protection has to be more far-reaching. Finally, a maximum protection is necessary for those bacteria which have settled in host organisms, since the defence mechanisms of the host, directed against the bacteria, have to be conquered. A great part of these protective functions is brought about by the cell wall. The protective function thus represents one of the most essential tasks of the bacterial cell wall. The importance of the cell wall is indicated by the fact that in Gram-positive bacteria it may make up to 30% of the total cell mass and that the biosynthesis of its components consumes quite considerable quantities of cellular substance and energy.

The enormous protective effectiveness of the bacterial cell walls is illustrated by a comparison between the mechanical stability of erythrocytes with that of bacteria.

The significance of the individual cell wall components for protection and stability of the bacterial cell can be proved by examining bacteria in which it is lacking. For this purpose, two possibilities exist. Firstly and predominantly, mutants can be used lacking the ability for biosynthesis or incorporation of the component. Secondly, it is sometimes possible to detach single substances from the cell wall by using cautious procedures. Even if both methods seem to be simple and easily comprehensible, difficulties may arise in applying them, due to mutual influencing of the components.

Some of the protective mechanisms caused by individual cell wall components are discussed later in the context of their functions. In this section, predominantly the protection given by the entire cell wall is regarded in detail, i.e. mainly the protective and the barrier function.

The mechanical and osmotic stability of the cell wall is essentially caused by the peptidoglycan. This can easily be proven by interrupting the biosynthesis of this substance, for example by addition of penicillin to the culture medium. The spheroplasts, formed in this case under loss of the bacterial cell shape, are stable only in isotonic solutions. After transmission into a hypotonic milieu they burst and release the cell interior.

The barrier functions of both the Gram-negative and the Gram-positive cell wall differ considerably; the latter is much more permeable. The outer membrane of the Gram-negative cell wall is of crucial importance for the low permeability. Its outer monolayer contains significant amounts of LPS which are decisive for the very effective barrier function of this membrane. LPS contain five to seven fatty acid residues per molecule as against only two in a phospholipid molecule. This makes the molecule more compact and, thus, interactions with adjacent molecules are reinforced. From this a higher packing density in the membrane plane results, and there are less temporary, hydrophobic gaps in the membrane caused by the lateral movement of the lipid molecules. A reduction in the lateral movement is also caused by the elevated percentage of saturated fatty acids in the LPS which also contributes to the high packing density of the membrane monolayer. All these factors make the membrane hardly penetrable for hydrophilic substances. In addition, the LPS molecules are held together not only by hydrophobic bonds in the membrane plane but also by the high number of Me^{2+} bridges between the anionic groups of the inner core region as well as by hydrogen bridges between the polysaccharide chains of the LPS. Both types of bridges create barriers against especially hydrophobic substances; the hydrogen bridges act as feltings in the polysaccharide layer.

Nevertheless, the outer membrane is not unassailable. The high density of negative charges on its surface makes it rather sensitive towards polycations. The reduction of the number of the negative charges, for example by integrating 4-amino-4-deoxy-L-arabinopyranose into the lipid A-Kdo region of the LPS (Sect. 2.2.4.3) lowers this sensitivity. Chelating agents like EDTA also may destroy the Me^{2+} bridges, remove part of the LPS, and thus lead to destructions in the cell wall.

As already mentioned, the "typical" Gram-positive cell wall basically consists of a thick peptidoglycan layer, in which polysaccharides (teichoic acids) and proteins are embedded, and in many cases S layers. Both layers may actually be penetrated by larger molecules, but nevertheless show some barrier functions.

Mycobacteria, which also belong to the Gram-positive bacteria, have an "untypical" and very penetration-resistant cell wall containing a thick layer of unusual lipids, but also some pore-forming proteins (Sect. 6.2.5).

The barrier function of the cell wall is associated to the vital processes of the bacteria. After their death, distinct changes in the permeability appear.

The periplasmic space of the Gram-negative bacteria or the periplasm-like space postulated for the Gram-positive ones between cytoplasmic membrane and peptidoglycan or S-layer (see Sect. 3.2.1.4) also cause certain protective functions. In the periplasm, the membrane-derived oligosaccharides (MDO, Sect. 3.1), for example, are located, contributing to the osmotic stabilisation. Degrading enzymes are especially enriched in the periplasm-like space which may degrade penetrated noxious substances.

8.3 The Importance of the Cell Wall for Pathogenicity and Virulence

During the contact of bacteria with a host organism, it depends on the nature and character of both whether or not a colonisation takes place. The bacteria have to multiply so fast that those killed by the defence mechanisms of the host can be more than just compensated. The hosts, on the other hand, must kill more bacteria than are able to grow. The course of the bacterial titre in the host represents the resultant of both processes operating against each other.

In the case of bacteria, the ability to colonise a host is not caused by a single property but by a whole complex of most miscellaneous performances. If one of these is missing, the germ may become incapable of colonising even though all others are present. This fact is sometimes overlooked.

The colonisation of a host mostly occurs via the following steps:
- adherence at the surfaces (skin, mucosae),
- survival at the surfaces and/or penetration into the tissue,
- multiplication within the host,
- undermining, repulse, or elimination of the defence mechanisms of the host.

Not all of these preconditions have in each case to be fulfilled. *Clostridium tetani*, for example, is not able to penetrate into the tissue. It mostly colonises in anaerobic areas of superficial wounds. Some bacteria succeed in colonising only if other microorganisms have been effective as pioneers (e.g. Staphylococci as followers of influenza viruses).

In this context, a further barrier against bacteria is worth mentioning which makes an infection much more difficult though it has nothing to do with the struggle host - microbe: the normal flora of the host. Many bacterial populations suppress the growth of others, a phenomenon called bacterial interference. The most frequent form of interference is the secretion of substances having toxic effects on other bacterial species, for example of bacteriocins. Deficiency of nutrients (inclusively oxygen) caused by the normal flora and competition for an attachment site, for example at the mucosa, represent additional possibilities.

The colonisation by bacteria may be beneficial, without greater meaning, or harmful for the host. Many bacteria of the normal intestinal flora are necessary for the hosts, and host and germ live in symbiosis. The bacteria may digest nutrients which are primarily indigestible into a form usable for the host or they may produce substances it needs (vitamins). These bacteria are called commensals; they differ from pathogens, i.e. from those causing damage to the host while colonising it, which may manifest as a disease. In this case, the process of colonisation is called infection, the ability of the bacteria for infection is termed pathogenicity or virulence.

Pathogenicity is a species- or serotype-specific, qualitative characteristic, giving the information whether the respective microbes basically have the capacity to cause disease in a defined macro-organism (according to a more recent definition: to cause damage in a host). A pathogenicity in itself does not exist, it is related to a particular host organism (e.g. a human pathogen).

Virulence is the quantitative expression of pathogenicity, it may distinctly vary from strain to strain within a serotype and even from population to population within one strain. Virulence is often indicated as the lethal dose, i.e. the number of microorganisms being sufficient to kill 50% of a host population within a relatively short period of time (LD_{50}). It is not to be confounded with the infective dose, i.e. the amount of bacteria being able to induce an infection.

Pathogenicity and virulence represent an expression for the degree of adaptation of the respective bacterial strain to the conditions of the host. They are indications of the interactions on the molecular level between bacteria and hosts and thus must be explained on this level.

To cause an infection, bacteria need the same preconditions as for   .
Another criterion, however, is added: the bacteria have to produce substances harming the hosts. The aim of this is not so much to strongly damage the macroorganism as a whole or even to kill it, but to reduce its defence forces to facilitate colonisation of the bacteria and thus to overcome the balance between bacteria and host. Therefore the production of substances which are noxious for the host is not a criterion exceeding the others mentioned but only a sideeffect, even though a very efficient one. There are cases in which strong (entero-)toxin producers multiply in a nutrient and generate so much toxin before uptake of the nutrient by the host that severe toxic reactions are caused even if a colonisation does not take place (e.g. botulism).

To cause infections, bacterial strains must have optimal combinations of specific properties. Formerly, it was supposed that these combinations and thus pathogenicity and virulence are serotype-dependent qualities since clear correlations between serotype and pathogenicity had been found. Certain combinations of O- and K-antigens thus seem to provide *E. coli* strains with an elevated aggressiveness. Meanwhile, it is known, however, that the epidemiological virulence of strains of the same serotype may be very different. Identity exists only within the clones, i.e. the progeny of one bacterial cell through asexual reproduction. Bacteria of virulent clones have acquired such optimal combinations of properties, presumably by gene transfer, and thus differ from strains of other clones from the same serotype. They are to a great extent qualified to spread in large territories thus to cause epidemics or even pandemics (clone conception).

Especially for energetic reasons, it is not appropriate that bacteria simultaneously produce all the cell (wall) components necessary for creation of optimal virulence properties, since the biosynthesis of most of these components is highly energy-consuming. This fact can easily be demonstrated by in vitro cultivation of bacteria in optimally composed nutrient media. Under such conditions the bacteria have deactivated the biosynthesis of a number of virulence-promoting components. During growth in vivo the individual environmental conditions change in the process of progressive infection. In this case, the bacteria produce only such components as they need in this special phase for survival and multiplication. Deviations from the correct order of biosyntheses may be noxious for the bacteria; thus fimbriae for example, though being necessary for the attachment process, are known to severely hamper the bacterial penetration into the host cells. On the other hand, components necessary for the protection against phagocytosis may cover substances needed for attachment and thus prevent adhesion. For this reason, bacteria must be able to sense their microenvironment and to produce nothing but all the required cell (wall) components in a given growth phase by means of switch-on/switch-off mechanisms.

Due to their exposed location, the cell walls and their components play an essential role in the development of pathogenicity and virulence. It is emphasised, however, that it would be wrong to suppose that the presence of an optimally developed cell wall would render the bacteria pathogenic and/or virulent. This is a necessary, but not a sufficient precondition.

Although, mainly due to genetic methods, it is often known which of the cell wall components are responsible for the respective virulence mechanism, only little information exists about the mechanisms themselves. Several techniques have been developed to study the infection processes. Theoretically, one has to distinguish between the cultivation of more or less completely developed bacteria and the test system itself. However, in practise both sides are often integrated in one system. In the past, several in vitro systems have been used. They suffer from the fact that the composition of bacteria grown in vitro and in vivo very often are not identical and that the tests cannot accurately reproduce all aspects of the host–pathogen interaction. Thus, physiological conditions have been created simulating the situation in vivo. For instance, tissue cultures have been used for this purpose. These are better than mere in vitro techniques but are hampered by the fact that pathogens may encounter different environments in the host body during the infection process and therefore the composition of its cell wall must not be constant over the whole time. Such a situation cannot be simulated in tissue cultures. Therefore, techniques have been developed to cultivate bacteria in animals and to test them at different phases; however, these techniques often are very labour-intensive, expensive, often highly animal-specific, and cause much pain for the animals. Several techniques have been developed recently that allow the in vivo examination of many virulence genes simultaneously. For this purpose, inbred strains have often been used. In addition, transgenic animals have been constructed that harbour human tissues.

8.3.1 Adherence of Bacteria at Surfaces

Basically, bacteria are able to attach to all inner and outer surfaces of the host, but infections mostly occur via colonisation of mucosae or via skin lesions. If the primary attachment sites are not situated on the body surface, it may be difficult for the bacteria to reach these sites at all. In the case of intestinal bacteria the host organisms have developed effective defence mechanisms, for instance extreme pH values in stomach and duodenum and production of bile salts.

In many cases, the attachment of the bacterial cells starts with a loose adhesion, which is relatively unspecific and may occur mechanically or via low-energy chemical bonds. For this purpose viscous substances or those with hydrophobic surfaces are suitable. Due to such loose attachment, the germs are easily removable and can be washed out, for instance by means of an elevated slime production or by fluids (e.g. urine).

The stronger adhesion occurring shortly afterwards is to a great extent both host- and tissue specific. Here, chemical bonds which may be connected and dissolved by relatively low activation energies play an essential role. The host-cell receptors are very often residues that are part of glycoproteins or glycolipids.

The substances involved in bacterial adhesion are of variable nature. However, they often enable reactions of specific receptors with complementary molecules called adhesins. These may be located on appendices (e.g. fimbriae) or immediately on the cell surface (AFA, NFA, see Sect. 6.4.2). The chemical nature of the adhesion-promoting substances is quite different. Apart from the proteins, bacterial surface polysaccharides like capsules, LPS or teichoic acids are also important for adhesion.

Fimbriae may either consist completely of lectin moieties or carry one or a few lectin molecules on their tip. Since lectins are complementary to an oligosaccharidic structure present at the binding site, the adhesion a very specific one and can be cleaved by the determinant monosaccharides. For a number of fimbriae of Salmonella and *E. coli* (for example the widespread type I-fimbriae), D-mannose represents the determinant monosaccharide, the adhesion mediated by it belongs to the large group of the so-called mannose-sensitive adhesions. On the other hand, one example of the mannose-resistant adhesion is the adhesion caused by the colonisation factor antigen I (CFA/I, fimbriae as well), in which the ganglioside GM_2 represents its receptor.

Adhesion may also occur via hydrophobic or electrostatic linkages. The bacterial cell wall as well as the fimbriae for this purpose often contain larger hydrophobic surface areas. The intermediate step of lipoteichoic acid release of Gram-positive bacteria described in Section 4.1.1.4 is recalled in this respect where the hydrophobic moieties of the acid points outwards, enabling the bacteria to bind to hydrophobic host surfaces. While blocking hydrophobic linkages, diarrhea caused by enterotoxigenic *E. coli* can be distinctly reduced in the animal model by administrating hydrophobic particles such as palmitoyl Sepharose.

Binding of the basic K88 and K99 antigens of *E. coli* to acidic polysaccharides of the intestinal mucosa occurs through specific electrostatic forces.

Many bacteria are able to express more than one type of adhesins, and it is the concerted action of these, often expressed at different stages in the course of infection, which is crucial for virulence.

Host and tissue specificity are mostly very high, even though not always absolute. Examples of the specificity are demonstrated by the fact that K88-carrying *E. coli* cause diarrhea predominantly in piglets, whereas those carrying K99 induce it in calves and lambs. On the other hand, the mouse-pathogenic Salmonella Typhimurium causes infections in humans as well (however, quite differently localised: bacteraemia in mice, gastro-enteritis in humans), and many urinary tract infections, especially in women, are induced by their own fecal *E. coli*.

Already in the moment of adhesion, the bacteria are often confronted with the defence mechanisms of the host, like a strong slime production for washing out the bacteria, the rejection of the uppermost cell layers, or the secretion of bacteriolytic enzymes, for example lysozyme in nasal secretions (which destroys the peptidoglycan network by cleaving the glycosidic linkage between *N*-acetylmuraminic acid and *N*-acetylglucosamine). To overcome these mechanisms and to start the infection process the bacteria from the very beginning must have an optimally composed cell wall.

Sublethal doses of antibiotics may prevent the adhesion of both Gram-positive and Gram-negative bacteria. This may occur by means of the prevention or decrease of adhesin formation (as described for *E. coli*), or by reduced incorporation and thus the reinforced secretion of cell wall components, such as that of lipoteichoic acids in Streptococci. The doses required for this lie below the bacteriostatic or the bactericidal concentrations. Low doses are beneficial since they reduce problems of antibiotic toxicity.

8.3.2 Survival at Surfaces and Penetration Into or Across the Tissue

The surface of the intact skin is very well shielded against attacks from the outside and therefore not a favourable place for bacterial growth. In most regions this is mainly due to the lack of water. Therefore the normal flora of the skin consists of mostly harmless Gram-positive bacteria belonging to only a few groups. Harmful bacteria are in most cases unable to survive and die. Therefore colonisation of the intact skin by such bacteria usually does not cause a disease. Injuries like, for instance, skin abrasions, burns, or insect bites (as for *Yersinia pestis*), however, represent gateways.

The inner surfaces of a macroorganism are easier assailable since they mostly consist of mucosae which are destined for the exchange of substances (gastrointestinal tract and respiratory tract). On their epithelial cells they carry protective layers predominantly composed of glycoproteins. The protection, however, is only a relative one since bacteria may also colonise on or in these layers. The epithelial cells respond to bacterial adherence in several ways, an important one of which is the production of cytokines. These substances are involved in inflammation and immunity processes with the aim to kill bacteria. An additional mechanism to remove (harmful) bacteria is the constant bathing of intact mucosal surfaces by secretions containing antibacterial enzymes and antibodies. On the other hand, the attachment may trigger a signal transduction occurring via the integrins of the eukaryotic cells and mediating the penetration of the bacteria into the tissue. Integrins represent transmembrane receptors.

Penetration of bacteria into the tissue is mostly an active process. Two possibilities can be differentiated:

1. The tissue is superficially destroyed, and the bacteria colonise only in the uppermost part (e.g. *Shigella flexneri*).

2. A passage of the bacteria into the tissue takes place (e.g. Salmonella Typhi).

There are bacteria that are able to alternatively use both possibilities (e.g. *Staphylococcus aureus*).

The penetration into the tissue is mostly not a single-step process, but facilitated by a cascade of events. In the case of *Shigella flexneri*, a secretory apparatus encoded by a large (220-kb) virulence plasmid is activated after the attachment at the cell surface, and subsequently formed Ipa-invasins make a rearrangement of the host cell cytoskeleton possible by means of interaction with the host membrane, including in particular a polymerisation of F-actin filaments that form bundles supporting the membrane projections which achieve the bacterial entry by endocytosis. The actin filaments are attached to one side of the bacterial cell and mediate the intracellular and intercellular migration of the bacteria. Although the invasins are constitutively produced, their secretion from the bacterial cell is dependent on a signal given by the intestinal cell of the host (Sansonetti 1992, 1998). The actin polymerisation is of importance not only in the case of Shigellae, but can be observed in many bacterial species. It is generated by a protein (ActA) located in the surface of older cell poles which is phosphorylated and thus activated after the entry of bacteria into the host.

In the case of Salmonella-induced enteritis, the invasion of the bacteria into the tissue also represents a very complex and multifactorial process in which Salmonella invasion proteins (SIP) and Salmonella outer proteins (SOP), secreted with the aid of the so-called type-III secretion system (TTSS), play a crucial role (for details see Wallis and Galyov 2000).

The further course of infection depends on the bacterial species. The germs may remain at the penetration site as a focus, or they may either reach a localised target via lymph or blood system or provoke a generalised infection.

To improve their dissemination in the macroorganism, the bacteria frequently excrete lytic enzymes that destroy the barriers of the body (cell walls, cementing substances). Worth mentioning are lipases (lecithinase, phospholipases, hemolysins, leucocidin) which attack the membranes of blood and tissue cells, as well as the hyaluronidase (also called spreading factor) which breaks up intercellular substances. Collagenase and streptokinase act similarly, the latter dissolving fibrin clots.

Those bacteria that penetrate into a tissue may be divided into two groups: i.e. the extracellular ones and the facultatively intracellular ones. The intracellular growth offers several advantages, e.g. the immune system is not effective there, the supply of nutrients is better in the cells, and other organisms are seldom present.

Penetration of pathogens into epithelial cells may occur in different ways. Frequently, an exchange of information (cross-talk) with the host cells takes place, for example by developing molecular trigger signals, which occurs by means of the phosphorylation of eukaryotic proteins and facilitates the penetration of the bacteria either directly or via influencing the signal transduction between the eukaryotic cells. If penetration does not succeed in this way, the bacteria have to destroy the surfaces of the epithelial cells by means of e.g. lytic enzymes.

Not in all cases do the bacteria have to invade into the epithelial tissue to provoke diseases. If bacteria excrete an effective toxin, its superficial colonisation may be sufficient to cause severe pathologic reactions in the host. In some cases bacterial growth and excretion of toxin do not occur in the host but in an appropriate nutrient. Although in this case no real infection takes place the bacteria may damage the host severely, even lethally. Typical representatives of bacteria which cause severe diseases via toxin production and excretion are *Vibrio cholerae*, *Neisseria gonorrhoeae*, *Corynebacterium diphtheriae*, *Clostridium tetani* and *C. botulinum*, and the enterotoxigenic *E. coli*.

8.3.3 Growth in the Host

Bacteria have to multiply so fast that the number of those eliminated by the host defence mechanisms is more than just replaced. A rapid growth and a high division rate are thus decisive for the bacteria. Optimal nutrient and environmental preconditions (pH, temperature, redox potential) at the site of infection are therefore advantageous; but even then, the in vivo growth rate seldom amounts to more than 10% of that measured in vitro.

The optimal utilisation of nutrients being present in the host is the precondition for a rapid growth of the bacteria. First of all, this requires a fast nutrient transport across the cell wall. As already mentioned (Sect. 8.1.2.1), Gram-negative bacteria dispose of several independent systems for this purpose. The unspecific porin pores allow the penetration of small molecules by diffusion without energy consumption, a way that is optimal in the presence of sufficient quantities of nutrients. If their concentration is too low or if the substances to be transported are too large, the synthesis of corresponding transport proteins takes place. While consuming energy, they pump the required substrates into the periplasmic space, eventually against a concentration gradient. From the energetic point of view, they are by far less effective than the diffusion pores. For rapid bacterial growth, however, they are indispensable.

The much higher permeability of the Gram-positive cell wall facilitates the diffusion of higher-molecular substances as well and seems to make active transport processes less important. However, these substances are frequently too large for transport across the cytoplasmic membrane. Therefore the bacteria secrete enzymes degrading them into transportable fragments. It is possible that some cell wall polysaccharides form a kind of channel between host tissue and the cell surface to prevent both the enzymes and the fragments from escaping into the surrounding medium and to provide sufficiently high concentrations of nutrients at the bacterial surface.

The hosts, on the other hand, tend to reduce the amount of nutrients suitable for and attainable by the bacteria. To give an example: among the trace elements that are necessary for the bacteria, iron is of crucial importance (see Sect. 8.1.2.1), and one mechanism of host defence against bacteria is to reduce the quantities of accessible iron ions present in the body fluids. Both hosts and bacterium have developed very effective mechanisms for the struggle for iron, in particular by the formation of highly efficient siderophores which can release the complex-bound Fe^{3+} of the respective antagonist and bind it in their own chelates. In addition, specific proteases may be synthesised which destroy the iron chelates and thus release Fe^{3+}. Finally, the strongly complex-forming Fe^{3+} may be reduced into Fe^{2+}, which forms only weak complexes.

The advantages of the ability of some pathogens to invade into the host cells and live there have already been discussed. For this ability a set of complex biological properties is necessary about which little is known at present.

8.3.4 Undermining, Repulse or Elimination of the Defence Mechanisms of the Host

There are various defence mechanisms of the host which consist of a number of unspecific and specific systems directed against the infecting bacterial agent. In both cases, cellular and humoral mechanisms are distinguished, however, combinations of both may occur.

Among the unspecific mechanisms are the natural resistance (which may have very different origins), the normal flora of the host, extreme pH values (in the stomach approximately pH 2), the production of bacteriolytic enzymes (e.g. lysozyme), inflammation, fever and many others.

The specific mechanisms are mainly represented by immune reactions. The major components of this response are antibodies and the complement system (complement represents a rather complex system of proteins present in serum or on surfaces of body cells which by means of several mechanisms may inactivate among others bacteria that have invaded into the macroorganism). Antibodies may be both humoral and cellular. They specifically bind the bacterial antigens under formation of antigen-antibody complexes. These complexes activate complement on the so-called classical pathway. Besides this, complement may be activated by the mere presence of bacteria on the alternative pathway (via surface components and the properdin system). The activated complement then kills and lyses the bacteria.

The transduction of the complement component C3 into the activated C3b, which is important in this process, is possible via both pathways. Many cell wall components (LPS, peptidoglycan, teichoic acids and others) may be activators. The activation of complement may induce phagocytosis (by the uptake and killing of bacteria by phagocytes via complement receptors) and/or via bacteriolysis (by the formation of lytic pores in the cell membrane by means of the membrane-attack complex; MAC).

Among the target cells representing components of the cellular immune system are the following:

- Polymorph-nuclear (PMN) leukocytes which for instance are able to enzymatically de-*O*-acetylate and de-*O*-phosphorylate the LPS and thus to detoxify them, as well as to secrete basic proteins for the neutralisation of LPS.
- B-Lymphocytes; they produce specific antibodies.
- T-Lymphocytes which produce lymphokines, especially γ-interferon.
- Monocytes and macrophages which secrete interleukins and the tumornecrosis factor α, and are able to produce secondary mediators such as oxygen radicals (including NO), prostaglandins, leukotrienes, and also proteases.
- Vascular cells; they produce interleukins as well as interferons.

To enable survival and growth of the bacteria, the defence mechanisms of the host must be hindered to come for effect. Several possibilities exist to achieve this goal, which all increase the efficiency of infections and make them more dangerous.

Bacteria are able to disguise themselves in such a way that the host does not recognise them as foreign and, besides others, does not form specific antibodies. The covering of the bacterial surface with polysaccharides, especially with anionic ones (mainly capsules), serves this intention. Another possibility is molecular mimicry, i.e. the biosynthesis of cell surface components which the host recognises as "self" (for instance, hyaluronic acid).

Usually, particles that are phagocytosed by a macrophage are present after ingestion in a compartment called the phagosome which fuses rather quickly with another compartment, the lysosome, to furnish the phagolysosome. Its interior is acidified, which contributes to the destruction of particles, e.g. bacteria. Pathogenic mycobacteria which have a remarkable capacity for infecting mammalian cells, primarily the macrophages, have developed tools that are not yet understood to prevent the fusion of phagosome and lysosome and, thus, to circumvent destruction.

In addition, bacteria try to escape the action of the immune system. This may be done by periodic changing their surface antigens (antigenic variation) or using the mentioned molecular mimicry. Another way is the use of the host´s own substances as a shield. Proteins, for instance, may be precipitated by means of staphylococcal coagulase, enclose the germs, thus disguise them, and fail to attack them.

A certain form of camouflage is represented by the growth within vacuoles; this possibility is practised, for example, by Chlamydia or Mycobacterium. In contrast to these bacteria, ingested Shigellae rapidly lyse endocytic vacuoles and multiply freely within the cytoplasm of infected mammalian cells.

Also, bacteria may produce particular substances, so-called aggressins, which inhibit the immune defence of the host. The term aggressin comprises members of the most diverse substance classes which must have at least one of the following qualities:
- They provide the bacteria with resistance against humoral antimicrobial agents (e.g. LPS, the polyglutaminic acid-capsule of *Bacillus anthracis*, capsular polysaccharides).
- They block the signal transduction between the macrophages (e.g. dephosphorylation of signal proteins by Yop, Sect. 8.1.2.2).

- They stop the mobilisation of phagocytes initiated by inflammation (e.g. peptidoglycan).
- They hinder the contact with the phagocytes and inhibit chemotaxis (e.g. capsular polysaccharides).
- They prevent the uptake (ingestion) and the subsequent destruction (phagocytosis) of the bacteria by phagocytes (e.g. polysaccharides).
- They inhibit the immune response or reduce its effectiveness (e.g. polysaccharides).

8.3.5 Damaging of the Host

The production of substances that generate severe damage in the host is not a prerequisite for the capacity of the bacteria to colonise a macroorganism. The damage rather represents side effects of these substances; it is primarily necessary to fulfil the criteria described in Sections 8.3.1-8.3.4 and may be induced by their directly toxic effects or via immunopathologic reactions.

It is conceivable that the incorporation of such substances into the bacterial cell wall occurs with the primary aim to resist the host´s own defence mechanisms, i.e. it represents rather a means of defence than of attack. Even the production of some exotoxins could exclusively serve this purpose. Due to their extreme toxicity, however, even the low secreted amounts have drastic effects on the host organism.

Some effector proteins excreted by several Gram-negative animal and plant pathogenic bacteria using the mentioned TTSS, for instance the SOPs, can also be considered to be toxins though they lack receptor-binding domains. They promote the invasion of the bacteria.

Most of the cell wall components are less toxic than the real exotoxins. In their bound form, they exhibit no direct toxic effects at all, but they develop toxic properties after having been released. Lysis of the bacteria, for example after administration of antibiotics, may cause such a release and render the substances toxic. In such a case the substances may become life-threatening. This may be illustrated by the toxicity of the mycobacterial cord factor and by LPS-caused septic shock.

It is even discussed that at least the LPS are not primarily toxic at all, but that their effect is induced by chaos reactions of the host organism which in themselves are unnecessary and uncontrolled. They thus belong to the category of immunopathologic reactions caused by cell wall components.

8.3.6 The Roles of Particular Cell Wall Components

The determination of the role of particular cell wall components in the fight between microbe and host is difficult for several reasons. As already mentioned, one single component of the bacteria or their cell walls can never be effective during the whole infection process; there are always several compounds acting in cooperation. The composition of the bacteria is not constant during the infection process, since these are able to optimise themselves using switch-on/switch-off mechanisms for the biosynthesis of individual cell wall components in dependence on the single phases of the infection process. In addition, false conclusions may be drawn from the fact that bacterial components which are chemically very similar but not identical can trigger rather different mechanisms.

8.3.6.1 Polysaccharides and Related Compounds

In many bacteria, substances containing oligo- or polysaccharide regions represent an essential part of the outermost layer. Capsular polysaccharides and (lipo)-teichoic acids are widespread in Gram-positive bacteria, capsular polysaccharides and lipopolysaccharides in Gram-negative ones. They may be required for a successful adherence to the host, represent protective agents for the bacteria, or may play a role as information carriers.

The protection of bacteria mediated by polysaccharides may be a passive or an active one. Envelopes consisting of acidic polysaccharides are especially efficient, in particular capsules. A passive protection is, e.g., the prevention of their identification as foreign by the host or of the accessibility by protective antibodies or complement of the cell surface. An active protection, for instance, is the blockade of the complement activation. This blockade occurs especially by inhibiting the formation of C3b by means of interaction with various complement components (e.g. C2bC4b) and may prevent uptake and killing of the bacteria by phagocytes occurring via complement receptors or by means of formation of lytic pores in the cell membrane by the MAC.

The complement resistance of encapsulated bacteria frequently cannot be conquered by antibodies against capsular polysaccharides. This probably depends on thickness and viscosity of the capsules which hinder the complement components to reach the cell surface. As an example, in *E. coli* and *Klebsiella* a clear correlation has been found between the magnitude of the capsule on the one hand and the serum sensitivity and virulence of the bacteria on the other. Highly encapsulated bacteria are often able to live and multiply within macrophages. Even the addition of acidic polysaccharides, for example mucin, to unencapsulated bacteria may increase their virulence.

The protection against phagocytosis caused by capsular polysaccharides can be conquered by opsonins. These generally represent components of the blood plasma which modify exogenous substances (for instance by covering them) in such a way that they may be more easily recognised by phagocytic cells and other cells of the immune system.

Many bacterial polysaccharides contain monosaccharide species which are quite unusual in nature. Therefore the host organisms contain no enzymes to split them off and/or to degrade them. This, too, is a reason why polysaccharide envelopes are difficult to detach and may provide a good protection for the bacteria.

The immunogenicity of polysaccharides, i.e. their ability to stimulate the production of specific antibodies in the host organism, depends on a number of parameters: the molecular size (enlarged by formation of homo- and/or heteroaggregates), the physical state (charge, hydrophobicity), the resistance against the host´s own enzymes, the given dose, but also nature and age of the host. One example: the O-specific polysaccharides of the LPS with a molecular mass of 10-20 kDa solely are not immunogenic, whereas complete LPS frequently induce antibody formation, especially due to their tendency to form high-molecular aggregates. The reason for the lacking ability of several capsular polysaccharides (e.g. *E. coli* K1 and K5) to induce antibody production will also be mentioned.

The nature of the produced antibodies depends on several parameters. Antiprotein antibodies are mostly of the IgG type, antipolysaccharide antibodies mostly of the IgM type. This, however, is not generally the case. Injection of encapsulated bacteria into host organisms in general induces the formation of protective anticapsule IgG-antibodies. On the contrary, antibodies produced by immunisation with purified CPS are of the IgM type. Finally, in the case of some animal species (rabbits) or of human babies, formation of anticapsule antibodies may not take place at all.

Polysaccharides tend much more to create immunologic paralysis (immunotolerance) than proteins. Small doses (1 µg per mouse) of pneumococcal polysaccharides induce a specific vaccination protection; however, the much higher dose of 500 µg does not. One reason for this is the "treadmill" effect which depends on the fact that polysaccharides are difficult to degrade. The "treadmill" effect means that polysaccharides initially react with the antibodies produced against them and the antigen-antibody complexes formed thus are subsequently phagocytosed by the host. Because of the stability of the polysaccharides against enzymatic degradation, only the antibodies are digested by the macrophages; the polysaccharides are released and able to bind antibodies again. Due to the thus possible complete exhaustion of the antibody-producing cells, the immune apparatus finally no longer regards the polysaccharides as foreign and stops antibody production.

The rather common assumption that polysaccharide surfaces should generally be hydrophilic is not correct. In the case of the helix of the O-specific polysaccharides from *Salmonella enterica* strains of groups A, B or D1, which has already been described in Section 2.2.4.1, the deoxy groups of the 3,6-dideoxyhexoses and of rhamnose form large hydrophobic surface areas. These areas facilitate the unspecific attachment at hydrophobic host structures or possibly the penetration across hydrophobic membranes of the host cells. A similar role is attributed to the sequence GlcNAc-Rha-Rha-Rha for the penetration of *Shigella flexneri* into the uppermost cell layer of the intestinal mucosa. After placing the *wba* locus of *S. flexneri* that determines this sequence into Ra mutants of *E. coli*, the recombinants achieve the ability to penetrate. Finally the O-specific polysaccharide of the LPS from *Legionella pneumophila* is worth mentioning which consists of an α-(2→4)-interlinked 5-acetamidino-7-acetamido-8-*O*-acetyl-3,5,7,9-tetradeoxy-L-*glycero*-D-*galacto*-non-2-ulosonic acid (legionaminic acid) homopolymer, which completely lacks free hydroxyl groups and thus is very hydrophobic.

8.3.6.2 Lipids (Including Lipid A)

As essential components of many bacterial cell walls, lipids are of great importance for pathogenicity and virulence. They develop their activity in different ways: as protecting components, by means of their toxicity, or by inducing hypersensitivity reactions.

The incorporation of lipids renders the cell wall more resistant against hydrophilic substances from the defence system of the host organisms. However, the penetration of nutrients across this cell wall may likewise be reduced. This proves also to be disadvantageous, the extremely low growth rate of many Mycobacteria caused in this way represents an example.

Many bacterial lipids are toxic. Mycolic acids, especially in the form of the already mentioned cord factor of, e.g., *Mycobacterium tuberculosis* are well known. They develop their activity by interaction with mitochondrial membranes and by specific inhibition of the oxidative phosphorylation. Though the cord factor may cause haemorraghiae and even death, it is of no influence on the bacterial virulence.

The reactions proceeding after application of LPS (endotoxins) into host organisms are particularly well examined. Endotoxins are regarded as not primarily toxic by some authors. Small amounts released from the bacteria may induce beneficial reactions in the host. They represent an important and very sensitive indicator for an infection and may initiate early protective mechanisms thus preventing disastrous consequences. Larger quantities, however, induce immunologic chaos reactions leading to severe damage in the host up to its death. In this process lipid A plays the crucial role.

Two groups of factors reacting with LPS are present in the serum. The first group reduces the effects of the LPS, the second group increases them and induces subsequent reactions. Within the second group, especially two host proteins are to be mentioned: the LPS-binding protein (LBP) and the leukocyte receptor protein CD14. LBP was discovered as a 60-kDa acute phase serum glycoprotein. CD14 occurs both as a 55-kDa glycosylphosphatidylinositol (GPI) anchored cell surface receptor glycoprotein named membrane CD14 (m-CD14) and a soluble form that usually exists in serum (sCD14) which is involved in LPS stimulation of cells lacking mCD14, e.g. endothelial cells.

Binding of LPS to mCD14 on monocytes is thought to be a prerequisite for their stimulation and the release of immune modulators. Shortly described, LPS present as autaggregates are bound to LBP, thus deaggregated, and the LBP/LPS adducts are passed on to the target cells, where they interact with mCD14 (note that recently the presence of additional receptors on different cell types have been postulated). The mechanism of cell activation after this binding is still not understood. Since mCD14 is a GPI-anchored protein without a transmembrane domain, it is not able to transmit a signal through the membrane. A good candidate for this process is the Toll-like receptor TLR-2 which very recently has been shown to be involved in LPS-induced cell activation. Its function depends on LBP and is enhanced by CD14.

In small amounts, the mediators induce resistance against the infections, in larger quantities, however, severe clinical complications up to septic shock.

In a recent publication (Khan et al. 1998) it is shown that LPS bound in the bacterial cell wall on the one hand and released LPS on the other are not comparable to each other with regard to their biological effect. It is also proven that lipid A is unambiguously responsible for the lethality of Salmonella infections and that the loss (msbB mutation) of 3-hydroxymyristic acid (3-OH-C14)-bound myristic acid does not, in fact, restrain the bacterial growth in the hosts, but drastically reduces the lethality of the infection.

A lot of bacterial lipids do not show any primary toxicity. Nevertheless, if having entered the macroorganism they are able to sensitise it. A second dose may trigger immunologic reactions which are possibly life-threatening. Such reactions which may be induced in the same way by proteins or other, even low-molecular, substances are able to essentially dramatise the course of diseases.

Braun´s lipoprotein also represents a very effective mitogen, its smallest active fragment being tripalmitoyl-cysteine (Seltmann and Rietschel 1988).

Hopanoids are believed to be the most abundant natural products of defined structure and have been recognised as the molecular fossils of a broad variety of bacteria, including Gram-negative genera like Acetobacter and Zymomonas, and several species from Rhodospirillaceae. Hopanoids are amphiphilic molecules. They are present in bacterial membranes, where they contribute to membrane stability. At least in one case, a hopanoid was coextracted with LPS, indicating their possible presence also in the outer membrane. For bacteria that contain hopanoids, their presence is essential for survival.

Mycobacterial lipoarabinomannan may, depending on its structure, represent an immunologically active substance which blocks a number of host defense mechanisms, e.g. T-cell activation. Thus, it may be important for the infection process.

8.3.6.3 Proteins

In the absence of S-layers, the outermost layer of mainly Gram-negative bacteria is much more seldom composed of proteins than of polysaccharides or lipids. They may be rather easily recognised as foreign by the immunological defence and readily degraded by proteases. Therefore protein envelopes are unsuitable as protective substances. A certain protection against proteases may be achieved by the incorporation of non-proteinogenic amino acids or by formation of unusual linkages. The capsule of *Bacillus anthracis* consisting of γ-D-glutamic acid-polypeptides may serve as an example. Another kind of protection is caused by the embedding of proteins into (glyco-)lipid layers, thus, enterotoxins are often excreted into the intestine in vesicles formed by LPS (vesicles as vehicles).

Nevertheless, cell wall proteins play an important role for the infection process mainly in "typical" Gram-positive bacteria. Protein A of *Staphylococcus aureus* and protein G of *Streptococcus pyogenes* react with the F_c fragment of IgG and thus block the region activating the complement system. The coagulase of *Staphylococcus aureus* causes the formation of a fibrin network around the focus and thus shields the bacteria from the host´s defence systems. In the same way, the fibronectin-binding protein of Streptococci creates a fibronectin layer for the bacteria on their surface. The streptococcal M-protein prevents the activation of complement by a blocking interaction with one of its components (factor H). In addition, it is believed to play a role during the bacterial attachment.

The S-proteins, e.g. of *Campylobacter fetus*, are examples in the case of Gram-negative bacteria. They prevent binding of the complement component C3b to the cell surface.

Besides proteins with protective functions there are others which actively support the infection process. Among them are the lytic enzymes (Sect. 8.3.2) and the invasins. Both are frequently found in invasive bacteria. They are mostly secreted by the bacterial cell, but may also be bound in the cell wall. The invasins switch on signal transduction pathways in intestinal cells that cause takeup of the bacteria.

The fimbriae and the "non-fimbrial adhesins" with their attaching functions are cell surface proteins of both Gram-positive and Gram-negative bacteria. In some fimbrial types, they may be formed completely of subunit proteins (fimbrins) of lectin character (major subunit adhesin); in other types, there are a few lectin molecules localised merely on their tip (minor subunit adhesin). In both cases, fimbriae mediate the binding of bacteria to oligosaccharide structures of the host surfaces. Examples for non-fimbrial adhesins are the M-protein of the streptococcal cell wall, the a-fimbrial adhesins of *E. coli*, and a number of adhesins (Inv, Ail and others) in the case of Yersiniae. The already mentioned Yersinia outer proteins (Yop) represent adhesins, but are only for a small part cell wall-bound. The Ptx-protein of *Bordetella pertussis* has a double function, serving as an adhesin in the cell wall-bound form and as a toxin when released.

In mycobacteria, the thick mycolic acid layer acts as an effective penetration barrier with the same function as the outer membrane in Gram-negative bacteria. To solve the transport problems, there should also be proteins enabling the penetration of low molecular nutrients and/or waste products, i.e. carrying out the same functions as the porin proteins. Such proteins were indeed found. They are mainly present in fast-growing mycobacteria such as *M. tuberculosis*.

A 23-kDa protein is associated with the peptidoglycan and shows a strong similarity to OmpF of *E. coli*. In the case of *Mycobacterium chelonae*, a 59-kDa protein with a channel function (2.2 nm in diameter) and a slight selectivity for cations has been described. In contrast to the porin proteins of Gram-negative bacteria, it is only present in small amounts and at least one tenth less permeable. The presence of a similar protein is also supposed for *M. tuberculosis*.

Tuberculin, which formerly had been regarded as a mycobacterial cell wall protein, proved to be a mixture of toxins and cell wall components. Even the PPD (purified protein derivative) obtained from it by precipitation is not homogeneous. It is able to induce the formation of antibodies after parenteral injection. Its importance, however, is due to the fact that it induces an allergic cutaneous reaction in sensitised organisms on intracutaneous application, which may indicate an existing tuberculosis.

8.3.6.4 Peptidoglycan

The macromolecular murein sacculus represents an efficient mechanical protection for the cell. Moreover, the peptidoglycan or its soluble cleavage products may show a considerable immunologic activity. They may represent a mitogen, be polyclonal activators of B-cells, or induce the induction of cytokine secretion in macrophages. Like LPS, they bind to CD14 and induce secretion of interleukins and the tumornecrosis factor α. As immunoadjuvants they are able to replace Mycobacteriae (e.g. in Freund´s adjuvant).

During the degradation of the peptidoglycan molecules in the course of cell enlargement and multiplication (see Sect. 3.2.2.5), only a minor part of the low-molecular material is released by the bacteria. One of the cleavage products, namely β-GlcNAc-(1→4)-MurNAc-L-Ala-D-isoGlu, was isolated from both the pathogenic *Staphylococcus aureus* and the apathogenic *Bacillus subtilis*. The isolates turned out to be an amplifier of NO formation induced by lipoteichoic acids, of septic shock and of organ failure. The smaller molecule MurNAc-L-Ala-D-isoGlu (muramyl-dipeptide) represents an immunoadjuvant.

Bibliography

El-Samalouti VT, Hamann L, Flad H-D, Ulmer AJ (2000) The biology of endotoxin. In: Holst O (ed) Methods in molecular biology, vol 145: Bacterial toxins: methods and protocols. Humana Press, Totowa, pp 287-309

Kartmann B, Stengler S, Niederweis M (1999) Porins in the cell wall of *Mycobacterium tuberculosis*. J Bacteriol 181:6543-6546

Khan SA, Everest P, Servos S, Foxwell N, Zähringer U, Brade H, Rietschel ET, Dougan G, Charles IG, Maskell DJ (1998) A lethal role for lipid A in *Salmonella* infections. Mol Microbiol 29:571-579

Klemm P, Schembri MA (2000) Bacterial adhesins: function and structure. Int J Med Microbiol 290:27-35

Molinari G, Chhatwal GS (1999) Streptococcal invasion. Curr Opin Microbiol 2:56-61

Passador L, Iglewski BH (1996). Bacterial pathogenesis. In: Myers RA (ed) Encycylopedia of molecular biology and molecular medicine. VCH, Weinheim, pp 103-108

Sansonetti PJ (1992) Molecular and cellular biology of *Shigella flexneri* invasiveness: from cell assay systems to shigellosis. Curr Top Microbiol Immunol 180:1-19

Sansonetti PJ (1998) Molecular bases of epithelial cell invasion by *Shigella flexneri*. Antonie van Leeuwenhoek Int J Gen Mol Microbiol 74:191-197

Seltmann G, Rietschel ET (1988) The outer membrane of Gram-negative bacteria with emphasis on its lipopolysaccharide and lipoprotein components. Wirkstofforsch/Wissenschaftspubl, Martin Luther Universität, Halle, pp 77-112

Wallis TS, Galyov EE (2000) Molecular basis of Salmonella-induced enteritis. Mol Microbiol 36:997-1005

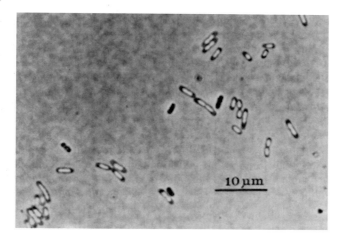

Fig. 8.5. Micrograph of bacteria from a growing *E. coli* culture. (Donachie WA, Robinson AC (1987) Cell division: Parameter values and the process. In: Neidhart FC, Ingraham IL, Low KB, Magasanik B, Schaechter M, Umbarger ME (eds) *Escherichia coli* and *Salmonella*: cellular and molecular biology. ASM Press, Washington, pp 1578-1593, with permission). Cells were grown in minimal medium in the presence of chloramphenicol (to arrest cell growth and induce nucleoid condensation), nalidixic acid (to prevent completion of chromosome replication) and penicillin G (to prevent cell division)

8.4 Significance of the Cell Wall for the Maintenance of the Bacterial Shape

Each bacterial species has its characteristic shape which has been phylogenetically developed and is genetically determined. However, it is not completely constant. Examination under a microscope of samples taken from a growing and multiplying bacterial monoculture shows a mixture of organisms of diverse sizes.

In the bacterial kingdom, basically two forms can be found, i.e. the spherical (coccoid) and the elongated (rod-shaped) one. Though the coccoid form is mechanically more stable and/or energetically more probable, many bacteria grow in elongated forms. Possibly this is advantageous to the bacterium, due to the resulting increase in the surface to volume ratio.

8.4.1 Rod-Shaped Bacteria

Rod-shaped bacteria, such as normally growing *E. coli* cells, form cylindrical particles with rounded poles. Their diameter is largely constant; however, their length varies between a minimum L and its double value. In addition, dividing cells of an approximate length of 2 L can be found (Fig. 8.5). To define the bacterial shape, either an average obtained from cells of all sizes has to be calculated or it must be related to a defined growth phase. Thus, for instance, cells immediately before or immediately after division could be characteristic for this definition. Sometimes the ratio length to diameter is used as a characteristic, but this is also of limited significance. Its value may change as there exist dependencies of the bacterial shape on the velocity of their growth and division. When entering the stationary phase, for instance, *E. coli* cells become much smaller and almost spherical.

Maintenance of the bacterial shape has two different aspects, namely its conservation and passing on during the multiplication process as well as its constancy and mechanical stability of the mature form. One precondition for shape maintenance is that all cell components are synthesised in the required quantity at the same time and after their biosynthesis are transported exclusively to their destination site. For this purpose, the cells need regulating and controlling mechanisms, which are especially active during the multiplication process; however, the biochemical basis of these mechanisms is still unclear to a great extent.

Essentially one component of the cell wall is important for shape and mechanical stabilisation of the form, namely the peptidoglycan (murein) of the bacteria and the similarly constructed pseudomurein of some archaea. Both are to a high degree appropriate for this task due to their stable spatial structure. In the case of Gram-positive bacteria they possibly act in cooperation with teichoic acids. In the case of the less robust peptidoglycan of the Gram-negative bacteria some OMP (Braun´s lipoprotein, OmpA; see for example Sect. 2.3.2.4.1) could have an additionally stabilising effect.

The maintenance of the shape during multiplication requires a coordinated and controlled sequence of growth and division while keeping to the assigned spatial pattern (template). The enzyme system responsible for this function may be influenced by drastic changes in the bacterial environment (stress), and temporary modifications in the bacterial morphology may be found induced by these changes.

The multiplication process of the bacteria starts with a length growth from L to 2 L at a constant diameter. After its termination, the cell division begins with the replication of the chromosome and the spatial separation of the two daughter chromosomes. The precondition for this process is the arrival of a defined minimal value of the ratio cell length to cell diameter or of a defined minimal mass (initiation mass). Subsequently, in the middle of the cell, a coordinated ingrowth of two chemically interlinked peptidoglycan layers and of a cytoplasmic membrane each on both sides of them occurs under formation of a cross-wall in the septum region (for details see Sect. 3.2.2.5). After its completion, both peptidoglycan layers are gradually enzymatically separated from each other, starting from the outside of the septum region. The murein in this region is simultaneously stretched by the bacterial inner pressure and the poles thus become bowed out. In the case of Gram-negative bacteria, outer membranes are formed and anchored in parallel to the separation process onto the peptidoglycan. In this way, the new poles of the two daughter cells are formed.

Cell division is a complicated collaborative interplay of several mechanisms with control systems independent of each other which up to now are understood only in few cases. For division, the cells produce specific regulatory proteins encoded by a single operon (*mra*). The individual mechanisms can be switched off without blocking the process of growth and multiplication on the whole. If the cross-wall formation is inhibited by addition of the antibiotic cephalexin, for example, long filaments with 2^n (n = number of the prevented cell divisions) clearly separated nucleoids are formed. On the other hand, spherical (coccoid) cells can be generated in *E. coli* by preventing the growth of the cylindrical region. Finally, it is possible to prevent the spatial separation of the daughter chromosomes and to induce septum formation in the middle between the double chromosome (generated by simple division of the chromosome) and the poles.

One model for septum formation in *E. coli* proposed by Höltje (1993) postulates that the polar caps and the cylindrical region are surrounded by peptidoglycan monolayers whereas the septal regions are trilaminar. Starting from the in-between theory (Sect. 3.2.2.5) he developed a model of the septum formation (three for one mechanism, Fig. 8.6). Prefabricated triplets are stepwise bound from the inside of the bacteria to a template strand present exclusively at the site of future cell division. In this way, some kind of cross-linked peptidoglycan bilayer is formed which expands as a cross-wall into the interior of the cell. After completion of the cross-wall synthesis the template strands are removed by cooperative action of both a muramidase and an endopeptidase step by step from the outside to inside and, thus, in the septum plane the two peptidoglycan layers are separated from each other. After formation of both membranes the pole caps of the daughter cells are completed.

To determine the location of septum formation, a set of proteins exists. The regulation obviously occurs in such a way that the septum formation does not take place too proximate either to the DNA or to the poles. According to one model, the distance of the site of septum formation has to be half of the distance between the two daughter chromosomes (i.e. about ½ L). In all other regions of the cell, the septum formation is presumably prevented by regulator proteins (encoded by the *min* locus). The above-mentioned septum formations between double chromosome and pole can also be explained on the basis of this model.

It may be speculated that in rod-shaped bacteria the murein sacculus at the septa or the poles on the one hand and the cylindrical regions on the other are chemically differently composed. This, however, seems not to be the case. The only difference is that in zones of length growth, predominantly tetra-tri bridges are present (see Sect. 3.2.2.4); in zones of division, however, predominantly tetra-tetra bridges. Obviously, the enzymatic regulation is in this case rather clearly arranged: PBP2 (see Sect. 3.2.2.5) is responsible for cell elongation, perhaps in cooperation with RodA, PBP1B for the initiation of septum formation and PBP3 for septum formation itself, possibly in collaboration with other membrane proteins.

8.4.2 Spherical (Coccoid) Bacteria

By means of electron microscopic examination of the division of coccoid bacteria, three different types have been detected:
- *Streptococcus* type: as in the case of rod-shaped bacteria, only one division under formation of two daughter cells.

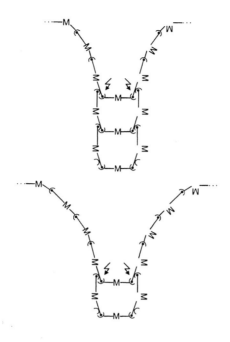

Donor peptide, ⟶•; Acceptor peptide, ⟶⊂;
Cleavage site of enzyme specially recognizing
trimeric crossbridges, ⟩⟨.

Fig. 8.6. Model of septum formation according to the „three for one" mechanism for Gram-negative bacteria (Labischinski H, Maidhof H (1994) Bacterial peptidoglycan: overview and evolving concepts. In: Ghuysen JM, Hakenbeck R (eds) Bacterial cell wall. New comprehensive biochemistry, vol 27. Elsevier, Amsterdam, pp 23-38, with permission). A projection along the glycan strands (*M*) running perpendicular to the plane of the drawing

- *Pediococcus* type; successive divisions in two spatial directions vertical to each other under formation of four daughter cells.
- *Sarcina* type; successive divisions in three spatial directions vertical to each other under formation of eight daughter cells.

These three divisional types logically involve three types of cell growth as well; i.e. in one, two or three spatial directions.

Staphylococcus aureus belongs (Fig. 8.7) to the latter type. Its cell division is initiated asymmetrically at one single starting point beneath the peripheral cell wall, followed by a centripetal growth of the cross-wall. After closure of this wall, the peripheral cell wall appears in most cases as a rather compact, homogeneous-looking structure. Nevertheless, it consists of two layers: the outer primary wall and the inner secondary wall (Fig. 8.8). The latter continues into the cross-wall.

Between both walls the so-called stripping system is located which is involved in the cell wall turnover. The cross-wall contains in addition the splitting system which is responsible for cell separation. It consists of concentrically arranged ring-shaped tubuli. Its chemical composition is not completely elucidated. It is supposed that lipoteichoic acids are associated with this system. Minute wall organelles, the murosomes, are located in two circumferential rows above the closed cross-wall. They may play an important role in lytic processes during cross-wall formation and initiation of cell separation.

The molecular mechanisms occurring especially in peptidoglycan during cellular growth, cross-wall formation, and separation of the daughter cells essentially more or less correspond to those described for rod-shaped bacteria. Thus, for example, endo-β-N-acetylglucosaminidase and N-acetylmuramyl-L-alanine amidase have been detected as murosome-associated autolysins in the case of *Staphylococcus aureus*. The diverse spatial orientations of the glycan strands of the single peptidoglycan layers in Gram-positive coccoids (Fig. 7.1) makes the stretching possible and thus the growth into more than one spatial direction (difference to that of rods).

Cell division does not in all cases result in a complete separation of the cells from each other. Some Gram-positive bacteria thus grow as clusters (staphylococci) or as strings of cells (streptococci).

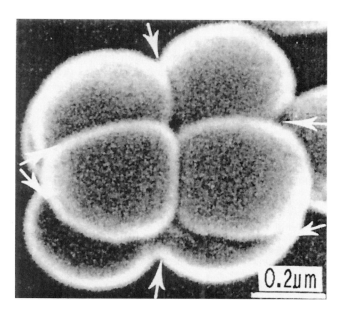

Fig. 8.7. Scanning electron micrograph of dividing *S. aureus* cells. *Arrows* Division planes (courtesy J. Wecke)

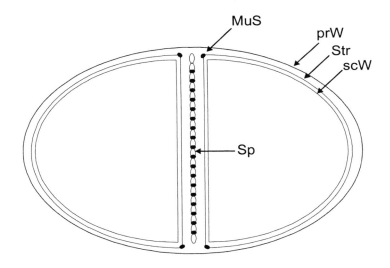

Fig. 8.8. Model of a dividing *S. aureus* cell (Giesbrecht P, Kersten T, Maidhof H, Wecke J (1998) Staphylococcal cell wall: morphogenesis and fatal variations in the presence of penicillin. Microbiol Mol Biol Rev *62*,1371-1414, with permission). *prW* Primary wall; *scW* secondary wall; *Str* stripping system; *Sp* splitting system; *MuS* murosome

In some cases, the cell division process is even more complicated than described. In the case of *Bacillus*, a sporulation may take place under certain circumstances, and in the case of *Caulobacter crescentes* two non-identical daughter cells are formed.

Bibliography

Ayala JA, Garrido T, de Pedro MA, Vicente M (1994) Molecular biology of bacterial septation. In: Ghuysen J-M, Hakenbeck R (eds) Bacterial cell wall. New comprehensive biochemistry, vol 27. Elsevier, Amsterdam, pp 73-101

Donachie WA (1993) The cell cycle of *Escherichia coli*. Annu Rev Microbiol 47:199-230

Donachie WA, Robinson AC (1987) Cell division: Parameter values and the process. In: Neidhart FC, Ingraham IL, Low KB, Magasanik B, Schaechter M, Umbarger ME (eds) *Escherichia coli* and *Salmonella*: cellular and molecular biology. ASM Press, Washington, pp 1578-1593

Giesbrecht P, Kersten T, Maidhof H, Wecke J (1998) Staphylococcal cell wall: morphogenesis and fatal variations in the presence of penicillin. Microbiol Mol Biol Rev 62:1371-1414

Höltje JV (1993) "Three for one" – a simple growth mechanism that guarantees a precise copy of the thin, rod-shaped murein sacculus of *Escherichia coli*. In: dePedro MA, Höltje JV, Löffelhardt W (eds) Bacterial growth and lysis. Plenum Press, New York, pp 419-426

Labischinski H, Maidhof H (1994) Bacterial peptidoglycan: overview and evolving concepts. In: Ghuysen JM, Hakenbeck R (eds) Bacterial cell wall. New comprehensive biochemistry, vol 27. Elsevier, Amsterdam, pp 23-38

8.5 Interactions Between the Bacterial Cell Wall and Bacteriophages

Bacteriophages (phages) represent DNA- or RNA-containing viruses which infect bacteria and multiply within them. Frequently, they consist of a head containing the nucleic acid and a tail being important for its transmission; however, the tail is not present in all phage types. The head envelope of the phages is often furnished by a protein, the tail is tubular and consists also of protein-containing materials. A basal plate with spikes and fibres is often located at the end of the tail. The phages attach at a receptor present on the bacterial surface and inject their nucleic acid into the bacterial interior. In the case of virulent phages, subsequently a change of the bacterial metabolism occurs, resulting in the formation of new phages within the bacteria. After their maturation, the bacteria burst and release the phages which then infect further bacteria. On the contrary, in the case of temperent phages, bacterial metabolism is not changed after nucleic acid injection, but the latter is integrated in the genetic material (chromosome) of the bacteria. This increase in genetic information partially leads to distinct alterations in bacterial activities (see below), especially to development of resistance towards a secondary infection by the same phage. After integration of the phage, the bacteria are called lysogenic, and the phages, which are latently present in the bacteria, are termed prophages. Together with the genetic material of the bacteria, prophages are transmitted to all descendants during cell division. Under certain conditions, such as UV irradiation, the prophage may become virulent (lytic) and initiate its multiplication process.

To inject its nucleic acid into the bacterium, the phage must attach specifically to the cell wall. This often occurs in a two-step process. In the first step, as a result of an occasional collision, an unspecific and reversible adsorption of the phage occurs at the bacterial surface. For this process, the distal ends of the tail fibres are often important. During the second step, the binding of the phage turns over into a specific irreversible one at a receptor that is characteristic for each phage. This binding involves the spikes of the basal plates. Tail-less phages have their attachment instruments at exposed surface sites. The presence of cations is necessary for the binding, e.g. T2-phages require monovalent metal cations, T4-phages Ca^{2+}. Finally, the injection of the nucleic acid takes place.

A number of phages have their receptor at the tip of pili or flagellae. Since injection of nucleic acid into the bacterial body is hardly possible from this site, it is supposed that binding to a second receptor present directly at the bacterial surface occurs beforehand.

The receptors are extraordinarily characteristic and specific for individual phages. Phage and receptor are sterically adapted to each other. Therefore the phage can be characterised by the receptor as well as the receptor by the phage. This is of both scientific and of practical importance. By means of phages it can often be tested much more easily than by chemical-analytical procedures whether or not a particular substance is present in the cell wall. For this purpose, suspensions of the bacteria in question are spread as a lawn on nutrient agar surfaces and the phage solution is dropped onto it. After incubation for 18 h, the bacteria containing the substance as receptor have not grown at the dropping site of the phage solution, which appears as a dark spot on the light bacterial lawn. As a precondition for this the bacteria must not contain substances, such as restriction endonucleases, hindering the phage nucleic acid from developing its effectiveness and the receptor may not be covered by other substances, e.g. capsules. The possible effect of restriction endonucleases may be excluded by testing the binding capacity of the bacterial cells or cell walls for the phage.

A further characterisation of the receptor substance may occur in two different ways. Firstly, the relevant substances may be isolated and added to the system bacteria/phages. Due to the competition for the phage between the added receptor and that present on the bacterial surface, an apparent decrease in the efficiency of the phage results. The method is limited, since frequently not a single component but rather a complex consisting of several, partly very different components arranged in a sterically adequate assembly are a prerequisite for the specific and irreversible binding of the phage. The second way is to test mutants lacking either the receptor or parts of it for their lysis by the phage. A variant of the second way is the chemical or genetic modification of the receptors and its addition to the system bacteria/phages. In this way, it is possible to localise the receptor site in the molecule. Like most methods, those described also have their limits and should be used in combination.

Table 8.1. Localization and structure of phage receptors

Bacteria	Phage	Receptor	Receptor site
Salmonella Anatum	ε^{15}	LPS region I	α-D-Gal-$(1{\rightarrow}6)$-β-D-Man
Salmonella Newington	ε^{34}	LPS region I	β-D-Gal-$(1{\rightarrow}6)$-β-D-Man
Salmonella	Felix O (=O$_1$)	LPS core	α-D-GlcNAc-$(1{\rightarrow}$
Salmonella/Shigella	RFfm	LPS core	α-D-Gal-$(1{\rightarrow}3)$-α-D-Glc
E. coli	T3, T7	LPS core	
E. coli	T4[1]	LPS core	Glc-Glc
E. coli	T4[1]	OmpC	
E. coli	K3	OmpA	
E. coli	K20	OmpF	
E. coli	T6	OMP Tsx	
E. coli	BF23	Vitamin B12 receptor	
E. coli	λ	LamB	
E. coli	K1	Capsule	$\rightarrow$8)-α-NeuNAc-(2$\rightarrow$
B. subtilis	Φ3T	Teichoic acid	Glc-GalNAc-1-$\circledP$-
S. aureus	52A	Teichoic acid	α-D-GlcNAc-$(1{\rightarrow}$
E. coli	MS2; R17	F-Pilus	
P. aeruginosa	Pf1	PAK-Pilus	
Enterobacteriaceae	χ1	Flagellum	
B. subtilis	PBS1	Flagellum	

[1] Details see text.

A further important practical application of the procedure allows the subdifferentiation of bacteria for epidemiological purposes. Even though molecular methods have recently been used to an increasing degree for this intention, the method called phage typing (German: Lysotypie) has hardly lost its importance. For this purpose, different phages are dropped onto different spots of the suspended lawn of bacteria. From the resulting pattern of lysed and non-lysed bacteria a conclusion can be drawn on the so-called phage types (German: Lysotypen) which are generally very constant and characteristic.

Table 8.1 shows a selection of phage receptors elucidated so far, partially including the structure of the receptor site. Besides the components of the outer membrane of Gram-negative bacteria and the cell wall of Gram-positive bacteria, also those localised in the capsules, flagellae and pili may be relevant as receptors.

In the case of phages having their receptor in the LPS, the "smooth phages" (e.g. ε^{15}, G341 or P22) attaching to the O-specific polysaccharide (Fig. 2.16) and the "rough phages" (e.g. RFfm, F1 or C21) attaching to the core region are distinguished. With rough phages it can easily be detected which rough type (Ra, Rb, etc.) is present. For the epidemiologically important typing phage Felix O, the receptor site has been localised in the outermost part of the core region.

An example of the requirement of more than one receptor component is give by phage T4, in the attachment of which both protein and LPS are involved. In the presence of the membrane protein OmpC, T4 needs an LPS for binding that contains a terminal heptose (Rd_2 chemotype), and in the absence of OmpC an LPS possessing terminal glucose in the core region (Rc chemotype). A second example is given by the *E. coli* K-12-specific phage 20. Alterations in the biosynthesis either of porin OmpF (removal of eight amino acid residues in loop 5 of the β-barrel, not influencing the channel activity; see Sect. 2.3.2.3.1) or of the LPS core (preventing the addition of galactose to the first glucose residue; the structure of this core is similar but not identical to that of the Salmonella R_a-LPS shown in Table 2.8) render the strains resistant to the action of this phage.

After its binding to the receptor, the phage must permeabilise the bacterial cell wall to make the injection of nucleic acid possible. For this purpose, phage T4 utilises lysozyme to locally destroy the peptidoglycan; other phages use other specific glycosidases. A misappropriated application of bacterium-specific transport systems has been observed: the capsule-specific phage ViII destroys its own receptor, the Vi-antigen, by deacetylation. It may be that the phage prevents in this way a secondary infection; however, it may also be that the phage uses this degradation to move inwards.

The investigation of the infection of Salmonella Anatum with phages ε^{15} and ε^{34} presents a nice example of the influence of bacteriophages on composition and structure of the bacterial cell wall. Salmonella Anatum is characterised by the O-antigen factors 3 and 10. It can be lysed and lysogenised by the phage ε^{15} and is resistant towards phage ε^{34}. After lysogenization by phage ε^{15}, the O-antigen factor 10 has disappeared, it is replaced by factor 15. In this way, the strain has been converted to Salmonella Newington (lysogenic conversion) and is now sensitive towards phage ε^{34}. After the infection with this phage, the O-antigen factors (3), (15) and 34 are detectable (those in brackets weakly), and the strain is now Salmonella Illinois.

The molecular background of this conversion is known (Fig. 8.9). The *O*-acetyl galactose-bound α-(1→6) to mannose is characteristic for the repeating units of S. Anatum. The lysogenisation by the phage ε[15] leads to the repression of the O10-transacetylase and thus to the suppression of the acetylation of the galactose. At the same time, the biosynthesis of a new-type polymerase is determined, forming a β-bond between galactose and mannose instead of the former α-bond. Simultaneously, the α-polymerase is inhibited. It has been found that both alterations are not induced by the same phage, but that the phage ε[15] represents a mixture of two mutants, one of which (ε[15a]) induces the α→β-change of the Gal→Man-linkage and the other (ε[15b]) the suppression of the acetylation. The phage ε[15] is no longer able to adsorb at the new repeating unit, its receptor site has disappeared. The newly synthesised repeating units are characteristic for S. Newington and are appropriate as a receptor for phage ε[34]. After lysogenisation by this phage, the biosynthesis of a UDP-glucosyl transferase is determined by substituting the galactose in 4 position with α-glucose. Now the repeating units have the structure typical for S. Illinois.

After the formation of new phages, they have to be released. Since the cell wall and especially the peptidoglycan in it are impeding this process due to their stability, they are enzymatically degraded from inwards (lysozyme, endolysines). A degradation of the phospholipids or, in some cases, a modification of the phospholipid head groups may also occur, e.g. such as from phosphatidyl ethanolamine to phosphatidyl glycerol. If the phages are not able to achieve this, bacterial cells firmly filled with phages can be observed under the electron microscope.

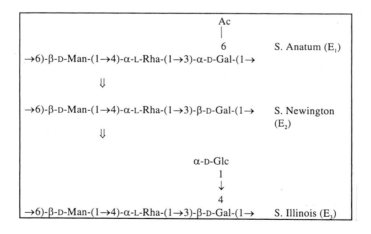

Fig. 8.9. Structural changes in LPS region I during lysogenic conversion in the Salmonella group E

Bibliography

Bennett PM, Howe TGB (1998) Bacterial and bacteriophage genetics. In: Coll A, Sussman M (eds) Topley & Wilson´s microbiology and microbial infe........ Balows A, Duerden BI (eds) Systematic Bacteriology, 9th edn. Arnold London, pp 231-294
Kemper B (1999) Bacteriophages as models for differentiation. In: Lengeler JW, Drews G, Schlegel HG (eds) Biology of the procaryotes. G Thieme, Stuttgart, pp 602-626
Lindberg AA (1973) Bacteriophage receptors. Annu Rev Microbiol 27:205-241
Taurig M, Misra R (1999) Identification of bacteriophage K20 binding regions of OmpF and lipopolysaccharide in *Escherichia coli* K-12. FEMS Microbiol Lett 181:101-108

8.6 Interactions Between the Bacterial Cell Wall and Antibiotics

Antibiotics are low-molecular substances mostly produced by microorganisms, which are able to enter bacteria on different pathways to kill them (bactericidal) or to inhibit their growth (bacteriostatic). Antibiotics can be classified according to both their chemical structure and their (very diverse) action mechanisms. Targets of the antibiotics may be components of the cell wall, of the cytoplasmic membrane or of the biosynthetic apparatus of the bacteria.

As a countermove, the bacteria have developed resistance mechanisms against antibiotics. These mechanisms may also be quite diverse and may be both specific or unspecific. Besides mechanisms which are naturally (primary resistance) present in bacteria, acquired (secondary) resistance mechanisms play an important role. Plasmids are often essential for formation and transmission of the latter.

In most cases, resistance to antibiotics imposes on the fitness of bacteria. However, as due to the resistance-determining mutations and accessory elements the bacteria remain able to grow in presence of the antibiotics, the loss in fitness is more than compensated.

One of the primary and unspecific cell wall-dependent resistance mechanisms of the Gram-negative bacteria is caused by the low permeability of their outer membrane. Due to this, *E. coli* wild-type strains are resistant to more than 90% of the antibiotics which effectively inhibit *S. aureus* strains. This permeability can be further reduced, for instance by mutation in the porin-encoding genes. In this case, a secondary resistance is additionally present. The distinct primary resistance of mycobacteria due to a strong barrier function of their cell wall has already been mentioned.

The effect of β-lactam antibiotics (penicillins, cephalosporins, carbapenems and monobactams) on peptidoglycan biosynthesis and a resistance mechanism of the bacteria directed against them have been discussed in Section 3.2.2.6. Due to the low permeability of their outer membrane, many Gram-negative bacteria have a primary resistance to β-lactam antibiotics. A second, basically different resistance mechanism depends on the formation of bacterial β-lactamases inactivating many, but not all β-lactam antibiotics. In Gram-negative bacteria they are localised in the periplasmic space, in Gram-positive ones they are partly situated in the periplasm-like space, partly they are excreted.

Glycopeptide antibiotics (e.g. vancomycin or teicoplanin) also inhibit the peptidoglycan biosynthesis by binding to the D-Ala-D-Ala-region of the UDP-muramyl pentapeptide and thus blocking both transglycosylation and transpeptidation. Replacement of the terminal D-alanine by D-hydroxybutyric acid or D-lactic acid renders the bacteria resistant by preventing the binding of the antibiotic, but not preventing the application of the D-Ala-D-Ala-like dipeptide as energy supplier for cross-linking. Mutants resistant towards such antibiotics were found to contain drastically increased amounts of intact D-Ala-D-Ala-carboxy termini in their peptidoglycan layer. These dipeptides bind the antibiotics, trap them in the layer, and thus prevent the antibiotic molecules from reaching sites of wall biosynthesis at the cytoplasmic membrane.

The action of polycationic antibiotics like aminoglycosides (gentamicin, kanamycin) and the polymyxins has been mentioned in Section 8.1.1. Their actual target is located in the cell interior. For example the basic polymyxin B (PmB), a decapeptide containing five positive charges, consists of a heptapeptidic ring and a tripeptidic tail possessing a hydrophobic tip formed by long-chain fatty acids (Fig. 8.10). Due to its structure, the molecule cannot pass via porin pores or the hydrophobic pathway. For adhesion to the cell wall it therefore needs a primary target, which is represented by the accumulation of negative charges (phosphate and carboxyl groups) in the lipid A-core region of membrane-bound LPS. In this region, PmB binding displaces the bivalent metal ions from their binding sites. In a second step the hydrophobic moiety of the antibiotic contacts the lipophilic part of lipid A, i.e. its fatty acid chains. This induces a breaking up of the LPS aggregates which, besides others, may result in a release of less-toxic LPS. The antibiotic integrates into the lipid matrix, induce lesions in the membrane structure due to its bulky head group, creates large channels, and thus destroys the barrier function of the membrane, at least partially. Via these channels both antibiotic and uninvolved molecules may enter the cell interior (self-promoted pathway). This mechanism is schematically shown in Fig. 8.11.

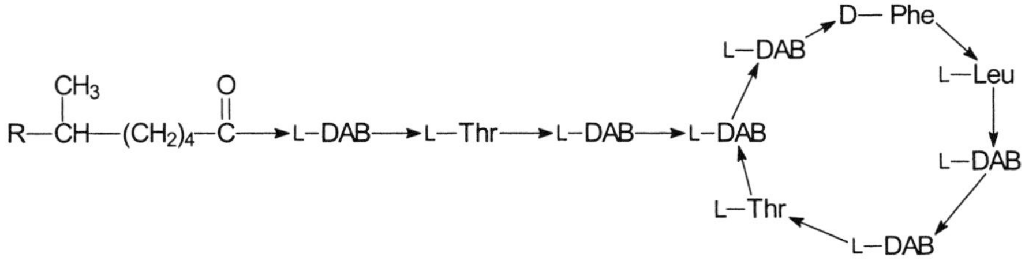

Fig. 8.10. Structure of Polymyxin B. For polymyxin B_1, R C_2H_5, and for polymyxin B_2, R CH_3

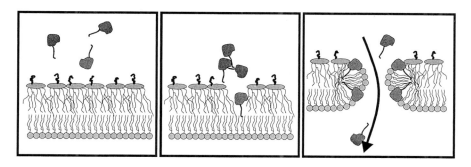

Fig. 8.11. Model for the interaction of polymyxin B (PMB) with the outer membrane. From left to right *1* Accumulation of PMB at the negatively charged LPS surface. *2* Binding to and integration into the lipid matrix. *3* Formation of membrane lesions and „self-promoted pathway". (Jahresbericht 1998, Research Center Borstel, Germany)

A general protective mechanism of the bacteria against the action of such antibiotics is the reduction in the density of negative charges at the cell surface and, thus, the impediment for attachment of such antibiotics. In the case of polymyxin B, the effect is achieved by the substitution of the 4´-phosphate residues in lipid A with 4-amino-4-deoxy-L-arabinose, which renders the molecule, and thus the bacteria, distinctly more resistant. However, examples have been described (*Pseudomonas aeruginosa*) which do not confirm such a correlation.

The antibiotic efficiency of aminoglycoside antibiotics may also be drastically reduced by their phosphorylation, acetylation or adenylylation by means of periplasmic enzymes. However, this is of only little influence on the penetration of the antibiotic across the cell wall.

Finally, bacteria are able to remove antibiotics which have penetrated into the cell by "pumping" them out. Such a mechanism was first discovered as specific in the case of tetracyclines. Later, an ATP-dependent system for removal of macrolides out of *Staphylococcus aureus* was described which resembles more general cell-own transport mechanisms. Besides systems removing only one type of antibiotics, "multidrug" efflux pumping systems exist. In Gram-negative bacteria basically two pumping systems may be present: those of the first pump the antibiotics into the periplasmic space, those of the second to the exterior. The latter is advantageous for Gram-negative bacteria because, in order to come again, the expelled antibiotic have to cross the low-permeability outer membrane.

The transporter system AcrB of *E. coli* represents an example for the second type, it is able to transport a broad spectrum of antibiotics and other harmful substances outwards across both membranes and the periplasmic space (see Table 8.2). It consists of three components: the transporter protein localised in the cytoplasmic membrane, a channel-forming protein in the outer membrane (presumably TolC), and a periplasmic accessory protein (a member of the MFP family) connecting both proteins. The system slightly resembles ABC transporters (Sect. 8.1.2.2) in structure, energy supply and involvement of the Tol complex. As shown in Fig. 8.12 for the example of an amphiphilic antibiotic, the transporter

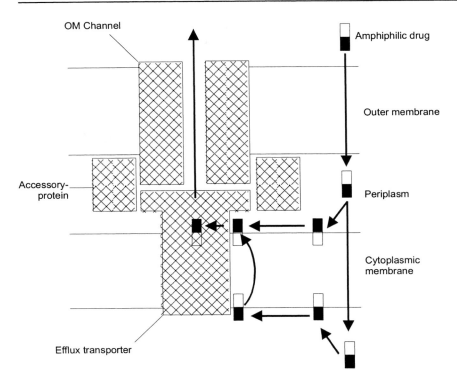

Fig. 8.12. Schematic construction of the AcrB-transporter system. (Nikaido H (1998) Antibiotic resistance caused by Gram-negative multidrug efflux pumps. Clin Infect Dis 27:S32-S41, with permission)

Table 8.2. Substances exported by the AcrB-transporter system

Group	Substances
Antibiotics	Chloramphenicol, erythromycin, fluorchinolone, fusidic acid, β-lactam antibiotics, novobiocin, rifampicin, tetracycline
Further substrates	Acriflavin, deoxycholate, ethidiumbromide, crystal violet, Na-dodecylsulfate

proteins are able to take up antibiotics both from the cytoplasm and/or the cytoplasmic membrane, and to transport them under energy consumption into the above-situated outer-membrane channels, through which they diffuse to the exterior.

It is supposed that a cone-like drug-binding pocket in the interior of the bacteria selects the molecules with regard to hydrophobicity, charge and shape. Recent data suggest a possible role of the periplasmic protein in causing membrane adhesion sites between the cytoplasmic and the outer membrane thus bringing the AcrB and TolC into contact (not shown in Fig. 8.12).

There are indications that this system represents a constituent of a common natural protective mechanism of the bacteria towards mostly lipophilic and amphiphilic noxious substances from the environment.

Bibliography

Amyes SGB, Gemmel CG (eds) (1992) Antibiotic resistance in bacteria. J Med Microbiol 36:4-29

Nikaido H (1998) Antibiotic resistance caused by Gram-negative multidrug efflux pumps. Clin Infect Dis 27:S32-S41

Zgurskaya HI, Nikaido H (2000) Multidrug resistance mechanisms: drug efflux across two membranes. Mol Microbiol 37:219-225

Uncommon Abbreviations

ABC	ATP-binding cassette
ACP	acyl carrier protein
ACL	antigen carrier lipid
AFA,	a-fimbrial adhesin
AG,	arabinogalactan
CFA	colonization factor antigen
CM	cytoplasmic membrane
COSY	correlation spectroscopy
CPS	capsular polysaccharide
DAG	2,3-diamino-2,3-dideoxy-D-glucopyranose
DAP	diaminopimelic acid
DOC	deoxycholate
ECA	enterobacterial common antigen
EDTA	ethylene diamine tetraacetic acid
EI	electron impact
ESI	electrospray ionisation
FAB	fast atom bombardment
FnBP	fibronectin binding protein
GalNN	2-acetamido-4-amino-2,4,6-trideoxy-D-galactose (Gal*p*NAc4N)
GEP	signal peptide-dependent general export pathway
GLC	gas-liquid chromatography
GPI	glycosylphosphatidylinositol
GPL	glycopeptidolipid
GSP	general secretory pathway
HPLC	high-performance liquid chromatography
HPTLC	high-performance thin-layer chromatography
Ipa	invasion plasmid antigen
Kdo	3-deoxy-D-*manno*-oct-2-ulosonic acid, old name 2-<u>k</u>eto-3-<u>d</u>eoxy-D-*mannc* -<u>o</u>ctonic acid
Ko	D-*glycero*-D-*talo*-oct-2-ulosonic acid
LamB	receptor of λ-phage
LAMMA	laser microprobe mass analyser

LBP	LPS-binding protein
LD_{50}	lethal dose for 50% of the animals
LDI	laser desorption ionization
LOS	lipooligosaccharides
LPS	lipopolysaccharide
LTA	lipoteichoic acid
MAC	membrane-attack complex
MALDI	matrix-assisted laser desorption/ionization
MD	membrane domain
MDO	membrane-derived oligosaccharides
MFP	membrane fusion protein
MS	mass spectrometry
NCAM	neural cell adhesion molecule
NFA	non-fimbrial adhesin
NMR	nuclear magnetic resonance
NOESY	nuclear Overhauser enhancement spectroscopy
OF	opacity-factor
OM	outer membrane
OMA	outer membrane auxillary (protein)
OMP	outer membrane protein
P	phosphate
PAGE	polyacrylamide gel electrophoresis
PBP	periplasmic-binding protein
PBP	penicillin-binding protein
PEP	phosphoenolpyruvate
PGL	phenolic glycolipid
PL	phospholipid
PP	periplasma/periplasmic space
PPD	purified protein derivative
SDS	sodium dodecyl sulfate
SLH	S-layer homology
TLC	thin-layer chromatography
TOCSY	total correlation spectroscopy
undec	undecaprenol
Yop	Yersinia outer protein

Index

Druck: Strauss Offsetdruck, Mörlenbach
Verarbeitung: Schäffer, Grünstadt